STUD MANAGERS' HANDBOOK

International Stockmen's School Handbooks

Stud Managers' Handbook
Volume 19

edited by Frank H. Baker and Mason E. Miller

The 1984 International Stockmen's School *Handbooks* include more than 200 technical papers presented at this year's Stockmen's School, sponsored by Winrock International. The authors of these papers are outstanding animal scientists, agribusiness leaders, and livestock producers who are expert in animal technology, animal management, and general fields relevant to animal agriculture.

The *Handbooks* present advanced technology in a problem-oriented form readily accessible to livestock producers, operators of family farms, managers of agribusinesses, scholars, and students of animal agriculture. The *Beef Cattle Science Handbook*, the *Dairy Science Handbook*, the *Sheep and Goat Handbook*, and the *Stud Managers' Handbook* each include papers on such general topics as genetics and selection; general anatomy and physiology; reproduction; behavior and animal welfare; feeds and nutrition; pastures, ranges, and forests; health, diseases, and parasites; buildings, equipment, and environment; animal management; marketing and economics (including product processing, when relevant); farm and ranch business management and economics; computer use in animal enterprises; and production systems. The four *Handbooks* also contain papers specifically related to the type of animal considered.

Frank H. Baker, director of the International Stockmen's School at Winrock International, is also program officer of the National Program. An animal production and nutrition specialist, Dr. Baker has served as dean of the School of Agriculture at Oklahoma State University, president of the American Society of Animal Science, president of the Council on Agricultural Science and Technology, and executive secretary of the National Beef Improvement Federation.

Mason E. Miller is communications officer at Winrock International. A communications specialist, Dr. Miller served as communication scientist with the U.S. Department of Agriculture; taught, conducted research, and developed agricultural communications training programs at Washington State University and Michigan State University; and produced informational and educational materials using a wide variety of media and methods for many different audiences, including livestock producers.

A Winrock International Project

Serving People Through Animal Agriculture

This *Handbook* is composed of papers presented at the
International Stockmen's School
January 8–13, 1984, San Antonio, Texas
sponsored by Winrock International

A worldwide need exists to more productively exploit animal agriculture in the efficient use of natural and human resources. It is in filling this need and carrying out the public service aspirations of the late Winthrop Rockefeller, Governor of Arkansas, that Winrock International bases its mission to advance agriculture for the benefit of people. Winrock's focus is to help generate income, supply employment, and provide food through the use of animals.

1984 INTERNATIONAL STOCKMEN'S SCHOOL HANDBOOKS

STUD MANAGERS' HANDBOOK Volume 19

edited by Frank H. Baker and Mason E. Miller

A WINROCK INTERNATIONAL PROJECT

LONDON AND NEW YORK

First published 1984 by Westview Press, Inc.

Published 2019 by Routledge
52 Vanderbilt Avenue, New York, NY 10017
2 Park Square, Milton Park, Abingdon, Oxon OX14 4RN

Routledge is an imprint of the Taylor & Francis Group, an informa business

ISBN 13: 978-0-367-28904-1 (hbk)
ISBN 13: 978-0-367-30450-8 (pbk)

CONTENTS

Part 13. FACILITIES AND BEHAVIOR

Part 14. COMPUTER TECHNOLOGY AND OTHER TOPICS

PREFACE

The *Stud Managers' Handbook* includes presentations made at the International Stockmen's School, January 8-13, 1984. The faculty members of the School who authored this fourth volume of the *Handbook*, along with books on beef cattle, dairy cattle, and sheep and goats, are scholars, stockmen, and agribusiness leaders with national and international reputations. The papers are a mixture of technology and practice that presents new concepts from the latest research results of experiments in all parts of the world. Relevant information and concepts from many related disciplines are included.

The School was held annually from 1963 to 1981 under Agriservices Foundation sponsorship; before that it was held for 20 years at Washington State University. Dr. M. E. Ensminger, the School's founder, is now Chairman Emeritus. Transfer of the School to sponsorship by Winrock International with Dr. Frank H. Baker as Director occurred late in 1981. The 1983 School was the first under Winrock International's sponsorship after a one-year hiatus to transfer sponsorship from one organization to the other.

The five basic aims of the School are to:

1. Address needs identified by commercial livestock producers and industries of the United States and other countries.
2. Serve as an educational bridge between the livestock industry and its technical base in the universities.
3. Mobilize and interact with the livestock industry's best minds and most experienced workers.
4. Incorporate new livestock industry audiences into the technology transfer process on a continuing basis.
5. Improve the teaching of animal science technology.

Wide dissemination of the technology to livestock producers throughout the world is an important purpose of the *Handbooks* and the School. Improvement of animal production and management is vital to the ultimate solution of hunger problems of many nations. The subject matter, the style of presentation, and the opinions expressed in the papers are those of the authors and do not necessarily reflect the opinions of Winrock International.

ACKNOWLEDGMENTS

Winrock International expresses special appreciation to the individual authors, staff members, and all others who contributed to the preparation of the *Stud Managers' Handbook*. Each of the papers (lectures) was prepared by the individual authors. The following editorial, secretarial, and word processing staff of Winrock International assisted in reading and editing the papers for delivery to the publishers:

Editorial Assistance

Jim Bemis, Editor
Essie Raun, Assistant editor
Paula Gerstmann, Research assistant
Randy Smith, Illustration editor
Venetta Vaughn, Illustration editor and proofer
Melonee Baker, Proofer
Elizabeth Getz, Proofer
Joan Hart, Proofer
Beverly Miller, Proofer
Mazie Tillman, Proofer

Secretarial Assistance and Word Processing

Patty Allison, General coordinator
Ann Swartzel, Secretary
Tammy Henderson, Secretary
Shirley Zimmerman, Coordinator of word processing
Darlene Galloway, Word processing
Tammie Chism, Word processing
Jamie Whittington, Word processing

Part 1

GLOBAL AND NATIONAL ISSUES

1

APPLYING AGRICULTURAL SCIENCE AND TECHNOLOGY TO WORLD HUNGER PROBLEMS

Norman E. Borlaug

Agriculture and food production have been my primary concerns in research, but by necessity, I have developed interest in the broad fields of land use--or misuse--and demography.

If one is involved in food production, it naturally follows that one must be concerned about the land base upon which we depend for food production and the number of people that land base must feed.

In the total plan of things, our earth is very small. On the surface, more than three-quarters of it, approaching 78% or 79%, is water, most of it salt water or ocean. Some is inland water, sweet waters, and lakes. Less than one-quarter of the earth's surface is land, but 98% of worldwide food production was produced on the land in 1975.

When we examine it, some of the land in the world is bad real estate (table 1). As far as arable land is concerned, only 11% of the total land area is classified as suitable for agriculture. Another 22% is classified as

TABLE 1. LAND RESOURCES OF THE EARTH

Land type	Area, ha (millions)	% of total land area
Arable land (annual and permanent crops)[a]	1,457	11
Permanent meadows and pastures[b]	2,987	22
Forest and woodland	4,041	30
Other (tundras, subarctic wastes, deserts, rocky mountainous wastes, cities, highways)	4,908	37

Source: FAO Production Yearbook (1972).
[a]Of the arable land area, 48% (698 million hectares) is cultivated to cereal grains.
[b]Total agricultural land, therefore, is about 33%.

suitable for grazing and animal industry. Both of these agricultural uses account for about 33% of our total land area. An additional 30% is classified as forestland and woodlots. The remaining 37% is called "other." "Other" means mostly wasteland, arctic tundra, deserts, rocky mountain slopes with very little soil on them, or good agricultural land that has been covered by cities, pavements, and highways. We continue to cover this good land at an appalling rate in many parts of the world, not only in the U.S. Several million acres of good land go out of production each year because it is easier and less costly, apparently, to build on flat land than on sloping land. On the surface, at least, this seems to be the case.

FEEDING FOUR BILLION PEOPLE

When we talk about food, we need to have some concept of how much food is needed to feed this population of 4.6 billion and about the possibilities of producing enough to maintain stability--social, economic, and political--in the next 4 decades.

When we consider food, we must consider it from three standpoints:

1. From the standpoint of biological need, which should be self-evident, for without food you can live only a few weeks at most, assuming you entered the famine or starvation situation in good health.
2. From an economic standpoint, the worth of food depends entirely on how long it has been since you had your last food and what your expectancies are for food in the future.
3. From the political standpoint, the importance of food can be observed when stomachs are empty. It makes no difference whether it is a socialistic or communistic system or whether it is a free enterprise system. To illustrate, think back several years ago to the devastating drought in the Sahel. You saw the consequences on your television screens--the misery and poverty and hunger. Six governments fell as a result of the shortages of food and the misery and suffering of their masses.

Anyone engaged in attempting to increase world food production soon comes to realize that human misery resulting from world food shortages and world population growth are part of the same problem. In effect, they are two different sides of the same coin. Unless these two interrelated problems and the energy problem are brought into better balance within the next several decades, the world will become increasingly more chaotic. The social, economic and political pressures, and strife are building at different rates in different countries of the world, depending upon human population density and growth rate and upon the natural resource base that sustains the different economies. The poverty in many of the developing nations will become unbearable, standards of living in many of the affluent nations may stagnate, or even retrogress. The terrifying human population pressures will adversely affect the quality of life, if not the actual survival, of the bald eagle, stork, robin, crocodile, wildebeest, wolf, moose, caribou, lion, tiger, elephant, whale, monkey, ape, and many other species. In fact, world civilization will be in jeopardy.

Unfortunately, in privileged, affluent, well-educated nations such as the U.S., we have concerned ourselves with symptoms of the complex malaise that threatens civilization, rather than with the basic underlying causes. In recent years, we have been attacking these ugly symptoms by passing new legislation or filing lawsuits against companies, individuals, or various government agencies for polluting the environment. Most of these lawsuits just fatten the incomes of lawyers without solving the basic problems.

THE HUMAN POPULATION MONSTER

Most of us are either afraid, or are unwilling, to fight the underlying cause of most of this malaise...The Human Population Monster. The longer we wait before attacking the primary cause of this worldwide problem--with an intelligent, unemotional, effective, and humane approach--the fewer of our present species of fauna and flora will survive.

About 12,000 yr ago, the humans who had been roaming the earth for at least 3 million yr, invented agriculture and learned how to domesticate animals. World population then is estimated to have been approximately 15 million. With a stable food supply, the population growth rate accelerated. It doubled four times to arrive at a total of 250 million by the time of Christ. Since the time of Christ, the first doubling (to 500 million) occurred in 1,650 yr. The second doubling required only 200 yr to arrive at a population of 1 billion in 1850. That was about the time of the discovery of the nature and cause of infectious diseases and the dawn of modern medicine--which soon began to reduce the death rate. The third doubling of human population since the time of Christ, to 2 billion, occurred by 1930...only 80 yr after the second doubling. Then, sulfa drugs, antibiotics, and improved vaccines were discovered. They reduced infant deaths spectacularly and prolonged life expectancy.

World population doubled again...to 4 billion people in 1975. That took only 45 yr and represents an increase of 256 fold--or eight (8) doublings since the discovery of agriculture. Currently, it has reached 4.7 billion.

It is obvious that the food/population ratio and competition between species is getting worse dramatically as the numbers of humans increase so frighteningly. And the interval between doublings of human population continues to shorten. At the current world rate of population growth, population will double again, reaching 8 billion souls by 2015 (figure 1)!

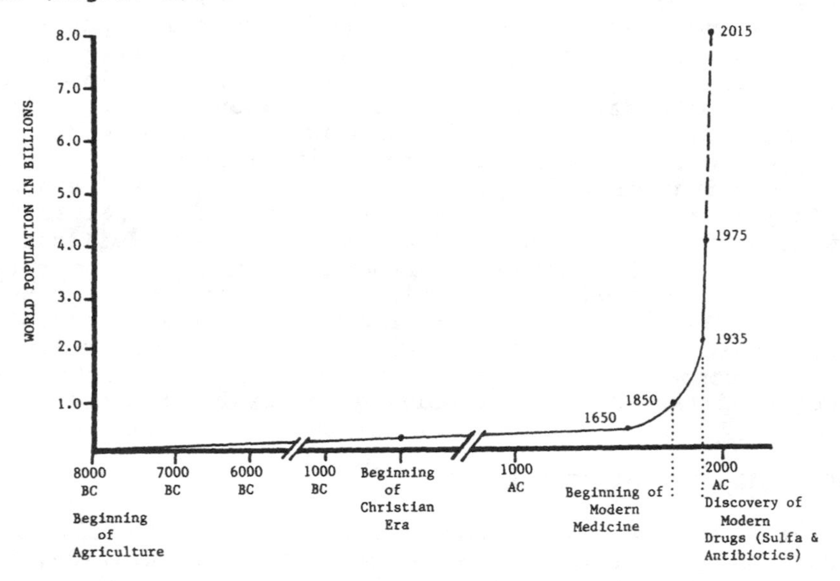

Figure 1. World demographic growth

TWO WORLDS

It is a sad fact that on our planet Earth, at this late date, there are two different worlds as far as food production and availability are concerned--namely, the "privileged world" and the "forgotten world" (Borlaug, 1978). The privileged world consists of the affluent, developed nations comprising about 33% of the world population. In these nations agriculture is efficient--and industrialization is well advanced--with only 5% to 20% of the population engaged in agriculture but capable of producing sufficient food for their own nation's needs as well as surpluses for export. The consumer in these nations has an abundant and diverse food supply available at a low price; his entire food budget represents only 17% to 30% of his income after taxes. Most of the people in these nations live in a luxury never before experienced by man. The vast proportion of the population (70 to 80%) in these countries is urban. They take the abundance of cheap food for granted. Many of them think it comes from the supermarkets and fail to understand the large investments in land and machinery required, the management skills, toil, struggle, risks, and frustrations on the ranches and farms that are required to produce the abundance they take for granted.

The "forgotten world" is made up of the developing nations where most of the people, comprising 50% of the world's population, live in poverty with hunger a frequent companion and fear of famine a constant menace. In these nations, a vast segment of the total population--ranging from 60% to 80%--is tied to a small plot of land in an inefficient subsistence agriculture. In these nations, food, and especially animal protein, is always in short supply and expensive. The urban consumer in such countries expends 60 to 80% of his income on food in normal times, and when droughts, floods, diseases, or pests reduce the harvests, all of his earnings go for food--and even then he is unable to buy what he needs. Many of the subsistence farmers themselves are often short of food, and even a larger proportion are suffering from protein malnutrition.

Why does this great discrepancy exist between the privileged and the forgotten nations in food production? Although many factors are involved, the four major causes are: 1) the difference in per capita endowment of natural resources, i.e., good arable land; 2) the availability or nonavailability of proper modern technology developed by research for increasing yields; 3) the presence or absence of strong economic and extension infrastructures; 4) and adequate or inadequate visionary policy supported by government. Of these, the two greatest problems of the developing countries are the small amount of arable land available on a per capita basis coupled with low and stagnant per hectare yields.

Table 2 illustrates the comparative food production capabilities of land exploited under hunting and various

TABLE 2. COMPARATIVE FOOD PRODUCTION CAPABILITIES OF LAND EXPLOITED UNDER HUNTING AND VARIOUS TYPES OF AGRICULTURE

System of exploitation	Area required, ha	No. of people fed
Hunting[a]	2,500	1
Foraging[b]	250	1
Hoe agriculture[c]	250	3
Plow agriculture[d]	250	750
Modern agriculture[e]	250	2,000[f]

Source: Storck and Teague (1952).
[a]Indians of the North American plains before European influence.
[b]California Indians before European influence.
[c]Eastern woodland Indians of North America before European influence.
[d]Ancient Egyptian agriculture.
[e]Highly developed modern agriculture of the U.S., based on 1950 yields.
[f]If 1980 yields were used, this figure would increase by 80% to 90%.

types of agriculture. It is apparent that modern American agriculture employing advanced technology is capable of producing much more food per unit of land than are other methods of exploitation.

Our Obligations

"Human rights" is a utopian issue and a noble goal to work toward. But it can never be achieved as long as hundreds of millions of poverty-stricken people in the world lack the necessities of life. The "right to dissent" doesn't mean much to a person with an empty stomach, a shirtless back, a roofless dwelling, the frustrations and fear of unemployment and poverty, the lack of education and opportunity, and the pain, misery, and loneliness of sickness without medical care. My work has brought me into close contact with such people and I have come to believe that all who are born into the world have the moral right to the basic ingredients for a decent humane life. How many should be born and how fast they should come on stage is another matter. This latter question requires the best thinking and efforts of all of us if, in my opinion, we are to survive as a world in which our children and their children will want to live and, more important, be able to live in.

Those of us who work on the food-production front, I believe, have the moral obligation to warn the political, religious, and educational leaders of the world of the magnitude and seriousness of the food/population problem that looms ahead. If we fail to do so in a forthright unemotional manner, we will be negligent in our duty and inadvertently through our irresponsibility will contribute to the pending chaos.

In the next 30 to 50 yrs, depending on how the world population continues to grow, world food and fiber production must be increased more than it was increased in the 12,000-yr period from the discovery of agriculture up to 1975. This is a tremendous undertaking and of vital importance to the future of civilization. Failure will plunge the world into economic, social, and political chaos. Can the production of food and fiber reach the necessary level in the next 31 yr? I believe it can, providing world governments give high enough priority and continuing support to agriculture and forestry. It cannot be achieved with the miserly and discontinuous support that has been given to agriculture and forestry during the past 50 yr.

We are all aware, from history and archaeology, of the disappearance of one civilization after another. We know that in some of the theocracies of recent times, the clergy as the privileged caste, lost contact with the masses, and their civilizations disintegrated. Time after time military dictatorships also have lost contact with the masses and their government--and sometimes the entire civilization--has perished.

APPROPRIATE USE OF SCIENCE AND TECHNOLOGY

Ours is the first civilization based on science and technology. Through the development and contributions of science and technology the present standard of living of much of the world has reached undreamed-of heights. But science and technology have also developed frighteningly powerful destructive forces...which if unleashed...are capable of annihilating civilization and much of today's population. To assure continued progress, we scientists must not lose contact with the needs of the masses of our own society nor of that of other societies of the world. Our own survival is at stake. We must recognize and meet the changing needs and demands of our fellow men. To do so we must strive for the proper balance between fundamental and applied research. And we must try our best to assist in training young scientists from the many developing countries who study in our universities so that their training will be useful to them when they return to their homeland. Moreover, our technical assistance programs to the developing countries should be organized so that they will be relevant to the needs of the host country. All too often, the approach is much too sophisticated...attempting to fit 1984 U.S. agricultural technology into areas of the world where 1920 technology would be more appropriate.

I have seen the consequences of oversophisticated approaches reflected in the research and education programs in many developing countries. Sometimes the irrelevant research being done and the expensive gadgets and equipment that one sees standing unused are the result of ideas brought back by students who had taken advanced degrees in foreign universities. At other times, it is the result of foreign consultants or scientists from one of the advanced countries promoting impractical, irrelevant research projects to unsuspecting government policy makers and with it the need for sophisticated equipment.

I also have my reservations about the highly specialized narrowness and lack of communication that has been creeping into our science during the past 2 decades. Dr. Thor Heyerdahl, in his book Aku-Aku, expresses this misgiving beautifully: "In order to penetrate even farther into their subjects, the host of specialists narrow their fields and dig down deeper and deeper till they can't see each other from hole to hole. But the treasures that their toil brings to light they place on the ground above. A different kind of specialist should be sitting there, the one still missing. He would not go down any hole, but would stay on top and piece all the facts together." To this thought I add: "He might even help decide where some of the digging should be done...perhaps more in the soil and less in space and in destructive powers."

With a team of scientific colleagues from many countries, I have spent the last 40 yr trying to help many developing nations increase the efficiency of food production of their agriculture. We have been working on total grain production or farming systems. The impact of our work is evident in data from India on land use before and after the wheat revolution (table 3).

POTENTIAL OF WORLD ANIMAL POPULATIONS

Research by Winrock International focuses on the potential of the world animal populations to contribute to solving world hunger problems. Approximately 75% of the animals in the world are maintained in mixed crop-animal farming systems. The animal populations in developing countries are at very low levels of productivity and output compared to developed countries (tables 4, 5, 6) (Spitzer, 1981).

The potential exists for changing the low animal productivity levels by developing a scientific approach to production similar to that used in the wheat and rice programs. An example of what can be done through application of science is the successful eradication of the screwworms in the U.S. during the past 25 yr.

There are vast grass savannahs in Central Africa where food production from livestock can be significantly improved through elimination or control of the tsetse fly. Similarly, the work at CIAT shows that improved livestock management systems can significantly improve the food output from the *llanos* of Colombia, Venezuela, and Brazil.

We must expand our scientific knowledge and improve and apply better technology if we are to make our finite land and water resources more productive. This must be done promptly and in an orderly way if we are to meet growing needs without, at the same time, unnecessarily degrading the environment and crowding many species into extinction. Producing more food and fiber, and protecting the environment can, at best, be only a holding operation while the population monster is being tamed. Moreover, we must recognize that in the transition period, unless we succeed in increasing the production of basic necessities to meet growing human needs, the world will become more and more chaotic, and our civilization may collapse.

REFERENCES

Borlaug, N. E. 1981. Using plants to meet world feed needs. In: R. G. Woods (Ed.) Future Dimensions of World Food and Population. Westview Press, Boulder, CO.

Borlaug, N. E. 1978. Food production in a fertile unstable world. World Food Institute Lecture. Iowa State University.

TABLE 3. LAND USE BEFORE AND AFTER THE WHEAT REVOLUTION

Year (harvest)	Area harvested, 1000 ha	Yield, MT/ha	Production, 1000 MT	Gross value of increased production, (million $)[b]	Adults provided with carbohydrate needs by increased wheat yield[c] 1961-66 period, (million persons)	Area required to produce crop at 1961-66 yield, 1000 ha	Area saved by yield increased over 1961-66 base, 1000 ha
1961-66[a]	13,191	.830	10,950	-	-	-	-
1967	12,838	.887	11,393	88	3	13,726	888
1968	14,998	1.103	16,540	1,118	41	19,928	4,928
1969	15,958	1.169	18,652	1,540	56	22,472	6,513
1970	16,626	1.209	20,093	1,828	67	24,208	7,582
1971	18,241	1.307	23,833	2,576	94	28,714	10,472
1972	19,154	1.382	26,471	3,092	113	31,893	12,738
1973	19,461	1.271	24,735	2,758	101	29,801	10,339
1974	18,583	1.172	21,778	2,166	79	26,238	7,655
1975	18,111	1.338	24,235	2,630	96	29,198	11,087
1976	20,458	1.410	28,846	3,580	131	34,754	14,295
1977	20,966	1.387	29,080	3,626	133	35,036	14,069
1978	20,946	1.480	31,000	4,010	147	37,349	16,402
1979	22,560	1.574	35,510	4,912	180	42,783	20,222
1980	21,962	1.437	31,560	4,122	151	38,024	16,061
1981	22,104	1.649	36,500	5,110	186	43,976	21,872
1982	-	-	37,800	5,370	196	-	-

Source: Ministry of Agriculture of India.
[a]Average for the six-year period 1961-66.
[b]Wheat value calculated at US $200/MT, similar to landed value of imported wheat in India.
[c]Calculation based on providing 375 g wheat/day or 65% of carbohydrate portion of a 2350 Kcal/day diet.

TABLE 4. ANIMAL DISTRIBUTION IN DEVELOPED AND DEVELOPING COUNTRIES

Animal	World	Developed Countries	Developing Countries
	---------(million head)---------		
Cattle	1,212	425	787
Buffalo	131	1	130
Swine	763	327	436
Chickens	6,706	3,022	3,684
Milk cows	214	114	100

Source: FAO Production Yearbook (1979). Adapted by R. Spitzer (1981).

TABLE 5. PERCENTAGE COMPARISON OF ANIMAL DISTRIBUTION AND PRODUCTION IN DEVELOPED AND DEVELOPING COUNTRIES

	% of distribution		% of production	
Animal	Developed countries	Developing countries	Developed countries	Developing countries
Cattle and buffalo	32	68	67	33
Swine	43	57	62	38
Chickens (hens)	45	55	66	34
Milk cows	53	47	84	16

Source: FAO Production Yearbook (1979). Adapted by R. Spitzer (1981).

TABLE 6. YEARLY PRODUCTION PER ANIMAL

Animal product	Developed countries	Developing countries
	--------(kg/animal)--------	
Beef and veal[a]	218	161
Milk	3,081	672
Pork[a]	78	58
Eggs	5.8	2.5

Source: FAO Production Yearbook (1979). Adapted by R. Spitzer (1981).
[a]Per animal slaughtered.

Borlaug, N. E. 1977. Forests for people: A challenge in world affairs. Society of Forestry Convention.

Borlaug, N. E. 1972. Human population food demands and wildlife needs. North American Wildlife Conference.

Borlaug, N. E. 1968. Wheat breeding and its impact on world food supply. Third International Wheat Genetics Symposium. Australian Academy of Science, Canberra.

FAO. 1979. Production Yearbook.

Raun, N. S. 1983. Beef cattle production on pastures in the American tropics. In: F. H. Baker (Ed.) Beef Cattle Science Handbook. Vol. 19. A Winrock International Project published by Westview Press, Boulder, CO.

Raun, N. S., R. D. Hart, J. De Boer, H. A. Fitzhugh and K. Young. 1981. Livestock program priority and strategy: U.S. Agency for International Development. Winrock International.

Spitzer, R. R. 1981. No Need for Hunger. The Interstate Printers and Publishers, Inc., Danville, IL.

2

FUTURE AGRICULTURAL POLICY CONSIDERATIONS THAT WILL INFLUENCE LIVESTOCK PRODUCERS

A. Barry Carr

American agriculture is in a state of decline. This decline can be measured in economic terms and in political terms; in fact, these economic factors and the political factors are related.

With each year that passes, the importance of the agricultural economy to the general economy is reduced--and this is true not only at the national level but also at the state and local levels. Recent studies have shown that even in the 100 most rural counties of the U.S., agriculture accounts for only 50% of the jobs and income. In the vast majority of rural counties, the portion of jobs and income derived from agriculture seldom exceeds 25% and is often less.

Not only is agriculture as a whole becoming a less important feature of our total economy, but Harold Breimyer, a prominent economist in Missouri, asserts that animal agriculture seems to be slipping in importance. He cites the declining share of total farm receipts earned by livestock and poultry. He also cites consumer touchiness about animal fats in their diets. For this and other reasons, a number of vegetable food products have been devised as substitutes for traditional foods of animal origin. On top of this is increased publicity from groups protesting the production methods used in animal agriculture. Another factor is the strong export market for feed grains and soybeans, which has caused many farmers to give up their livestock operations. Although Breimyer laments these changes, he sees little prospect that these trends will soon be reversed.

At the same time that agriculture's importance to the nonfarm economy is decreasing, farmers and ranchers are becoming increasingly dependent on earnings from nonagricultural sources. In 1981, farm families obtained 60% of their total family earnings from off-farm sources. This is one of the reasons why the condition of the nonfarm economy is increasingly important to the farm sector--but there are other reasons as well. Consumers' demand for agricultural commodities is determined by their purchasing power, which in turn is affected by their employment status, their wages, and even the size of their government assistance checks.

Interest costs paid by farmers and ranchers are affected by the monetary policy of the Federal Reserve System and the fiscal policy of the Administration and the Congress. Imports and exports of agricultural commodities are impacted by foreign policy decisions of various types.

Public policies with respect to land, water, and environmental quality frequently have direct consequences for farmers and ranchers. Currently, for example, various interest groups are working for policies or programs to preserve farmland from conversion to nonagricultural uses, to reduce soil erosion by eliminating crop program payments on certain lands (sodbuster bill), to eliminate water pollution, to protect wildlife, and to further restrict the use of pesticides.

As the influence of nonfarm factors has increased, some important changes have taken place in the farm sector. The average size of farms and ranches has increased each year. Agricultural units have gotten so big and so expensive to own and operate that few can afford them as hobbies any more. Agricultural operations have gotten so big, and so expensive to own and operate, that federal farm programs can do little by themselves to support the farm economy. If you do not believe this, consider that farm program costs went up $10 billion in 1983 while farm income remained at the same level as the year before. Agricultural prices and farm income have been very low for the past 3 yr. Large federal expenditures for commodity programs have not brought prosperity to the farm sector. And yet, 1/3 of the normal base acreage of grain and cotton are out of production. I do not need to tell you that the PIK program can put a real squeeze on livestock producers.

Over the past decade, policymakers have been gradually shifting farm programs toward more of a market orientation. Even in the face of this trend, however, farm groups have successfully worked for and received commodity price-support levels that exceed market equilibrium prices. This has occurred in spite of the fact that most of the benefits of support programs go to a small proportion of the farmers. In fact, they go almost entirely to wheat and feed grain producers, with lesser amounts to dairy and cotton producers. Although the next omnibus farm bill is not due until 1985, in 1983 we saw legislative attempts to change the price-support programs in ways that would reduce government involvement and costs.

CURRENT ISSUES OF INTEREST TO THE LIVESTOCK SECTOR

I would like to narrow the focus of my remarks to specifically address some of the current issues of interest to the livestock sector.

In 1982, Congress passed legislation to reduce the government's cost for the dairy price-support program. What *actually happened here*, for the first time, was that production-control incentives were included in this price-support program. Before this happened, milk was the only major program to operate without marketing quotas, production bases, or some other mechanism to encourage individual producers to keep supplies in reasonable balance with demand. This new, revised dairy program has the potential to become a model for other commodities.

Some are arguing that these changes have not come in time and are not nearly enough. In 1979, the Community Nutrition Institute (a consumer group), three milk drinkers, and a dairy operator petitioned the USDA to eliminate provisions of a milk-marketing agreement that made the use of reconstituted milk economically unfeasible. When the USDA stalled on the issue, the petitioners went to court. They won an important step in January 1983, when the U.S. Circuit Court of Appeals for the District of Columbia ruled that milk drinkers had standing to sue. In a sense, their suit is a challenge to the entire milk-marketing order system. But I would point out that if milk drinkers have standing to sue, what about beef eaters? Think about the possibilities.

Turning for a minute to beef, there is little good news. Per capita beef production has fallen from 94.4 lb in 1976 to 77.5 lb in 1982. Part of this decline is due to dietary concerns, and part is economic. Pork and chicken production is up over the same 6-yr period. Add too many packing plants and too many feed lots to this overcapacity in the beef industry. High interest rates and severe weather have also been problems. And on top of all of this, the PIK program promises to raise grain prices in the long run.

In the past 2 yr, we have seen a relaxed attitude at the Packers and Stockyards Administration toward regulation of feedlots and packers. It is too early to call the trend, but should packer-owned feedlots become commonplace, I believe some fundamental structural changes will take place in the cattle-feeding business--and in ranching as well.

The animal welfare movement is another factor in the current policy scene. H.R. 3170, a bill pending in the Congress at the present time, would establish a commission to investigate intensive farm animal husbandry to determine if these practices have any adverse effect on human health and to examine the economic, scientific, and ethical considerations with respect to the use of intensive farm animal husbandry. The more immediate goals of these groups include regulations on confinement feeding of poultry and livestock. However,if you examine the backgrounds of the activists in this movement and read their literature carefully, you will find a close connection with the vegetation movement and the antihunting movement as well.

When farmers and ranchers are confronted with the charges of the animal welfare movement, they tend to defend themselves by claiming that as livestock producers they naturally have the welfare of animals at heart. I just want you to know that you have to do better than that. Last year, George Stone, in his address at this school, stressed the need to keep the dialog open between both sides on this issue. This is important. But more importantly, you have to educate the large mass of the general public who, although largely uncommitted to either side at the moment, finds itself in sympathy with many positions of the animal welfare movement.

An example of such efforts was Senate Joint Resolution 77, which commemorated the 75th anniversary of the American Society of Animal Science by designating July 24 to July 31, 1983, as National Animal Agriculture Week. This resolution honored the tremendous progress in animal agriculture and the role of animal products in our lives. The resolution pointed out that foods from animal origins supply 70% of the protein, 35% of the energy, 80% of the calcium, and 60% of the phosphorus in the average American's diet.

The full list of policy issues of interest to livestock producers includes more than the few topics I have just discussed. I also would mention, as other topics on a long list, legislation to control the importation of meat and dairy products, proposals to alter the status of the Cooperative Farm Credit System, and even restrictions on the export of live horses for slaughter purposes. But I would hope that you, as members of the livestock industry, recognize the importance to your own welfare of all farm legislation.

Secretary Block and the Congress have already opened the debate on the 1985 omnibus farm bill. Secretary Block has stated that he believes farm policy is at a crossroad. He believes that the direction we set in the 1985 farm bill will largely determine the nature and scope of the U.S. agricultural system and its role in the world economy for years to come. If U.S. agricultural policy is at a crossroad, it is a choice between making further progress toward a market-based economy or reversing directions and returning to a rigidly controlled and heavily subsidized farm sector. I hope you will choose to be a part of that debate.

A LOOK AT THE FUTURE

Economists are projecting relatively stable demand conditions in agricultural markets for the remainder of the 1980s. At the same time, world production of food will continue to increase. After adjustment for inflation, the real prices of agricultural commodities probably will decline.

The producers who survive, and maybe even prosper, will be those who increase their productivity at a rate faster than the decline in real prices. The less-productive producers will not be able to stay in business. This is not a new phenomena. The economic pressures felt by these producers will create political pressure to provide price supports above market equilibrium levels or some other form of assistance. Of course, this is not a new phenomena either.

I expect in the future, however, that the political arena where agricultural policy is made will be less sympathetic to the cries of distress from the agricultural sector. The federal budget dilemma will be an important constraint when considering future agricultural programs.

It is possible that recent changes in the dairy and tobacco programs have created a precedent that may be applied to other commodities such as feed grains, cotton, and wheat. That precedent is to have producers finance their own programs, or, at least, to cover the losses so that the burden does not fall upon the taxpayer.

Two other new risk-sharing possibilities are also on the horizon--commodity-futures opinions and revenue insurance. Options markets will not do anything to alter fundamental supply and demand conditions, but they can provide producers with a mechanism to shift the risk of price movement. Both the USDA and the Congressional Budget Office are exploring the possibility of a revenue-insurance system that would give producers the opportunity to purchase a guarantee of a set revenue per acre of crop.

A CLOSING NOTE

Government intervention is usually justified when there is no other way to cope with a problem. The government is involved, and will continue to be involved, in agriculture, and all other sectors of our economy and society. But government involvement can be good or bad when viewed from the standpoint of agricultural producers. Much of this depends on how informed you are about the issues and how involved you, as producers, are in the process that shapes the laws and writes the regulations. If you do not take part in that process, the results may not be to your liking. Fortunately, you do have a voice, if you want to use it. But a lot of others have a voice as well.

It will be in your best interest to keep America's increasingly urban population informed of the importance of the agricultural sector and its problems. Overcoming the vast communication gap between rural and urban America will require your best effort and constant attention.

3

EFFECTIVE WAYS FOR LIVESTOCK PRODUCERS TO INFLUENCE THE POLICYMAKING PROCESS: PRACTICAL POLITICS

A. Barry Carr

This paper discusses the farm policymaking process; therefore this paper discusses politics. The purpose of this discussion is to help you understand how the agricultural policymaking process operates and how you might influence that process.

Farm policy is designed, enacted, and carried out in a political environment. The political arena is a place where conflicting viewpoints come together to be heard and to do battle. It is rarely a place where any one side achieves complete victory; most often it is a place of compromise. It is a place where those who will not bend will be broken and ineffective. It is a place where half a loaf is better than none. To repeat, the political arena is a place of conflict and compromise. There are a number of important political arenas for agricultural policy--the U.S. Congress, the U.S. Department of Agriculture, state legislatures, state agencies, and county governments. Each of these institutions, at the federal, state, or local levels, is designed to represent a collection of competing special interests.

AGRICULTURAL POLICYMAKERS

As little as 15 yr ago, the agricultural policymaking triangle was easy to describe. The triangle of power consisted of the Secretary of Agriculture, the powerful chairmen of the House and Senate Agriculture Committees, and the leadership of the major farm organizations. There was a sense of farm-sector control over farm policy. This is no longer true.

In the past several years we have had two changes in the chairmanship of the House Agriculture Committee and one change in the Senate committee. And although the House and Senate Agriculture Committees are the authorizing committees with primary jurisdiction over farm legislation, their power to shape farm policy has been reduced by recent changes in congressional procedures. Spending limits imposed by the new congressional budget process are doing as much to shape farm programs as are the needs and concerns of agricultural producers. In recent years, several important pieces of farm program legislation have been passed as part of the budget reconciliation process.

With each passing population census and the congressional redistricting process that follows, we have seen a steady decline in the number of congressional districts that are primarily agricultural. The current membership of the Congress reflects a predominately urban and suburban population whose concerns include lower food prices; diets that are nutritious, healthy, and easy to prepare; a clean and pleasant environment; and low taxes. These concerns can be translated into resistance to production controls; support of food labeling requirements; controls on agriculture to reduce soil erosion, noise, smells, and other disagreeable aspects of food production; and a reduction in outlays for commodity programs. We have even seen members representing urban districts or states seeking seats on the agriculture committees because of the growing importance of the food stamp and other feeding programs.

To the extent that political influence is related to numbers of voters and economic significance, the political clout of the farm sector is growing weaker. The numbers speak for themselves. There are only 2.4 million farms. The farm population is 5.8 million men, women, and children, only 2.6% of the total U.S. population. Only 100 counties out of 3,041 are totally rural, with no city over 2,500 residents. Income of the farm population is only 2.2% of the total national income. Even in rural counties, agricultural jobs account for only 23% of the total employment. Data show that agriculture is less important to the economic base of rural communities, states, and the entire nation than it has been in the past.

On the Executive side, we have seen farm policy taken out of the USDA and escalated to the White House level where the Secretary of Agriculture has only one vote and one voice, as do the Secretary of State, the Secretary of the Treasury, the Director of the Office of Management and Budget, and 10 other cabinet officers.

And finally, the farm organizations are no longer the only public voices heard in farm policy debates. Nonfarm special interest groups have become active in the agricultural policymaking process. Groups representing consumers, environmentalists, farm input suppliers, and food marketers are among the contenders for access and influence.

For example, consumer pressure recently caused the USDA to withdraw its proposal to revise the beef-grading system. Consumer pressure is an important force in the current attacks on marketing orders for milk and other commodities. A group of milk drinkers is even suing in court to have milk market order rules changed to facilitate the sale of reconstituted fluid milk produced from dried milk.

In most instances agribusiness interests are consistent with farm interests, but not always. Nonfarm business interests are an effective political force that is increasingly coming into conflict with farm interests. There is an inherent conflict between marketing or processing firms (such as meatpackers) that want low-priced farm products and producers who want higher prices for their crops, animals, and animal products. Food imports and the use of imitation food products are another source of friction. For example, access to lower-priced foreign supplies of sugar, tobacco, cheese, and meat is desired by the food industry; however, domestic producers of these commodities want price and market protection through import quotas and tariffs. The maritime industry wants guaranteed shipping rights for a portion of U.S. agricultural exports. Consequently, cargo preference rules are applied to shipments of donated U.S. foodstuffs and subsidized sales of U.S. farm products. Legislation is pending that would require at least 20% of all dry bulk U.S. exports and imports be carried in American vessels.

The general farm organizations and the more specific commodity groups are still important political forces. There are, however, serious divisions within the farm block--regionalism, conflicts among and within commodity groups, and an even greater range of differences between farmers and ranchers. When agricultural producers and their organizations cannot present a united front, their political effectiveness is diminished.

A case in point is the extended wrangling over ways to solve the dairy dilemma. Milk producers have been embroiled in a divisive internal conflict. With budget costs out of control and surplus production continuing, southern producers want to reduce the price support level; another group wants them left at $13.10, with an assessment placed against farmers to defray the cost of government purchases of dairy products; and a third segment wants a program to pay farmers to reduce production. While the USDA and members of the agriculture committees of Congress are anxious to move ahead, they are unable to do so without the support and agreement of the dairy industry. In the resulting confusion, no action was taken. I repeat, when farmers cannot present a united front, their political effectiveness is diminished.

FARM POLICY TIMETABLE

The next major battle in farm policy will be the writing of a new farm bill when the current 4-yr act expires in 1985. The opening skirmishes of that battle are under way. Secretary Block has stated that he believes farm policy is at a crossroads and that the direction taken in 1985 will determine agricultural policy for years to come. The Secretary has convened a summit conference of agricultural leaders to obtain their views. The Senate Committee on Agriculture, Nutrition, and Forestry has asked any and all interested parties to provide written comments on the farm bill that will be published as a committee print. The Joint Economic Committee has held several days of hearings in 1983 to obtain views on new farm legislation. But all of this is just the beginning.

The earnest efforts to shape a new farm bill will begin next year, after this fall's elections are safely out of the way. But the time for you to get involved is now. And you will have to stay involved if you expect to be effective. Now is the time to be thinking through and presenting program concepts, concepts--not nitty-gritty program details. Now is the time to be debating them in your farm organization at the local, and later at the state, levels. Now is the time to begin to develop consensus among yourselves--remember the value of a united front. Now is the time to begin building coalitions--alliances with other farm organizations, with other commodity groups, and with other political interests outside of agriculture. In doing this you will undoubtedly have to compromise some items on your agenda and set some priorities.

THE LEGISLATIVE PROCESS

Once the legislative process begins to deal with an issue, you will have to track that legislative action to be effective. Let me just briefly summarize the steps a bill normally takes in the process of becoming a law.

Bills are introduced by interested members of either the House or Senate. The bill is simply a draft of how that member thinks the law should be. Other members may decide at that time or at a later time to cosponsor a bill. Obviously, the more cosponsors a bill has, the more attention it attracts. After introduction, the bill will be referred to a committee for further consideration. In the case of agricultural legislation, the committee of referral will usually be the agriculture committee. If the committee decides to take action on the bill, it will usually schedule a public hearing where interested parties are invited to give testimony concerning the bill. At this time the witnesses may register support or opposition to the bill and may suggest changes in the bill to make it more acceptable. A committee may consider several related bills simultaneously in the same hearing. After the hearing process the committee may decide to proceed further with the bill. In that case it will "mark up" (rewrite the bill) and report it to the full House or Senate. At this point the bill is in the hands of the leadership of that body in terms of whether it will be "called up" (scheduled) for consideration by the full body. If the bill is considered on the floor of the full body, it may be amended and passed or defeated. If the bill is passed, it is sent to the other body where a similar course of action takes place. In the event that both houses take action on a bill, and in the unlikely event that the bill as passed by both houses has identical language, the bill is sent to the President for approval. In most cases there are differences in the language of the House and the Senate bills, and the measure is sent to a specially constituted committee (conference) of members of both houses to work out a compromise version that then must be accepted by both houses before being sent to the President. The President can sign a bill into law or can veto it.

The preceding steps may sound complicated to you, but they are actually a slightly simpler-than-real-life description of the course taken by most bills. I have provided some written material that gives a more detailed description of the process, and I would recommend that you spend some time reading it. The purpose of the above description is to show you all of the steps a bill goes through, because in each of these steps you have an opportunity to affect the legislation.

So, the first step is to know where the bill is in the process. Where can you obtain timely information on the status of a bill? Newspapers and television news are convenient, but they usually report events after they happen. Your farm and commodity organization newsletters often contain sections to alert you to upcoming legislative activity. The Congressional Record, often found in your local library, carries advance notice of committee hearings. If a particular piece of legislation is really important to you, you might ask the national staff of your organization or even your legislator's office to keep you notified.

PUTTING IN YOUR TWO CENTS

The second step is to register your position on the measure with your elected representatives. Obviously, if the bill is in the House, you contact your Representative; if it is in the Senate, you contact your Senators. If your own legislators are not on the agriculture committee of their body, you should contact the chairman of that committee and perhaps the chairman of the appropriate subcommittee of that committee. You can find the names of these individuals in the material I have distributed today. Here is where working through your farm and commodity organizations can pay off. When they speak to the committee, they speak with the collective voice of their membership.

After you have identified where a bill is in the process and are ready to express your position, how do you make the most effective contact? Letters have been the most common method of constituent communication; in fact, most Representatives get over 1,000 pieces of mail per day and the average Senator twice that number. If time is a factor, you might consider a telegram or a phone call. Many people do not know that their Representatives and Senators maintain offices in their home state with a full-time staff. A call or visit to these offices is a convenient and economical way to reach your legislator with your message. If you really want to become effective, become personally acquainted with your legislator's legislative assistant for agriculture. You can obtain this person's name by contacting the office, and it is possible to arrange a visit with that person in Washington or when he visits the home state. The legislative assistant is the most effective communication channel short of talking directly to the legislator. If your Senator or Representative is on the agriculture committee, he or she will also have staff working directly on the committee staff, and you should become acquainted with that person as well. Don't be bashful; these people are there to serve you and are, without exception, very constituent oriented.

When you make your contact with a congessional office, how should you structure your message? Use all of the rules of good common sense. Be friendly and considerate. Be informed about the issue; know the bill; know your position. Be specific and be brief. Remember that the measure under consideration must deal with a variety of conflicting viewpoints, so explain why you are opposed to particular provisions and suggest alternatives that would be acceptable to you. It is important to register your support of a measure as well as your opposition. Follow up with another contact at the next stage of action on the bill.

One additional point--most of what has just been said about the federal legislative process and how you participate in it is as applicable at the state level. Only the names and addresses are different.

Why is all of this necessary? Why is the government in your business in the first place? Why do you have to worry about it? Government intervention takes place when there appears to be no other way to solve a problem. Believe it or not, the government is involved in your business because somebody asked it to be. And if you do not take part in the process, the results may not be to your liking. Fortunately, you have a voice, if you want to use it.

Generally farmers and ranchers want a voice in decisions that affect them. But remember, there is a lot of difference between having a voice and having your way. Others, who are also affected by government policies, are exercising their voices as well. The answer lies in being heard. Are you speaking at the right time and to the right people? Are you stating your position clearly and intelligently? Are you offering fair alternatives and a willingness to compromise? Are you a good loser and a gracious winner? And most of all, are you willing to keep trying?

This paper represents the view of the author and should not be attributed to The Congressional Research Service, Library of Congress.

4

FACING THE FACTS IN GETTING STARTED AS A LIVESTOCK PRODUCER

Dixon D. Hubbard

The existence of humanity is unconditionally tied to the existence of green plants. Only green plants are equipped to use solar energy, carbon dioxide from the air, and nutrients from the soil in growth and reproduction processes. Directly or indirectly, these processes create our food supply, and there is no other way. Not only are these processes the basic source of man's food, but this process also yields the oxygen that man and other animals must breathe to sustain life.

American agriculture is the world's largest commercial industry. Its present assets approach $1 trillion, which is equal to 90% of the total assets of all manufacturing corporations in this country. It represents about one-fifth of our gross national product.

Agriculture accounts for the employment of more than 23 million people--22% of America's labor force. Approximately 15 million people work in some phase of agriculture--the growing, storing, transporting, processing, merchandising, and marketing of all farm commodities.

Productivity of farmers has outstripped all other segments of U.S. society. An hour of farm labor in 1981 produced 14 times as much food as it did only 60 yr ago. Thus, food in this country is modestly priced compared with the rest of the world, and the 16.5% of our income that we spend for food is the lowest of any country.

American farmers are also one of our nation's largest consumers. In 1981, farmers spent $141.6 billion on production goods and services. For example, they spent $14 billion for farm machinery, farm tractors, trucks, and other vehicles. They spent another $13 billion for fuel, lubricants, and maintenance of vehicles and equipment. They used 6.5 million tons of steel and approximately 33 billion kwh of electricity. In addition, they paid $8 billion in taxes, and farm exports exceeded imports by $28 billion--a major contribution to our nation's balance of payments.

Being a livestock producer, therefore, means being a part of the largest commercial industry in the world and providing mankind with vital services at modest prices. Simultaneously, it means being part of the most productive and one of the largest consuming segments of U.S. society, accounting for 22% of employment for the people in this country.

These statistics make producing livestock and being part of the agriculture industry sound very enticing. However, income from agriculture, including livestock production, has been very erratic in my lifetime, and the financial condition of most farmers and ranchers has been especially volatile over the past decade. This culminated in the farm protests in the winters of 1978 and 1979, when farmers drove their tractors to the nation's capitol and forecast bankruptcy for one-quarter of American farmers if steps were not taken to improve their prices and incomes.

We still hear a lot about "the" farm problem today. In reality, the problem is low income, which is basically the problem that has periodically confronted farmers and ranchers. Presently, this problem has four distinct aspects, each requiring different policy measures for relief: 1) deflated farm assets and the high interest rates that caused the deflation, 2) the depressed state of the whole U.S. economy, 3) domestic farm programs that are only partially effective, and 4) foreign restrictions on U.S. farm exports.

One or more aspects of the present problem have been part of every previous farm problem. This problem and the continuing discussion of the poor outlook for farming often cause young people to elect not to enter farming as a profession. In fact, parents often discourage their sons and daughters from becoming farmers. I will not attempt to provide solutions to the various aspects of the farm problem. This will have to be done by someone with greater wisdom than I. All I want to do is point out that periods of low income have always plagued farmers and ranchers--a major reason why they keep decreasing in numbers year after year. It appears the older farmers are the ones leaving the profession, because in the 10 yr from 1969 to 1978, the average age of farm operators decreased from 51.2 yr to 50.1 yr.

It's been my life-long ambition to produce livestock, and I do own cattle and land. However, I have paid for most of my livestock operation with income from other sources. Based on my experience, anyone who does not have a massive bank account, or the backing of someone who does, needs to study this lesson well before entering the livestock business.

My son was interested in farming, and I have been helping him get established in farming and beef production over the past decade. The cost has been equal to his obtaining several college degrees. The point is that my son and I are average or above when it comes to know-how and application of technology. However, if we had not been well-financed, expected some of the difficulties that we have had, and been solidly committed to sticking with our objective, we would have long since given up. Using the very best outlook information available, we were unable to predict what has happened to us in the way of increased cost of production coupled with the low prices we have received for our product. As an extension animal science specialist for the past 20 yr, I have not been able to recommend that other people go into beef cow-calf business, even in the best of times. This was particularly true if they had to borrow over 50% of the capital and pay interest and principal and take a living out of the business. Thus, I should have known better than to recommend it to my own son.

Based on the observation I have made during my lifetime, going into beef cow-calf production can be equated more to a disease for which a vaccine has not yet been developed than to a business. This is not to say that those who are already established in beef cow-calf production do not live a good life, even if they sell their labor cheap and obtain low returns on their investment. Land inflation has been good to them by increasing their net worth. If they do not over-leverage themselves to where they have to make major principal payments, they can do quite well. However, for those trying to establish themselves in this business, it is a different story.

There are exceptions to this rule. I know people who have succeeded in establishing themselves in agriculture and the livestock business. In most cases these people have good business savvy and have applied the principles of good business management. However, for the most part their success has been more a function of timing than of knowledge and skill. If someone starts at the right time, there is not much that can keep them from temporary success in the livestock business. If they apply good business management principles, this success can be extended into a lifetime. However, there are peaks and valleys in the livestock business, and I expect there always will be. It is not hard to ride the peaks, but the valleys can be devastating for anyone unprepared to deal with them. I have always said that the day a person is born is more of a factor in his succeeding in agriculture than of intelligence, because timing is a major factor in determining when he decides to go into agriculture. I do not think I can overemphasize the correlation between timing and success for anyone thinking about getting started in the livestock business. The only problem is that one seldom realizes whether they started at the right time until after they either succeed or fail. Probably a good rule of thumb for making a decision about timing is one that my grandfather used. He always said that when everyone else was trying to get out of the business was the time for those wanting to get started to be getting in, and when everyone else was getting in, that was the time for those trying to get started to stay out.

An important element in the future of animal agriculture is the introduction of new personnel in the form of young farmers. Without outside help, young people have great difficulty in entering farming as a profession. Potential sources of help are: 1) family members; 2) other individuals such as other farmers; 3) industry organizations; 4) financial institutions such as banks and lending agencies such as FmHA (Farmers Home Administration); and 5) international, philanthropic organizations that provide animals and other input to support limited resource farmers in strengthening their operations. An example of the latter is the work of Heifer Project International.

Examples of help from individuals are employee-incentive programs--such as feedlot operators, ranches, or pork producers may have--that permit employees to own groups of animals in the units in which they are employed. In certain geographic areas, there are cases where a single successful feeder has created an industry in the area by providing partnership feeding opportunities to employees or former employees. In the case of industry organizations, some can and have assisted aspiring young farmers to be placed in a protege relationship with successful producers who can help them get started. Some rural banks have a very positive program of attempting to go the extra mile to assist young farmers to get started and stay on the land. In fact, the founder of the Bank of America gave special attention to farmers and their problems during the years of the Great Depression. The loan program of FmHA for limited resource farmers has helped young farmers to get started. In the past, certain nonprofit groups have helped young people get started with animal projects in 4-H club or FFA projects. Later these youngsters became farmers. There should be other ways. I am thinking of something such as animal lending rings in which a young farmer might be loaned cows, sows, or ewes for 2 or 3 yr. At the end of the period, the animals or replacements of them would be returned to the nonprofit sponsoring group.

An observation I have made relative to farmers and ranchers is that most of us claim our primary purpose for being in business is profit. However, when the profit motive interferes with our independence, we usually sacrifice profit. American farmers and ranchers are among the most independent people in the world, and our actions demonstrate this independence is our number one priority. I'm not critical of this attitude, because it is what makes our country great. However, there is always a price to be paid for independence. Farmers and ranchers have paid a rather high price at times. Our inability to regulate production the way strictly profit-oriented industries do has been devastating on our prices and profits at times. The adoption of new technology is lower in the livestock business than in industries where the driving force is strictly profit. This has reduced our competitive advantage, since speed of adoption of new technology is the key to being competitive in any business.

The clothes I am wearing are indicative of this independent attitude. I have worn this type of clothing all my life, and I do not intend to change. I want to identify with my heritage. When I work at home, I get up in the morning and put on levis, boots, hat, and other appropriate attire.

This clothing brings with it an early 20th Century mentality that comes naturally to those who grew up and were part of the farming and ranching environment in which I was raised. This mentality could best be described as a cross between Puritan work ethic and frontiersmanship. It says with hard work, true grit and a little luck, everything will work out okay. This approach has served my family fairly well, and it has been fairly good for most farmers and ranchers in my lifetime. Also, this type of aggressiveness will be required for those who plan to survive in farming and ranching in the future. However, it will not be enough. Profitable farming and ranching will require incorporation of concepts like management by objectives, of strategic planning, and of key result areas. Can you imagine John Wayne tying up his horse and saying "Got to go up to the house and recap some cash flow projections." It surely does not fit the image I have always had of farming and ranching. However, with present interest rates, and input costs, and all the other changes taking place, it is going to require using every sound business tool available to survive in the farming and ranching business in the future.

Farmers and ranchers tend to equate activity with results--work hard and you will succeed. Hard work is necessary, but it is not activity that measures success. Results are the true test of success. Planning is a process that will focus your operation away from activity and toward desired results. Management by objective can not be separated from the planning process.

The point I am making is this--the attitude of an individual starting in the livestock business and the image they have of themselves in this business both have a great deal to do with success. Exceptions to this rule are those with an MBA and those who can make sufficient outside income to be able to do their thing independent of the profitability of their livestock operation. However, those starting out in the livestock business who must make a living from their livestock better have their head screwed on right as to their attitude and image.

It would be an understatement to say that if I knew then what I know now, I would have done a lot of things differently when I first started in the livestock business. By the way, I got caught in one of the most severe droughts ever recorded in western Oklahoma. Also, this drought occurred when cattle cycle numbers peaked in the 1950s. To say the least, I sold some cattle very cheap after I had the pleasure of feeding them for a year. As you may have guessed, I went broke. I learned a lot and have since been able to avoid many of the errors I made in my first endeavor in the livestock business.

When my son decided to go into the livestock business in the 1970s, I thought I would help him capitalize on a producer plight similar to what I had experienced in the 1950s. Cattlemen were once again confronted with drought and peak numbers of cattle, so I encouraged him to get into the business while cattle were cheap (Grandfather's advice) Little did I know what was going to happen but I can assure you it has been tough. He will make it, but only because I am able to help him. I could provide similar examples of people who have started in sheep, swine, and veal production. Most people who start in the livestock business will experience some major problems and hardships. It has been my experience that the ones who have succeeded are people who were solidly committed self-starters who were willing to work hard to accomplish their objectives, have studied their lesson before starting, were able to start when the timing was right, and have applied sound business and scientific principles in both production and marketing. Even though these people experience problems and hardships, most of them are able to stay in the livestock business, enjoy what they do, and make a living.

I have advised lots of people who were planning to go into the livestock business. Also, I get a number of letters and phone calls from people wanting information on how to get started in farming and ranching. Seldom do I feel that any of these people have much of a chance of success. They give little indication of an unbiased evaluation of available information on the type of livestock business they plan to enter They generally have visions of grandeur and a much higher regard for their knowledge and experience than is justified. Also, their opinion of life on the farm and(or) ranch is far different from anything I have experienced or observed of people who preceded them in the same type of endeavor.

My advice to most people planning to start in the livestock business is that before they commit themselves to a major action such as buying land and facilities, they should first consult a number of knowledgeable people who can provide them with sound advice. These include Extension agents and specialists in the county and state where they plan to start their livestock operation, producers who are willing to show them their records, and lenders with money loaned to people trying to make a living in the same kind of livestock business they are planning to go into. In addition, people interested in starting in the livestock business should answer the following questions to develop a more realistic view of what they are planning to do. If these questions are answered honestly and used as a guide, they provide a useful tool. Anyone who can answer "yes" to most of these questions will have a fairly good chance of success in the livestock business if the timing is right when they start.

BASIC DECISION QUESTIONS

1. Have you decided whether you want to produce livestock on a full-time or part-time basis?

2. Do you plan to get started in the livestock business with an established producer either as a partnership in a closely-held corporation, or if you have no prior experience, as a ranchhand or manager trainee?

Financial Questions

3. Have you chosen a farm or ranch location suited to your family, found out the rental charge per acre, or how much land costs if you want to buy it?
4. If you want to make livestock production a full-time occupation, can you get from $300,000 to $400,000 in loans and other assets--or, if you want to produce livestock on a part-time basis, can you raise $150,000 or more, or raise somewhat less money and start off more slowly and work your way up?
5. If you have a farm or ranch picked out, do you know how much property taxes you would have to pay, the overall cost of local government, and if local government is planning any new project which might increase your real estate taxes?
6. If you plan to be living on nonfarm income, will it support you in case you make no net profits from farming and ranching, and can you keep living on nonfarm income indefinitely if necessary?
7. If you do not want to farm or ranch full-time, are you willing to take an extra part-time job at prevailing rural pay rates as a way to make ends meet, assuming you do not have a steady income from other sources?
8. Assuming a "normal" farming or ranching year in the area where you plan to start, have you made out a thorough and honest budget of your expected sales, farming expenses, and net farm income, and found that you can make out to your satisfaction on farm or ranch income and income from other sources?
9. Have you looked ahead to when you might want to or have to leave farming or ranching, and have you thought about related tax matters, such as capital gains and(or) inheritance taxes?

Personal/Management Questions

10. Are you flexible and "tough" enough so you do not mind taking risks with your own money?

11. Are you a "self-starter", and can you plan and do your own work on schedule whether you want to or not?
12. Are the other members of your family interested in working together at chores or special projects and can you effectively manage their work?
13. Do you like to work with your hands and do not mind physical work outdoors in all kinds of weather?
14. Are you looking forward to farming and ranching to get away from indoor confinement, busy offices and crowds, and city noise and smog--in exchange for other kinds of problems such as greater isolation, limited flexibility to go places, and greater distances from facilities?
15. Do you realize that despite its placid image, farming and ranching is very stressful because farmers and ranchers are self-employed and under constant pressure to keep up with the work and cope with constant changes in weather, market conditions, technology, and uncertainty of income?
16. Have you thought about health care, and do you know how far you would be from the nearest doctor, medical specialist, dentist, ambulance service, and hospital?
17. Are you ready for the social life of country living, which includes substantially fewer and different recreational and social events than the city?
18. Would you enjoy getting involved in rural community activities and do you know about your prospective locality's civic organizations religious groups service clubs, extension clubs, or the groups you might find interesting?
19. Do you have mechanical ability and like doing odd jobs around your home, such as fixing faucets or broken water pipes, doing carpentry work, painting, replacing rusted-out gutters, or laying or repairing concrete?
20. Have you considered that producing livestock requires that someone be around the farm or ranch to do chores basically 7 days a week, month after month, and that it usually is not easy to find someone to relieve you on short notice?
21. Have you had what you feel is enough experience on a farm or ranch similar to the type you plan to operate to ensure that all the little things necessary to success get done?

22. Are you aware of the manure disposal and run-off problems in the areas where you plan to start producing livestock?
23. Would you be able to handle the stress of rebuilding your herd or flock if it is "wiped out" by disease?
24. Do you know how to obtain technical advice from specialists such as county agents, nutritionists, entomologists, soil scientists, geneticists, conservationists, engineers, and others?
25. Would you enjoy shopping around to get the best price and make the best deal for feed, equipment, fuel, seed, and fertilizer?
26. Do you know about market reports and would you enjoy searching out the best markets for your livestock?
27. Do you know the meaning of these initials: USDA (U.S. Department of Agriculture), CES (Cooperative Extension Service), ASCS (Agricultural Stabilization and Conservation Service), CCC (Commodity Credit Corporation), FCIC (Federal Crop Insurance Corporation), ACS (Agricultural Cooperative Service), ERS (Economic Research Service), SRS (Statistical Reporting Service), SCS (Soil Conservation Service), REA (Rural Electrification Administration), FmHA (Farmers Home Administration), SBA (Small Business Administration), PCA (Production Credit Administration), FLB (Federal Land Bank)? And do you know what each might have to do with your selected or potential farm or ranching operation?
28. If you need help on your farm, do you think you can hire farm workers and do you know whether custom hiring services are available?
29. If you need to hire labor, are you familiar with state and federal laws concerning the safety and well-being of your employees?
30. Do you know that a farm or ranch employer can be held liable for negligent acts of his or her employees and that you are liable for your own negligence regarding safety and health of your employees?
31. Are you aware of the local and state fencing laws in the area where you plan to produce livestock, and the cost of fence of the type your neighbors may insist upon to protect them from damage by your livestock?
32. Are you familiar with local and state laws in the area where you plan to produce livestock relative to abatement, prevention, and policing of air and water pollution and do

you know that court injunctions and dairy fines may be imposed on farmers and ranchers who violate court orders against pollution?

This is not an all-inclusive list of questions. However, it gives people thinking about starting in the livestock business a base for understanding that they can not just go out and crank-off without proper planning and evaluation and expect everything to work out according to some mental image they may have.

Hopefully, this paper will help people who are considering going into the livestock business to properly analyze all the facts before they start. The land is going to be farmed and livestock is going to be produced in this country to provide our people with food. However, those who are going to do the farming and ranching and produce the food are, for the most part, going to be knowledgeable of the facts that govern success. Within this context they will use good judgment and apply good business and scientific principles in their farming and ranching operations.

5
JOBS IN THE HORSE INDUSTRY

Cecile K. Hetzel

The horse population in the U.S. increased with the country's growth until it peaked in 1915 with a total number of 25,199,552 horses and mules recorded on farms and ranches and an additional 2,000,000 head in cities (Ensminger, 1977). Today we find some confusion in getting an accurate count of the U.S. horse population because the Census Bureau data on horses have always been limited to those on farms and ranches, with no count taken of horses owned and kept by suburbanites. However, it seems clear that the distribution of our horse population has taken a shift from farm to town.

In 1976, Lloyd Davis, a retired USDA extension director, estimated that the number of horses in the U.S. ranged from 8.5 million to 10 million head (Bradley, 1981). Other recent estimates show the number to be 10 million to 12 million. We seem to be in the middle of a really big horse "boom." People have more leisure time, more and more people reside in suburban and rural areas and, most important, today's population is deeply concerned with physical fitness and sports. Thus, the horse population should continue to grow to accommodate the supply and demand, with a predicted 14 million by the year 2000. Factors that can affect that growth include the nation's economy, local zoning regulations, and certainly the promotional efforts of horsemen and their organizations.

In 1980, it was reported that 64,499 horses participated in thoroughbred races. Other types of breed racing (Quarter Horses, Appaloosa, and Arabians) included over 22,000 horses (Wood, 1982).

The American Horse Shows Association reported 36,700 horse shows conducted in the U.S. in 1979 with a paid attendance of 24.5 million. Remember, this is only an account of paid spectators; many shows do not charge spectators an admission (Wood, 1982).

Registered horses in the country continue to grow in numbers. Since its beginning, the Jockey Club has registered approximately 600,000 thoroughbred horses and about 1.8 million Quarter Horses have been registered since the beginning of the association in 1940. The Appaloosa horse registry totaled 310,000 and the Arabian 202,743 (Wood, 1982).

Some of the leading states in terms of horse numbers are California, Texas, Oklahoma, Illinois, Ohio, Missouri, Tennessee, Michigan, Montana, Kentucky, Minnesota, and New York (Figure 1).

It is estimated that the worth of the horse business to the U.S. economy is somewhere in the area of 15 billion dollars, which is certainly big business.

Employment statistics from 1960 to 1975 showed nearly a quarter of a million full-time job equivalents in the industry. For every 35 horses, there was one full-time position. Today this figure is considerably higher; at the end of 1977, there were nearly one-third million full-time job equivalents in the U.S., again showing a full-time job equivalent for every 35 horses. As the number of horses increases, so does the number of jobs.

So many people today fail to realize that success in the horse industry is a constant struggle and nothing short of hard work gets you a job. It seems to be an industry that requires one to start at the bottom and work up.

Many colleges and universities are adding equestrian programs to their curriculum. There are more horse-related positions in research and teaching throughout the country. Educational requirements for these positions are demanding a doctorate or at least a master's degree with experience.

The most logical places for a job in the horse industry are in the area where the equine population is greatest (figure 1).

To locate a job, you can utilize the professional organizations and breed associations that have a job referral department. You can look in the breed publications and all-breed horse magazines for positions available. Word of mouth is one of the best ways to locate a position.

The same job in California might pay five times as much as that job in Missouri. Jobs in densely populated areas tend to pay more.

To get any job in the horse industry, it is of the greatest importance that we evaluate ourselves and know for sure what our individual goals will be. Honest answers to the questions listed here will give you a good insight into what is important in the kind of work that you select.

- List things that you do well.
- List things that you know you do not do well.
- Do you express yourself well in speaking to someone?
- Do you express yourself well when you write?
- Would you consider yourself a leader?
- Do you work well under pressure?
- Do you work well alone?
- Do you like supervision?
- Do you seek responsibility?

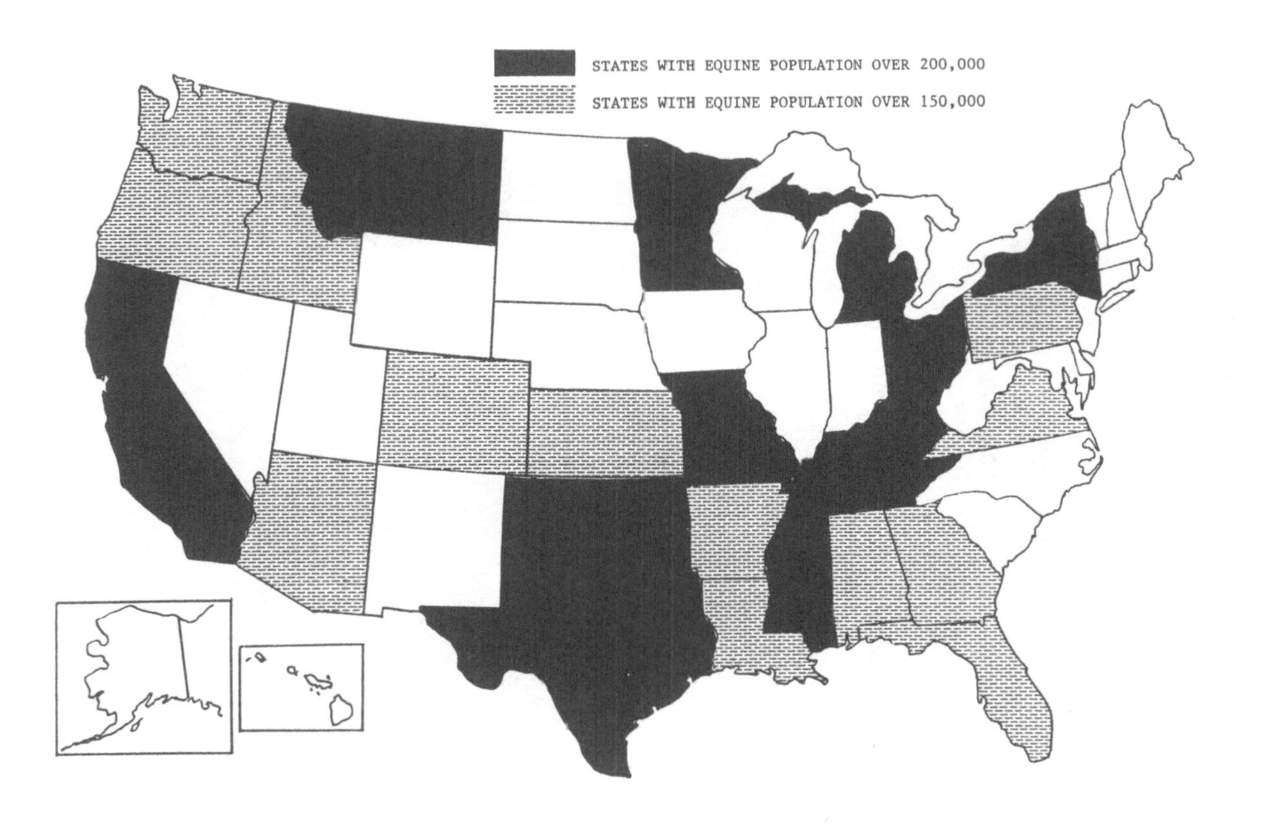
STATES WITH EQUINE POPULATION OVER 200,000
STATES WITH EQUINE POPULATION OVER 150,000

Figure 1.

- Will you follow directions?
- Do you like new ideas?
- Which is more important to you: 1) regular salary 2) profit sharing 3) commissions 4) commissions and salary?
- Do you prefer working 1) 9 to 5, 2) odd hours, 3) week-ends?
- Would you object to traveling half of your working time?
- What is important to you: 1) living in an urban populated area (one million or more) 2) living in an area that is 100,000 to 800,000, 20,000, 70,000 3) rural population under 15,000 4) working indoors; working outdoors, combination?
- Do you like fringe benefits and would you work for less money for good benefits?
- Would you move anywhere for a job?
- Would you prefer to work for a large company or a small company?

If you answered these questions honestly then you should be able to tell what is important to you in any kind of work that you might consider doing.

JOBS IN HORSE INDUSTRY

A "Guideline for the National Horse and Pony Youth Activities Council: Employment and Educational Opportunities for Youth in the Horse Industry" provides a list of over 100 jobs (Appendix) with a brief explanation of those positions.

We will review in depth several jobs from this list and evaluate them as to the geographic area where the job is more prevalent. Skills necessary to obtain the job will be discussed and the educational level needed to secure the position. In addition, the pay scale will be concluded as to low, average, or high.

(I have a videotape showing my interviewing of several people working in the horse industry. I ask them to answer questions that will cover how they became a part of the horse industry, what background they had before getting their position, what areas of the country they feel are best for their particular job, and finally, how satisfied they are with their pay scale.)

TITLE DESCRIPTION CODE

An explanation of the five-digit number preceding the job title:

FIRST DIGIT

1. Most current holders of jobs with this title have a high-school education or less. Work is relatively unskilled.
2. Most current holders of jobs with this title have a high-school education or less. Work is semi-skilled.
3. Most current holders of jobs with this title have a high-school education. Work is skilled.
4. Most current holders of jobs with this title have at least 1 yr of education beyond high school.
5. Most current holders of jobs with this title have at least 2 yr of education beyond high school.
6. Most current holders of jobs with this title have at least a baccalaureate degree from a recognized college or university.
7. Most current holders of jobs with this title have at least a baccalaureate degree from a recognized college or university and substantial additional training and(or) experience.
8. Most current holders of jobs with this title have at least a master's degree from a recognized college or university.
9. Most current holders of jobs with this title have either the Ph.D. degree or a professional degree from a recognized college or university.

This numbering system is employed only to indicate the general level of competence required for the various positions. It should not be interpreted to mean that one cannot obtain the position or serve in such a position well unless he has the indicated training or degree. Many positions are filled satisfactorily by those with less than the indicated training and education.

SECOND DIGIT

1. Administrative responsibilities
2. Government control and regulation
3. Training horses
4. Training personnel
5. Design and(or) manufacturing
6. Research and(or) laboratory work primarily
7. Sales
8. Services

THIRD DIGIT

1. Most jobs are in racing only.
2. Most jobs involve horse shows only.
3. Most jobs involve pleasure-horse activities only.
4. Most jobs involve racing and horse shows only.
5. Most jobs involve racing and pleasure-horse activities only.
6. Most jobs involve horse shows and pleasure-horse activities only.
7. Jobs in this classification involve all three branches of the horse industry: racing, horse showing, and pleasure horses.

FOURTH AND FIFTH DIGITS

These digits constitute a kind of serial number to make easy references to the particular job title. No two job titles have the same five-digit number. No attempt has been made to arrange the job titles bearing the same first three digits in any particular order. One should not assume, therefore, that job title number xxx32, for instance, requires more background or experience, or is more difficult to perform than job title xxx01.

REFERENCES

Bradley, M. 1981. Horses. McGraw Hill Book Co., New York.

Ensminger, M. E. 1977. Encyclopedia of Horses. A. S. Barnes and Co., Cranbury, NJ.

Wood, K. 1982. 1982 Tax Handbook for Horsemen. Wood Publications. Aurora Grande, CA.

APPENDIX

Employment and Educational Opportunities

NHPYAC Committee on Education and Employment

CODE	JOB TITLE	BRIEF EXPLANATION
13101	Hot-walker	Cools horses after racing or exercising.
13401	Exercise rider	Rides race horses during morning owrkouts.
17601	Clothing-store clerk	Sells riding habits and accessories.
17701	Tack-store clerk	Sells tack and saddlery.
18101	Silks-room attendant	Sees that jockeys receive shirts and caps of proper colors.
18102	Jockeys' Room attendant	Maintains proper conditions in jockeys' room.
18103	Pony-boy (girl)	Leads horses from paddock to starting gate.
18201	Jump-crewman	Sets up and removes jumps in horse-show arena.
18202	Rodeo laborer	Performs various jobs at rodeos, usually with rodeo stock.
18301	Trail-crewman	Builds and maintains riding trails.
18401	Gateman	Takes tickets and/or guards gates.
18402	Ticket-seller	Sells admissions.
18403	Bus-boy	Clears tables in restaurants.
18404	Dishwasher	Washes dishes.
18701	Parking attendnat	Aids in parking automobiles.
18702	Rest-room attendant	Keeps rest-rooms clean.
18703	Water-truck operator	Drives truck to water track or arena.
18704	Tractor operator	Drives tractor (to pull starting gate or trailer carrying jumps, etc.)

CODE	JOB TITLE	BRIEF EXPLANATION
18705	Grounds maintenance	Maintains grounds, often making minor repairs.
18706	Office clerk	Performs miscellaneous office jobs, such as mailing, filing, etc.
-----	-----	-----
21101	Paddock judge	Supervises activities in paddock before race.
21102	Patrol judge	Watches race from tower to detect fouls.
21103	Starter	Supervises loading of horses into starting gate and then starts race by opening gate stalls.
21104	Clocker	Times horses during morning work-outs.
21105	Timer	Times races manually, with stop-watch.
21701	Ground manager	Supervises maintenance of grounds.
27701	Advertising salesman	Sells advertising for program, magazines, etc.
27702	Tack salesman--wholesale	Sells tack to retailers and often also to professional horsemen.
27703	Tack salesman--wholeslae	Sells habits and other horse-related clothing to retailers.
27704	Feed salesman	Sells feed at track, horse shows, or feed store.
28101	Condition-book technician	Aids racing secretary in producing condition book.
28102	Photo-finish camera operator	Operates photo-finish camers.
28103	Film- or TV-patrol operator	Operates movie or TV camera to film races, so stewards can review action during race, in case of claim of foul, etc.
28104	Horse-identifier	Identifies horse in paddock before race or horse-show class, to be sure horse is actually the horse that is entered.

CODE	JOB TITLE	BRIEF EXPLANATION
28201	Premium-list technician	Aids horse-show secretary in preparing premium list.
28202	Rodeo cowboy	Usually a professional--competes in rodeo events or performs under contract (clowns, etc.)
28203	Rodeo pick-up man	Helps bareback and saddle-bronc riders dismount after ride. Herds stock and sometimes ropes horses or bulls.
28301	Packer/guide	Leads pack parties into recreational areas, usually mountains.
28401	Reservations clerk	Fills mail and telephone requests for box and/or seat reservations.
28402	Announcer	Announces races or horse shows, on grounds or on TV or radio.
28403	Program manager	Supervises preparation of printed program.
28404	Crewman--props	Takes charge of seeing the proper properties are in the right place at the right time.
28701	Guard	Patrols grounds and protects other personnel.
28702	Film-processing specialist	Processes film of race, horse show, or other sporting event.
28703	Painter	Paints buildings, etc.
28704	Landscaper	Plants and maintains flowers, etc., under direction of landscape architect.
2705	Gardener	Waters, prunes, and tends plants and grass.
28706	Farrier	Shoes horses.
28707	Sign-maker	Provides signs as needed.
28708	Groom	Cares for one or more horses, under direction of owner or trainer. Occasionally supervises hot-walkers.

CODE	JOB TITLE	BRIEF EXPLANATION
28709	Identification-file clerk	Maintains files of horses at track or on show grounds.
28710	Auction clerk	Maintains records of price, buyer, and horse at auction.
28711	Mail clerk	Collects and distributes mail, and often sends appropriate printed replies to requests for information.
*28712	Horse dentist	Floats teeth--general care of a horse's teeth.
31101	Trackman	Leads post parade and has other administrative responsibilities.
31201	Horse-show secretary	Has charge of entries, stall assignments, and often other activities.
31202	Rodeo secretary	Maintains records of rodeo entries, prize money won, and points won.
31203	Rodeo stock contractor	Furnishes stock for rodeos and usually supervises all arena activities.
31701	Director of parking	Supervises parking, preparation of parking lots, collections of fees, etc.
32701	Office personnel--regulatory	Handles filing, correspondence, etc., for government agencies regulating horse-oriented activities.
33701	Trainer	Trains horses and supervises grooms and other subordinate personnel.
33702	Stable manager	Handles stabling, feeding, and other physical necessities of horses.
34601	Riding instructor	Teaches equitation.
37701	Concession operator	Contracts for and operates food or novelty concessions.
37702	Feed-store operator	Operates retail feed store.
37703	Publications distributor	Distributes magazines and books to retail dealers.
38101	Jockey	Rides horses in races.

CODE	JOB TITLE	BRIEF EXPLANATION
38102	Driver	Drives horses in harness races.
38103	Jockey's agent	On behalf of jockey, contracts with trainers for mounts in races.
38104	Mutuel clerk	Sells and/or cashes mutuel tickets.
38105	Editor--scratch sheets	Determines likely winners, and prepares scratch sheets for sale at track and elsewhere.
38106	Clocker--scratch sheets	Clocks morning workouts and prepares information for scratch-sheet editor.
38201	Horse-show judge	Judges various horse-show classes.
38202	Horse-show steward	Represents American Horse Shows Assn. or other organization at show; reponsible for seeing that rules are enforced.
38203	Ringmaster	Lines up horses for judging in horse-show ring.
38204	Rodeo clown	Protects bull-riders by "fighting" bulls or working with barrels. Amuses audience.
*38205	Horse-show organist	Provides music for hose shows.
38401	Waiter	Waits table in track or horse-show restaurant.
38601	Saddlemaker--leather craftsman	Makes and repairs saddles, tack, and other leather goods.
38602	Harness-maker	Makes harnesses and allied equipment; repairs leather tack.
38701	Fireman	Fights fires and aids in fire-prevention activities.
38702	Printer	Prints programs, condition books, premium lists and other similar items.
38703	Office personnel	Handles more complicated details of office work.
38704	Carpenter	Performs carpentry work and often designs wooden equipment.

CODE	JOB TITLE	BRIEF EXPLANATION
38705	Leather dealer	Buys and sells leather for use in making tack. Often also supervises tanning and other preparatory processes.
38706	Veterinarian's Asst.	Helps veterinarian.
38707	Bookkeeper	Keeps track of accounts receivable and payable.
38708	Breeder	Owns or manages stallions and broodmares.
38709	Auctioneer	Conducts auctions.
38710	Horse-buyer	Buys horses for clients.
38711	Farrier--corrective shoeing	In cooperation with veterinarian, shoes horses to correct defects of hooves and lower legs.
38712	Publications production specialist	Makes up dummy for magazines and/or books. Determines placement of advertisements.
38713	Typesetter	Sets types for horse books, magazines, etc.
38714	Photographer	Takes photographs for sale or publication.
38715	Binder	Binds books, magazines, programs, etc.
38716	Identification maker	Marks horses with freeze brands or lip tattoo.
41401	Director of advertising	Selects media and supervises copywriters.
46701	Laboratory technicial	Performs various laboratory tests.
47401	Advertising copywriter	Prepares copy for advertisements.
47402	Commercial artist	Prepares artwork for advertisements.
47701	Advertising sales manager.	Supervises advertising salesmen.
47702	Publication circulation manager	Supervises subscription and newsstand sales effort.
48401	Technical-equipment installer.	Installs technical equipment.

CODE	JOB TITLE	BRIEF EXPLANATION
48701	Fire-prevention spec.	Makes surveys to find fire dangers.
48702	First-aid personnel	Administers first aid, often under direction of a registered nurse.
48703	Program coordinator	Has charge of printed program and often of sequence of arena events.
48074	Veterinary technican	Aids veterinarian in technical aspects of his work.
48705	Reporter	Covers equine events for magazines and/or newspapers.
48706	Motion-picture/television writer	Prepares script for movie or television.
48707	Motion-picture cameraman	Operates movie or television cameras.
48708	Transportation specialist	Arranges transportation for horses and/or personnel, usually by air, sometimes by van.
*51200	Horse-show secretary	Supervises horse show offices. Handles entries and sends in reports.
51201	Horse-show manager	Manages horse show and supervises other personnel.
51401	Fair or exposition manager	Manages fairs and/or expositions.
2201	Soring inspector	At horse shows, inspects horses to determine if they have been illegally sored.
52701	Drugging inspector	Determines if horses have been illegally drugged.
54701	Technical-school teacher	Teaches courses in technical school.
55701	Clothing designer	Designs riding habits and other clothing and accessories.
56701	Film editor	Edits motion-picture film.
56702	Blood-typing specialist	Aids horse-identification by typing blood.

CODE	JOB TITLE	BRIEF EXPLANATION
58101	Turf-club director	Supervises operation of turf club.
58401	Maitre d'hotel	Supervises dining services of turf club or clubhouse restaurants.
58402	Technical representative	Solves technical problems arising in equipment installed or used at track or show.
58701	Nurse	Supervises health services.
58702	Public - address system specialist	Installs and/or maintains public-address system.
58703	Artificial inseminator	Under direction of veterinarian, performs artificial insemination operations.
58704	Film-distribution specialist	Arranges for distribution of horse-oriented film or video tapes to theatres or television stations.
*58705	Computer programmer	Operates and programs computer.
61101	Racing steward	Controls racing activities at the track.
61102	Racing secretary	Arranges races and stabling at track.
61103	Handicapper	Determines weights horses will carry in handicap races, and determines morning-line odds in all races.
61104	Director of mutuels	Supervises betting and mutuel operations.
61701	Director of public relations	Supervises public-relations staff.
61702	Director of security	Supervises guards; responsible for protection of patrons and of track's property.
61703	Executive secretary of horse-oriented organization	Supervises all activities of organization.
61704	Field secretary for horse-oriented organization	Under supervision of executive secretary, coordinates activities across nation; usually involves much travel.

CODE	JOB TITLE	BRIEF EXPLANATION
61705	Youth director for horse-oriented organization	Develops and coordinates youth activities.
61706	Lobbyist	Represents organization in relations with various legislatures in capital cities.
62101	Racing commissioner	Exercises control over racing within a state.
62102	Secretary of racing commission	Serves as the executive of the racing commission and carries out the policies of the commissioners.
62701	State director of identification services	Supervises and executes state policy on marking and identification of horses.
64701	Director of technical school	Runs technical school; determines courses and hires instructors.
65101	Manufacturer of mutuel machines	Runs factory making mutuel machines.
65401	Manufacturer of timing equipment	Runs factory making timing equipment.
65701	Manufacturer of clothing	Runs factory making clothing.
65702	Manufacturer of agricultural equipment	Runs factory making agricultural equipment.
65703	Commercial feed manufacturer	Runs factory making and distributing commercial feeds.
66701	Laboratory technician-development	Develops chemicals, feed mixtures etc., usually under supervision of animal scientist of engineer.
67401	Director of advertising (for newspaper or magazine)	Supervises advertising salesmen and often determines advertising policies.
67402	Salesman of timing equipment	Sells timers and allied equipment.
67701	Salesman of agricultural equipment	Sells agricultural equipment relevant to horse operations (feeders, tractors, etc.)
67702	Insurance salesman	Sells rain insurance, life insurance on horses, etc.
67703	Bloodstock agent	Buys and/or sells horses for others; arranges breeding contracts, etc.
68101	Mutuel-machine maintenance specialist	Maintains and repairs mutuel machines and allied equipment.
68401	Advertising-space buyer	Determines advertising needs and contracts for space in the various media.
68701	Landscape architect	Determines landscaping needs and supervises gardeners and landscapers.
68702	Dietician/food-service specialist	Plans menus for restaurants and supervises restaurant personnel.
68703	Auditor	Audits books.
68704	Clothing buyer for retail store	Buys clothing and determines what brands and kinds of clothing store will sell.
68705	Horse-feed wholesale distributor	Finds retail outlets and distributes feed.
68706	Rehabilitation therapist	Under direction of veterinarian, operates equipment (hydrotherapy, etc.) to benefit injured or ill horses.
68707	Insurance investigator	Investigates insurance claims, to prevent fraud.
68708	Farm/ranch manager	Supervises operation of farm or ranch.
68709	Agricultural and research economist	Determines economic trends and predicts economic phenomena.
68710	Editor of horse magazine	Has charge of editorial content of magazine.
68711	Business manager of horse magazine	Has charge of finances of the magazine.
68712	Publisher of horse books	Selects manuscripts, arranges terms with authors, printers, binders, distributors.

CODE	JOB TITLE	BRIEF EXPLANATION
68713	Business manager of book-publishing house	Has charge of the financial aspects of publishing.
68714	Field man and acquisition specialist for book-publisher	Works with writers of books to be published by the publishing house, and seeks suitable new manuscripts for publication.
68715	Motion-picture producer	Produces and finances or arranges for financing or horse-oriented movies.
68716	Slide-series producer	Produces and finances or arranges for financing or horse-oriented slide series.
68717	Television producer	Produces and finances or arranges for financing of television shows about horses; also occasionally arranges distribution (stations carrying program).
*68718	Video-producer	Produces video tapes for shows, individuals, farms and ranches.
72101	Racing chemist	Analyzes blood, urine, and saliva samples to determine if horse has been illegally drugged.
72301	Trail engineer	Designs and supervises construction of trails, trail bridges, underpasses, etc.
*72701	Real estate agent	Specializes in the sale of land, farms, and homes for horse owners.
72601	Park and recreation administrator	Supervises administration of public parks; plans park and trail development and recreational programs.
72602	Recreation planner	Assists administrator in planning trails, facilities, and programs.
72701	County agricultural agent	Advises farmers and ranchers; serves as liaison between farmers and state agricultural college.
74701	High-school agricultural teacher	Teaches agricultural courses in high school; usually advises FFA and sometimes 4-H.

CODE	JOB TITLE	BRIEF EXPLANATION
75101	Mutuel-machine design engineer	Designs pari-mutuel wagering equipment.
75401	Timing-equipment design engineer	Designs timing equipment for race tracts and horse shows.
75701	Agricultural-equipment design engineer	Designs various kinds of agricultural equipment.
76701	Agricultural researcher	Performs various kinds of agricultural research (genetics, nutrition, etc.).
78701	Actuary	Develops actuarial tables for horse insurance and human health insurance.
*78702	CPA	Certified Public Account specalizes in horse business.
82701	Conservationist	Develops conservation programs and serves as inspector to prevent waste.
84701	College professor--agricultural	Teaches horse-science and other agricultural courses.
88701	Architect	Designs stable buildings, race tracks, horse-show arenas and stadiums, etc. Supervises construction.
88702	Pedigree analyst	Analyzes horse pedigrees; determines advisability of breeding mares to particular stallions.
88703	Agricultural engineer	Supervises building (and often design) of stable feeding and watering systems, sewage disposal systems, etc.
88704	Technical consultant	Advises in many fields, such as determining validity of motion-picture and television scripts.
91101	Track veterinarian	Determines fitness of horses to enter race; supervises animal-health and sanitary conditions at track.

CODE	JOB TITLE	BRIEF EXPLANATION
91201	Horse-show veterinarian	Passes on questions of unsoundness of horses, at request of judge; provides emergency care of ill horses at show.
92701	State veterinarian	Enforces animal-health regulations in the state; recommends to legislature advisable animal-health legislation.
94701	Extension horse specialist	Advises and instructs in horse questions within a state; serves on national committees concerned with equine activities.
94702	Extension animal scientist	Performs same duties as extension horse specialist, but also as concerned with other species of animals.
94703	Dean, College of Agriculture	Supervises various departments and determines policy.
98701	Veterinarian	Often specializes in equine practice; treats ill animals; occasionally performs research.
98702	Geneticist	Performs research on horse-genetics problems and often advises on pedigree problems; also often teaches courses on genetics.
98703	Horse-feed development specialist	Tests various kinds of feed, observes horses using feed, makes recommendations on feed production; often supervises production.
98704	Animal nutritionist	Often teaches animal-nutrition courses and delivers public lectures on animal nutrition; frequently performs nutrition research.

*Job titles listed after original list was made. These jobs from a list compiled by Hetzel.

An explanation of the five-digit number preceding the job title:

FIRST DIGIT

1. Most current holders of jobs with this title have a high-school education or less. Work is relatively unskilled.
2. Most current holders of jobs with this title have a high-school education or less. Work is semi-skilled.
3. Most current holders of jobs with this title have a high-school education. Work is skilled.
4. Most current holders of jobs with this title have at least one year of education beyond high school.
5. Most current holders of jobs with this title have at least two years of education beyond high school.
6. Most current holders of jobs with this title have at least a baccalaureate degree from a recognized college or university.
7. Most current holders of jobs with this title have at least a baccalaureate degree from a recognized college or university, and substantial additional training and/or experience.
8. Most current holders of jobs with this title have at least a master's degree from a recognized college or university.
9. Most current holders of jobs with this title have either the Ph.D. degree or a professional degree from a recognized college or university.

This numbering system is employed only to indicate the general level of competence required for the various positions. It should not be interpreted to mean that one cannot obtain the position or serve in such a position well unless he has the indicated training or degree. Many positions are filled satisfactorily by those with less than the indicated training and education.

SECOND DIGIT

1. Administrative responsibilities.
2. Government control and regulation.
3. Training horses.
4. Training personnel.
5. Design and/or manufacturing.
6. Research and/or laboratory work primarily.
7. Sales.
8. Services.

THIRD DIGIT

1. Most jobs are in racing only.
2. Most jobs involve horse shows only.
3. Most jobs involve pleasure-horse activities only.
4. Most jobs involve racing and horse shows only.
5. Most jobs involve racing and pleasure-horse activities only.
6. Most jobs involve horse shows and pleasure-horse activities only.
7. Jobs in this classification involve all three branches of the horse industry: racing, horse showing, and pleasure-horses.

FOURTH AND FIFTH DIGITS

These digits constitute a kind of serial number to make easy references to the particular job title. No two job titles have the same five-digit number. No attempt has been made to arrange thejcb titles bearing the same first three digits in any particular order. One should not assume, therefore, that job title number xxx32, for instance, requires more background or experience, or is more difficult to perform than job title xxx01.

6

ECONOMIC OUTLOOK FOR THE LIVESTOCK INDUSTRY

Robert V. Price

The domestic livestock industry in the U.S. is a large and complex business. During 1982, 1.6 million cattle producers marketed 39.3 million hd of cattle and calves, while 484,000 hog operations were responsible for providing 82.8 million hogs to slaughter, and 129,000 sheep operations produced 6.7 million hd of sheep and lambs. Although farm-level marketings amount to billions of dollars per year, this pales in comparison to the total scope of animal agriculture that includes not only the producers of dairy and dairy products, poultry, and other meats, but also the processors, marketers, retailers, and others in the complete marketing chain from the farm gate to the kitchen table.

The economic well-being of the livestock industry is dependent upon a multitude of complex factors that could threaten to disrupt the delicate balance between supplies and demand for meat products. The large number of uncoordinated producers, especially in the red meat sector, is, alone, sufficient to confound the pursuit of the elusive equilibrium point. But when the industry is buffeted by what economists refer to as "exogenous shocks," the resulting uncertainty and volatility can completely disrupt the economics of livestock production. While there are winners that are able to take advantage of this volatility, they are far outnumbered by the losers who do not survive.

This paper is an attempt to outline the current situation of the livestock industry in relationship to supply and demand, to discuss the economic environment in which the industry must operate, and to speculate on what the future may hold for the U.S. livestock industry.

Today we are operating in a relatively mature meat industry. Meat is no longer a growth business, such as it was in the 1950s, 1960s, and early 1970s. Average annual per capita consumption of all meat has leveled off at around 200 lb, retail weight. The total is not likely to change significantly in the years ahead. But the mixture can change in direct proportion to consumer's preference in buying decisions and production and marketing skills behind each of the commodities.

As shown in table 1 and figure 1, per capita meat consumption of all red meat and poultry has changed very little over the past decade or more. However, the relative shares held by each meat product have undergone some dramatic changes. Per capita beef supplies were in the neighborhood of 84 lb in 1970 and by 1983 had dropped below 77 lb. Pork supplies per person generally have been in the 55 lb to 62 lb range throughout the past two decades.

TABLE 1. PER CAPITA MEAT CONSUMPTION IN RETAIL WEIGHT (POUNDS)

Year	Beef	Pork	Broilers	Turkeys	Total poultry	Total red meat & poultry
1970	84.0	62.3	36.8	8.0	48.4	200.2
1971	83.4	68.3	36.5	8.3	48.6	205.5
1972	85.4	62.9	38.2	8.9	50.6	204.0
1973	80.5	57.3	37.2	8.5	48.9	190.9
1974	85.6	61.8	37.2	8.8	49.5	201.0
1975	87.9	50.7	36.7	8.5	48.5	192.6
1976	94.4	53.7	39.9	9.1	51.8	205.0
1977	91.8	55.8	41.1	9.1	53.2	205.8
1978	87.2	55.9	43.8	9.1	55.8	202.8
1979	78.0	63.8	47.7	9.9	60.4	205.4
1980	76.5	68.3	47.0	10.5	60.6	208.4
1981	77.2	65.0	48.6	10.8	62.4	207.8
1982	77.2	59.1	50.0	10.8	64.0	203.4
1983 (est.)	76.9	61.7	50.7	11.3	64.8	206.3

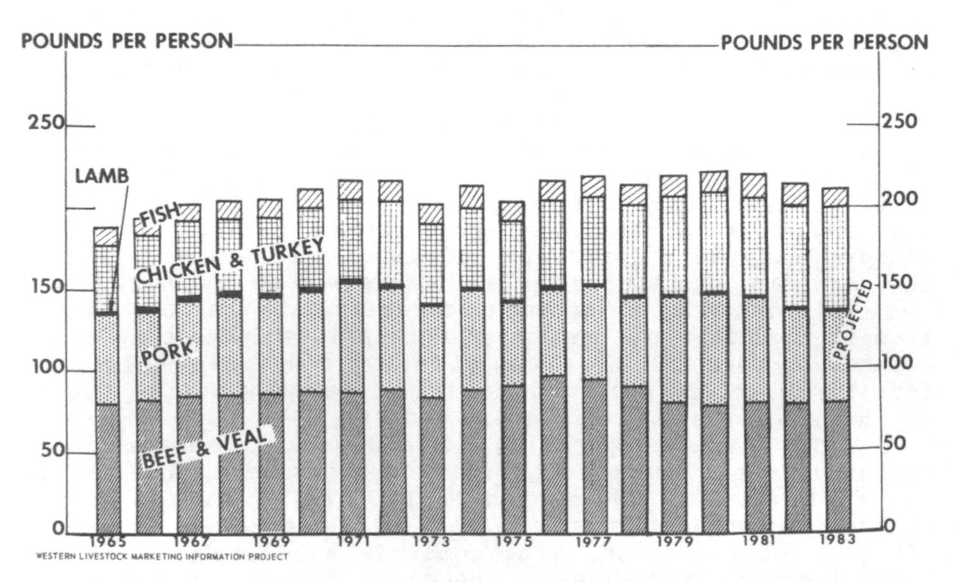

Figure 1. Consumption of meat, poultry, and fish - 1965 to 1983 (retail weight equivalent).

Total poultry production has increased dramatically, from around 48 lb/person in 1970 to nearly 65 lb/person in 1983. A good share of the increase in poultry production has resulted from the declining consumption of beef.

Consumer purchasing power has not improved significantly for a number of years. Although per capita disposable income (the purchasing power of an individual) has increased, smaller-sized families, the relatively greater percentage of single-person households, and other factors have kept total household purchasing power near 1970 levels (figure 2). Each household is faced with the cost of housing, energy, and other items that compete with the food dollar, and especially meat purchases. In addition, meat is purchased more often than other foods, which tends to keep the consumer aware of price differences among the various types of meat.

Future beef industry growth will depend on: population growth, which is expected to average about .8%/yr in the 1980s; expansion of exports, which now account for only 1% of U.S. beef production; and beef demand, the development of new beef uses and beef products; and improvements in the competitive relationship between beef and other meats. Keeping beef at the top of the protein mix, as it is today, will require the best efforts of the beef cattle industry as well as improved efficiency on the part of the individual cattleman.

The cattle industry has experienced a cycle in numbers in every decade of this century (figure 3). The current cycle seems to be somewhat abbreviated, with a mild liquidation having begun after only a 3-yr buildup in cattle numbers that started in 1979. Inventory numbers on January 1, 1983, (which will be reported later this month) are expected to show little change from total cattle inventories recorded last year at this time.

Hog producers also tend to adjust numbers in a cyclical pattern, although their cycle is adjusted much more rapidly and is of shorter duration than is the cycle for cattle. Hog producers now are in an uptrend in numbers for the current hog cycle (figure 4). The resulting low prices of the past several months will undoubtedly prompt liquidation in hog numbers sometime during 1984.

After 40 yr of steady declines, sheep numbers in the U.S. turned upward slightly in 1980, 1981, and 1982, providing some optimism in the industry that the declining trend in numbers had been arrested. However, sheep inventories were further liquidated during 1982 and 1983; thus the sheep inventory in January 1984 will be the lowest inventory on record in this country. It appears that further declines in the numbers of sheep and lambs may occur for the next few years (figure 5).

The economic environment for the livestock industry cannot be separated from the general economic climate in the country, or from that of the whole world. Decisions made by domestic policymakers can have profound effects on the

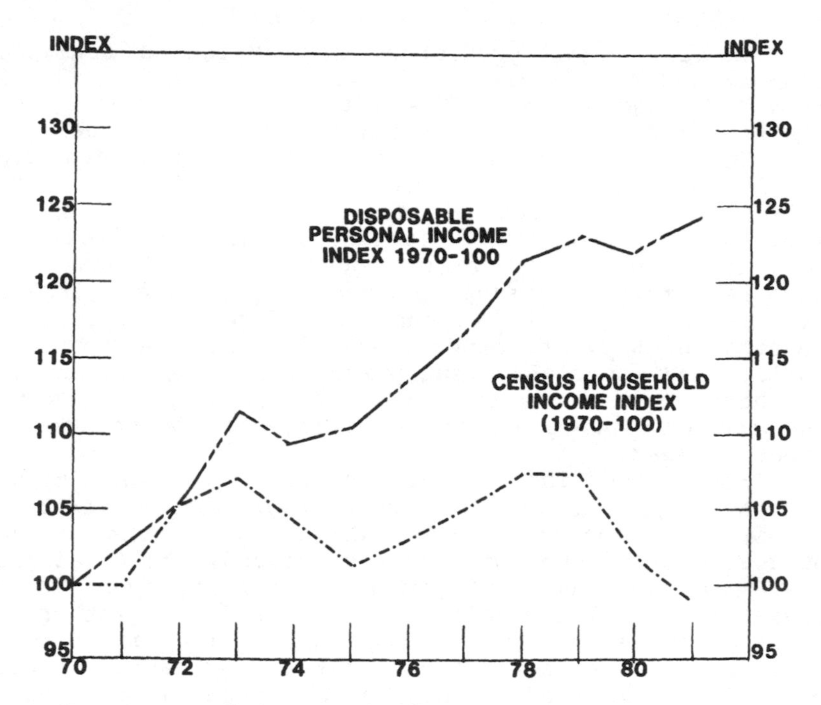

Figure 2. Personal income indices.

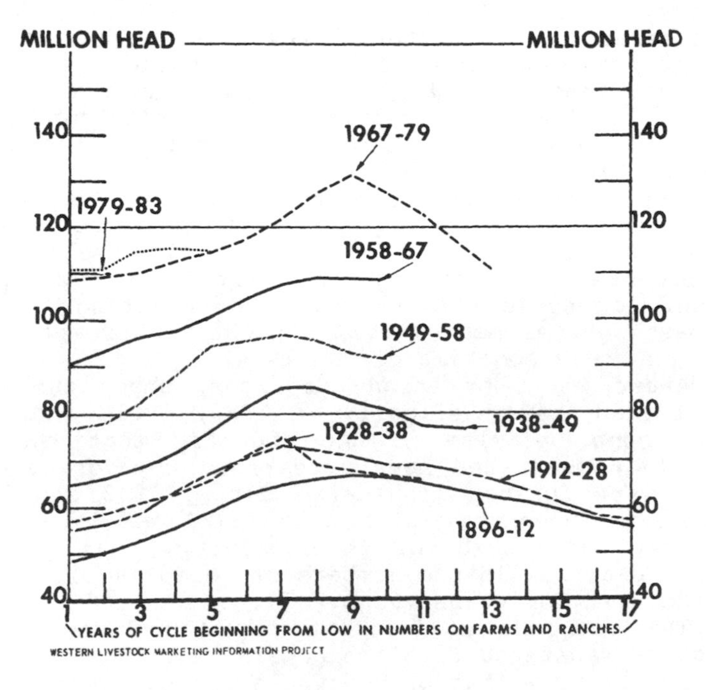

Figure 3. Cattle on farms by cycles.

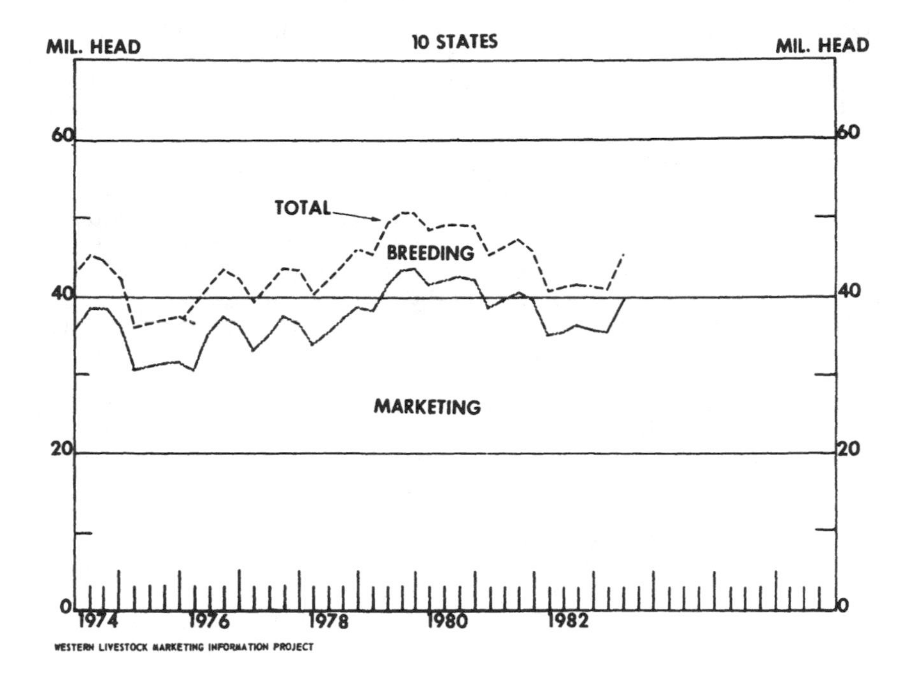

Figure 4. All hogs and pigs, quarterly 1974 to 1983.

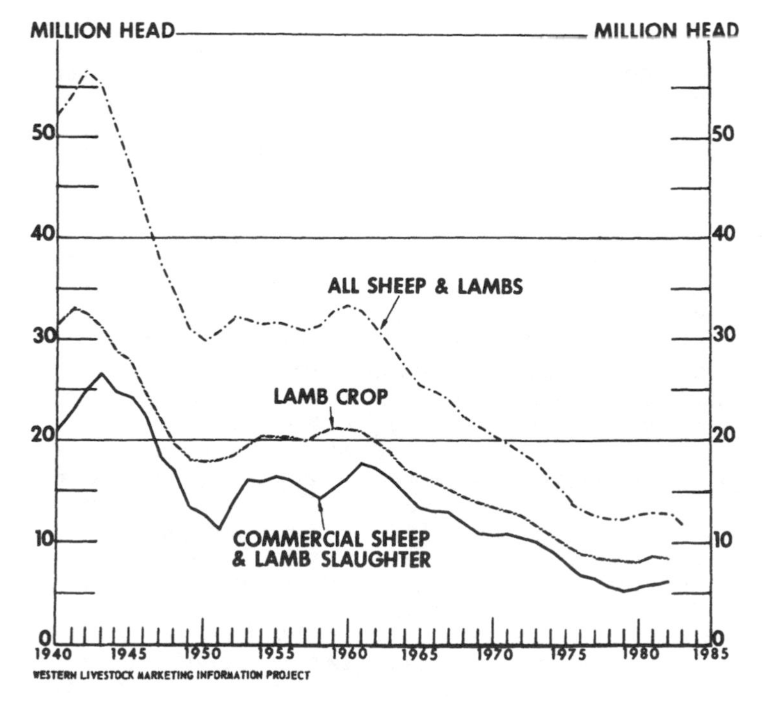

Figure 5. Annual data series, sheep and lambs.

industry. Deficits of the size currently seen in the federal government suggest that almost 100% of individual savings will need to be allocated simply to supply the credit demands of the federal government. This has resulted in record interest rates, and interest rates will probably continue to be high throughout this decade.

Domestic farm policies also can profoundly affect the livestock industry. Without a doubt, livestock producers bore a major portion of the cost of the Payment-in-Kind (PIK) program for 1983 crops in this country. Not only were livestock producers affected by the increases in feed prices, but also by the disruptions in feeder livestock supplies as grain producers clammered for stocker animals to run on acres set aside under the program.

Unfortunately, it also appears that the domestic livestock industry will shoulder a major portion of the cost of the new dairy legislation that has been enacted. The incentive for increased slaughter of dairy cows appears greater now than at any time during the past several years. These increased cow supplies may well come at a time when total supplies of red meat in the country are already running at high levels. These increased supplies of cow beef undoubtedly will be quite detrimental to livestock prices in general.

THE FUTURE OF THE LIVESTOCK INDUSTRY

Probably the greatest lesson of the past 10 yr in regard to economic policy is that some critical elements of policy are not in the hands of domestic policymakers. Instead, policies are affected by what economists call "exogenous shocks," whereby seemingly far-away events (such as war between Iran and Iraq) can impact the domestic economy just as much as does the destruction of a portion of the cereals crop by excessive rain or drought.

These uncontrollable events affect food and energy supplies, and like it or not, we are part of an interrelated international business environment. No nation, continent, or region is an island, and this interdependence will be the dominant key characteristic with which international policymakers must cope--and the success of their efforts will heavily condition the relative health of U.S. agriculture. If the choice is protectionism, we can expect a continuance of a relatively depressed domestic agricultural sector. On the other hand, we know that inflation is also costly. While agriculture may benefit in the short-term, resulting deficits place more and more strain on the impacted nations' ability, and will, to sustain and service the resulting debt load.

The picture is not as rosy as in the early 1970s when the world food shortage was expected to result in an indefinitely prosperous U.S. agriculture, and it may not be as simple as "feed them or fight them." But it is clear that the imbalance between producing and consuming countries will continue.

This dilemma will be dealt with in the international policy environment, and at this time it is unclear which direction U.S. policymakers will pursue in solving it. A key word describing the situation seems to be "instability," and it seems highly likely that the domestic livestock industry and its affected managers must learn to cope with this instability if they are to survive and prosper in the 1980s and beyond.

Since 1977, those of us who classify ourselves as "industry watchers" have been issuing warnings regarding the impact of changes in real incomes and relative prices on the consumption of livestock products. A large number of forces and actors are tugging at consumers' pocketbooks, urging them to consume everything from cars and electronic products to more beef and pork--and these forces are likely to continue in the future.

The issue is not one of convincing them they should consume more products, because we know that they consume the highest levels of products during times when the industry is in greatest trouble--during liquidations or herd reductions. Instead, the issue is that of developing a profitable level of production, over time. And this implies a knowledge of the consumers' future wants and needs, relative economic conditions, and types of products required, as well as smoothing out the gluts and shortages that now occur with prevailing industry management.

This knowledge will not come easily and will cause further erosion in the number of commercial operations in the U.S. This is not to say that commercial operations will necessarily be larger but, instead, that they will be better managed. The most pressure will be on the commercial land-intensive western operations because costs associated with owning and operating will continue to mount. We can expect crop-residue operations to continue to be successful, but cost pressures of different kinds will continue to plague them as well. In both cases, operations will be better managed and more business oriented. However, we can expect some rather large ownership turnovers (particularly during the next 3 yr to 5 yr) in those operations that have relied strictly on land equity to sustain them.

A look into this particular crystal ball shows that products will be produced more quickly and efficiently from existing operations. The economic environment seems likely to continue in which energy and feed supplies will probably continue swinging back and forth from shortage to surplus. Operators must possess flexible production, marketing, and financial programs in order to cope with this problem and others, particularly inflation and deflation.

Meat products will be leaner and will reach the marketplace more quickly, passing through fewer hands than has been the case during the past 25 yr. All of this reflects the market's answer to continuing competitive pressures.

Feedlot and packer numbers will continue to decline and successful operations in the future will maintain continuous coordination and communication. The extraordinary development of electronic communication and computation systems ensures that this will be an operational reality.

We might characterize successful managers of the future by a few useful generalities. First, they must accept the responsibility for marketing and market planning. Second, they must build an acceptable immediate, intermediate, and longer-term production and marketing plan that "builds-in" some of the above-mentioned contingencies and the means of dealing with them. Third, they must calculate the relative costs associated with their program and how and by whom they will be financed. And this latter choice is an extremely important one since extraordinary understanding and knowledge of the agribusiness sector from the financier's point of view is also necessary. Fourth, the individual must have a program to deal with risk associated with crop failure and market planning. This program will vary by individual and region, but it must be in place and may include a combination of hedging in the futures market, forward contracting, crop insurance, diversification of operations, or some other combination. But it will be in place and will be translated into a real, dollars and cents, cost-of-production framework.

Is there reason for real pessimism from an individual's point of view? Quite to the contrary, individuals can profit even more from the instability of the future. Proper timing of inventory accumulation and reduction is but one example of how an individual can prosper. The availability of low-cost electronic computers and communication programs will allow small to intermediate operations to be cost competitive in management if they choose to be--and this latter attribute is the name of the game in the 1980s and beyond.

Part 2

ENERGY SOURCES AND ANIMAL WASTES

7

PHOTOVOLTAIC SOLAR POWER FOR SMALL FARM AND RANCH USE

H. Joseph Ellen II

PHOTOVOLTAICS--A LAYMAN'S INTRODUCTION

Photovoltaic, or PV, solar energy is the direct conversion of sunlight to electricity. The panels that accomplish this are basically a series of silicon wafers that are interconnected by metal strips to conduct electrical current. These wired cells are then encapsulated by materials to weatherproof them and are covered with a resilient, transparent surface--usually glass--to allow the passage of light. As light strikes the cells, an electron flow is created and electricity is the end product. Assuming that the panel is properly constructed, a direct current, or a DC generator, will be created in an amount proportionate to the sunlight available. These devices generally last 20 yr.

THE ECONOMICS OF PHOTOVOLTAICS

With the introduction of the solid-state, long-life generation devices that require no fuel, a number of applications become available for their use. In effect, any electrical device can be powered, but the constraint of economics is to be considered. PV cells are still expensive and the storage of electricity for night use requires consideration of the inefficiency, maintenance requirements, and cost of batteries.

One should avoid the use of alternating current, AC, because a system must incorporate an inverter to change the DC current generated by the cells and stored in the batteries to AC to run the appliance. A good inverter can cost many thousands of dollars and have only a 75% efficiency. Thus, one is losing 25% of the electricity before it is even delivered to power equipment.

Generally, photovoltaics will suit any small need for lights; for fence charging as has been done for many years; for radio communications; or in some instances, for small dwellings. Projects by the U.S. Indian Health Service in Arizona and California have shown that for remote housing, supplying lights and enough electricity for communications (radio or television) can be cost effective using photovoltaics as opposed to using propane or diesel generation. When using photovoltaics for supplying electricity in large quantities, consider the following economic facts:

Remoteness. Will it be cheaper to bring in developed utility electricity to the site? If so, it is unlikely one would want to use photovoltaics. The quality of electricity being delivered by the utility and its service record of delivering electricity on a consistent basis should be considered.

Internal combustion power. The traditional alternative to photovoltaics is the use of a generator powered by a gasoline or diesel engine. In considering the alternatives, one should determine the maintenance costs including labor, transportation, availability of spare parts, and fuel.

Today, perhaps the most practical use for photovoltaics other than communications and small-lighting generation is water pumping. Water is essential for any sort of agriculture, be it farming or ranching. There are some 50 million farms in the world supporting 250 million people that need water for less than 2 or 3 acres. There are also untold numbers of ranches using windmills.

Water pumps are ideal for use of PV because they eliminate the need for batteries, regulators, and battery chargers, and because water can be pumped in the daytime and stored in storage tanks for use when needed. Small lift and small-volume consumption of water as used by most ranches and remote villages in the world can be provided economically by water pumping with PV solar. Presently there are systems up to 20 HP in the world with operating records under practical conditions for up to 6 yr. In the past 3 yr, the author has personally had experience with the installation and maintenance of over 60 systems that are all currently viable plus a close association with a few hundred other systems.

To use the windmill in many places in the world, one has to purchase and maintain it, and provide a standby internal combustion engine to pump during nonwindy periods. In southwestern U.S., quite often when wind is available water use is actually at its lowest. When use is at its highest there is very little wind available and one foregoes the "free power" provided by the windmill and utilizes a gasoline or diesel system to operate a pump, which is costly to maintenance.

With PV solar we have a system which is basically installed one time. It automatically begins pumping when the sun comes up and quits pumping when the sun goes down. The sizing constraint with photovoltaics to accommodate for cloudy weather or for climate using known solar insolation, is possible with data which has been collected by the U.S. Department of Energy for many years. A well-made solar pump will take into consideration the various amounts of sun energy striking the earth that is referred to as insolation, and will match its performance to the available insolation.

Interestingly enough, the curve of solar insolation almost directly matches the curve of evapotranspiration rates in plants, animals, and human beings. Therefore, on a day when one has 40% sun, and a solar device is producing 40% of its potential, the plants, animals, and people are using about 40% of the water.

Another advantage to solar pumping is that water is pumped over the entire daylight period, as opposed to using a gasoline engine, and pulling it up in 1 to 2 hr. This means that the jack pump itself, cylinder, and the well are all performing at a slower rate. This can allow slow-recovery wells to produce more water and should result in a longer life of the cylinder, the leather within the tube, and the pump itself.

In a remote area with no utility and where transportation is a logistical problem, to minimize the service and maintenance overview of water pumping, PV solar can be utilized today.

There are various solar pumping devices on the market. We have had great experience with the devices produced by Tri Solar Corporation of Massachusetts. They are the only company at present that are manufacturing a deep-well system for pump jacks that does not incorporate batteries to handle the electronics in motor fluctuations caused by different amounts of sunlight. Battery-less pumps reduce total system cost, eliminate maintenance of the batteries, and increase reliability considerably.

SIZING A PHOTOVOLTAIC SYSTEM

When purchasing a photovoltaic system, one must be certain that needs are clearly defined. First of all, look for an appliance that will fulfill the needs best with a minimum use of power. For reasons we spoke of earlier, it would be best to consider a DC appliance. A primary consideration is the amount of sunlight available in your area of the world. In the southwestern part of the U.S., one receives a little over 6 hr a day of peak production from a solar array. A rancher would determine his need, deciding how many hours a day he was going to use his appliance, divide by 6 and would have arrived at a preliminary array size. A preliminary formula would be the following:

$$\frac{\text{Use}}{\text{6 hr per day}} \div \underset{\text{\& system losses}}{.7\ \text{Battery}} \div \underset{\text{used)}}{.75\ \text{(if inverter}}$$

= Approximate array size in watts output

If the rancher is going to use a battery storage system, he should expect somewhere in the neighborhood of a 30% loss through the system, wires, batteries, and battery charges. Now he has a safe estimate for producing the power needed.

Water pumping avoids many of the above losses of power because of storage. However, we should plan to produce about 5% or 10% more water per day than we really need so that our storage tanks can give us a 3-day or 4-day period to carry us over in the event of foul weather or inevitable breakdown of any system that is man-made. This will ensure that cattle or humans will have water during time of repair or maintenance of the system. In sizing a pumping system, the total lift from static water level and daily volume required must be known. Because of the many factors involved, sizing is not a simple evaluation.

SIX STEPS IN PURCHASING A PHOTOVOLTAIC SOLAR POWER

1. As Packard Automobile Corporation used to say, "Ask the man who owns one." Does the company have a reputation? Can we talk to the people who have purchased a system? How did they like it?
2. (I think this step is very important.) Is there someone available who will service the system? Will the company that manufactured it stand behind the system with some sort of a warranty? A very quick way to discern the reliability of the company's warranty and reputation would be to see if the financing were available through some major financing or lending institution. Do they consider the product and the company reliable enough to lend money? It would be an indication that one would probably get service and maintenance as outlined in the agreement. If I were buying a system for myself, I would look for a minimum of a 5-yr warranty and would expect the components to be reliable enough to last that period of time so that I could amortize my investment in the system over that 5-yr period.
3. I would demand that the company have a full-coverage warranty covering every potentiality, and that 50% of the payment for the system be placed in escrow until I was assured that the system and the manufacturer performed properly.
4. I would take the time and effort to look at a solar system the manufacturer has already installed. Then I would take a close look at the internal portion of the control box and inspect the workmanship. Quite often the good workmanship indicates a good end product. Companies who take the extra effort of wiring a product with care, making sure it's painted and put together with care, will

probably have a good end product. There are people who can throw something together and end up with a good product, but I don't believe this is generally the case. Good workmanship is always a hallmark.

5. A buyer should inquire if there are spare parts that he can be trained to change, or if a system could have a built-in second circuit in the event of a failure; thus minimizing the potential of a person being without power. If a company cannot provide experience, warranty, finance, etc., then I would go to another company.
6. Bearing in mind that the low price is not always the best product, go to another company and get a second bid. The University of Arizona, Arizona State University, and the Arizona Solar Energy Commission have had considerable experience with photovoltaic systems within that state. I do not know what experience is available to buyers in other states and countries, but I am sure that Arizona would accept out of state inquiries and be happy to look over a system proposal. It is good to have someone who knows the business of PV to look a proposal over.

CONCLUSION

With careful definition of need and cautious purchasing of equipment, solar photovoltaics electrical systems can be highly reliable and most cost effective with a minimum amount of care and maintenance for a lifetime of use.

8

ANAEROBIC DIGESTION OF BEEF AND DAIRY MANURE FOR ENERGY AND FEED PRODUCTION

William A. Scheller

INTRODUCTION

Anaerobic digestion of waste materials has been a long standing method of treatment in municipal sewage plants. The purpose of digestion in such cases is to reduce sewage volume and to destroy certain undesirable microorganisms. Historically the sanitary engineers have not placed a high priority on the recovery of digester gases as a fuel.

American dairymen and cattlemen, along with the American farmer, have become very concerned about both the cost of energy and the availability of fuel in this time of inflation and political unrest. Anaerobic digestion offers those with large concentrations of animals (cattle, cows, poultry, swine) the opportunity to produce a portion (or sometimes all) of their energy needs from a renewable waste source, i.e., manure and to have as a coproduct a cattle feed component that can provide valuable fiber and cellular protein to the animals producing the waste. If the coproduct cannot be used as a feed component, it makes an excellent soil conditioner. The liquid phase separated from the digester effluent solids is usable as a fertilizer that contains a stable nitrogen in the form of cellular protein.

The plug-flow digester has been demonstrated to be very efficient in the production of fuel gas and coproducts. Its operation and energy balance, as well as the properties of its solid effluent as a feed component, are discussed in this paper.

PROCESS PRINCIPLES

Anaerobic digestion is a mixed culture biological process in microorganisms that converts organic matter into a gas mixture containing methane, carbon dioxide, hydrogen sulfide, and nitrogen. The heating value of this gas when saturated with water vapor at ambient temperatures is about 590 btu/cu ft. While all organic matter is probably digestible to some extent, materials such as cellulose, hemicellulose, starch, other carbohydrates, proteins, fats, etc., have digestion rates that are sufficiently high to permit their conversion into biogas in an economical period of time (5 to 25 days). The conversion process takes place in three stages:

1. Conversion of the digestible organic materials to sugars and long chain fatty acids. This process is usually referred to as hydrolysis.
2. Conversion of the product from #1 above into volatile fatty acids, alcohols, hydrogen, etc., by what is normally referred to as acidification.
3. Methanation of the materials from #2 above into methane and carbon dioxide.

The numerical distribution of types of microorganisms in the digester will depend on a number of factors including the nature of the organic material fed to the digester and the temperature of the digester. Three temperature ranges are usually defined: psychophilic (under 20C), mesophilic (20C to 45C), and thermophilic (45C to 65C). In this country and in Europe operating digesters are at either mesophilic or thermophilic conditions. Some small digesters that lack temperature control may operate from time to time at psychophilic conditions but gas production at this level, if any, is very low. Thermophilic digesters, on the other hand, may require a large portion of the gas produced to maintain the digester contents at the operating temperature, which results in only a small net gas production.

Many people consider mesophilic conditions the best since they normally use only 5% to 20% of the gas production to maintain the operating temperature. If the biogas is used in an engine to drive an electrical generator, there may be enough or more than enough waste heat in the engine exhaust gases to maintain mesophilic operating conditions in the digester.

DIGESTER TYPES

Anaerobic digesters are usually classified according to the type of flow in the digester and the operating temperture. These include plug flow, stirred tank, and packed bed (anaerobic filter). In terms of on-farm anaerobic digestion, there are more plug-flow digesters operating with long-term success than either of the other types.

The plug-flow digester is constructed as a horizontal, in-ground tank with a length-to-width ratio of about 6.5:1. A set of heating pipes is generally run lengthwise through the tank. Some kind of device to break bubbles and foam is mounted at or near the liquid surface. The open top of the tank is covered with a large plastic bag to hold the gas as it is generated. Manure is mixed with water or recycled effluent to adjust the solids content to 10% or 12% of the total solids. It is then pumped into the inlet end of the digester over a 1- or 2-hr period once a day. As the feed is added to the digester, effluent overflows from the opposite (outlet) end of the digester. This is sent directly to a lagoon if the effluent solids are not recovered or to a holding pit if the material is centrifuged to recover the solids. Liquid from the centrifuge is sent to a lagoon to be used at the appropriate time as liquid fertilizer. The solids may be dried or partially dried and sold or used as a feed component, as a soil conditioner, or as bedding for the cattle. Gas that has collected in the plastic bag is piped through appropriate safety equipment to a furnace, boiler, gas-driven electrical generator, or otherwise used. When there are hot exhaust gases, as from the furnaces or gas engine, these gases may be used to heat water to circulate through the digester heating pipes. If hot exhaust gases are not available, a portion of the biogas must be burned in a water heater. A plug-flow digester with electrical generating system and centrifuge requires about 1 to 1.5 hr per day of labor for operation and maintenance.

The stirred-tank digester and the anaerobic filter are vertical above-ground tanks. As the name indicates, the stirred-tank digester is equipped with a motor driven mixer. It has internal heating pipes and is fed once per day. The anaerobic filter is fed either on a batch basis or continuously. It too has internal heating coils and is filled with an inert packing material that serves as a medium to which the microorganisms can attach themselves. Stirred-tank digesters are usually run at thermophilic conditions and the other two systems at mesophilic conditions.

PLUG-FLOW-DIGESTER OPERATING EXPERIENCE

Energy Cycle, Inc., a subsidiary of Butler Manufacturing Co., and its predecessor company have built six plug-flow, mesophilic digesters at five large dairy farms in the U.S. since 1979. The digesters range in volume from 60,000 gallons (230 cu meters) to 240,000 gallons (910 cu meters). They have also completed a plug-flow digester system consisting of twelve 240,000 gallon (910 cu meters) unit at a feedlot for 50,000 head of beef near Lubbock, Texas.

A typical installation is the 180,000 gallon (680 cu meters) unit completed in March 1981 at Baum Dairy Farm near Jackson, Michigan. The plug-flow digester processes the waste from 650 cows. The biogas is used to generate electricity for use on the farm. The digester effluent is dewatered, with the solids used as bedding for the cows and the liquid as fertilizer after storage in a lagoon. Gas production is 41,600 cu ft (1180 cu meters) per day at design loading except in the winter when the rate is about 15% less. A 65 kw generator driven by a 1200 rpm Waukasha gas engine was installed at the time the plant was built. Actual gas production is higher than expected and plans are in progress to install a 90 kw generator and engine in place of the 65 kg unit. The current level of electric power generation has reduced electric costs by $30,000 per year.

About 900 tons per day (dry weight basis) of centrifuged solids are recovered from the digester effluent at a moisture level of 70% to 75%. This material is used as bedding for the cows, which eliminates an annual bedding bill of about $36,000. The owner claims that use of this material has reduced the occurrence of mastitis in the lactating cows by at least 80%. Liquid from the centrifuge contains 5% to 6% solids and is rich in single-cell protein. No value has been placed on it by the owner.

The owner of Baum Dairy believes that the installation of the digester system has actually reduced the labor required to handle and dispose of the manure from the cows. Maintenance to date has been routine and consists of such tasks as spark plug change every 600 hr, oil change, lubrication, etc. It is estimated that the annual cost of maintenance to date is not over $1,000.

CASH FLOW FROM THE INVESTMENT

The investment required for the digester, generator, centrifuge system, holding pits, and the buildings was $225,000 in 1981. This total investment was financed for 30 yr at an average interest rate of 10.93% per annum. Monthly payments of principal and interest total about $25,600 per year. In the early years most of the loan payment is for interest. If one assumes 1 hr per day of labor at $7 per hour, $1,000 per year maintenance, and $3,400 per year for property taxes and insurance on the digester system, then total expenses associated with the digester system are about $32,500 per year. This includes a portion of the loan principal. If no value is placed on the fertililzer derived from the centrifuge liquid or on the reduced losses from mastitis, then the economic benefit from the digester system is in total $66,000 per year from the bedding and electric power. This gives a cash flow of $33,500 per year or 14.9% per year of the initial investment. Continued increases in electrical costs and the general inflation can make this return larger in the future.

VALUE OF DIGESTER SOLIDS AS CATTLE FEED

In the Baum case the savings in electricity paid almost all of the digester expenses including loan repayment. The return on the project came from the value of the by-product. It is important to have a good market for the digester solids and, if possible, the liquid too. Cattle feeders do not need bedding for their animals. In a dairy operation, a steady state recycle of bedding will build up in the digester and recovery system--meaning that excess solids can be recovered for sale if an adequate recovery system is purchased and installed.

A number of investigators have studied digester solids as a cattle feed component (see reading list at end of this paper) and concluded that it has good potential and value. The values of feed components are constantly changing not only in absolute value but also relative to one another. It is reasonable to assume that the two dairy and beef cattle feed properties of greatest importance are the weight percent crude protein and the price per ton of total digestible nutrients (TDN).

To test this opinion prices were gathered for thirteen feed components from "Feedstuffs" and "The Wall Street Journal" (table 1). The NRC compositions of these components were also tabulated in table 1 and the prices per ton of total digestible nutrients were calculated. These were then plotted on semilog graph paper with the weight percentage of crude protein as the abscissa (figure 1). A least squares fit of the data was then calculated. The correlation coefficient (r) was .961. Eleven of the thirteen points fall within a range of 10% on either side of the regression line. The other two points are 11% and 12% from the line. The regression equation for the price line is:

$$\ln(\$/\text{ton TDN}) = 4.5566 + .0203(\%\ \text{crude protein})$$

A set of analyses was obtained for digester centrifuge cake and centrifuge liquid from Baum Dairy's digester system (table 2). On a dry matter basis the centrifuge cake contains 10.66% crude protein and the cake is 53.17% TDN. From the above equation the value of the centrifuge cake is $118 per ton of TDN or $63 per ton on a dry basis. The centrifuge liquid contains 54.80% crude protein on a dry basis and the dry matter is 83.80% TDN. From the equation, this dry matter is valued at $290 per ton of TDN or $243 per ton of dry matter. Therefore, an evaporative drying system might show an attractive rate of return.

The value of $63 per ton of dry matter for the centrifuge cake is consistent with other estimates published in the literature. No other estimates of the value of the dry matter from the centrifuge liquid have been found. These values are consistent with corn at $2.60 per bushel and SBM at $187 per ton. This same technique can be repeated as grain prices change. With sufficient historical data one might be able to develop a price index between corn and the centrifuge cake and between SBM and dry matter from the centrifuge liquid. The final answer to the question of value of these digester products will only come from the placement of the products into the feed marketplace along with an evaluation of their performance.

TABLE 1. COMPOSITIONS AND PRICES OF VARIOUS CATTLE FEED COMPONENTS

Feed component	%DM	Wt % of DM		-----Price $ Per-----	
		TDN	Cr Prt	Ton as Fed	Ton TDN
Molasses, cane	75	91	4.3	73.00	107
Corn	89	91	10.0	92.86	115
Milo	89	80	12.2	95.00	133
Wheat	89	88	14.3	108.83	139
Wheat bran	89	70	18.0	76.66	123
Alfalfa dehy.	93	92	19.2	112.00	131
Corn glut. feed	90	82	28.1	109.00	148
Brewers grains	92	66	28.1	92.00	152
Dist. dry gr.	92	84	29.5	145.00	188
Linseed meal	91	76	38.6	160.00	231
Soybeans	90	94	42.1	207.00	245
Cottonseed meal	91	75	44.8	160.00	234
Soybean meal	89	81	51.5	187.00	259

Note: Prices from either "Wall St. J." (6-24-82) or "Feedstuffs" (6-29-82). %DM, TDN, Crude Protein are NRC values as published by Church, D. C. 1977. Livestock Feeds and Feeding. p 292-295. O & D Books, Inc., Corvallis, OR. DM = Dry Matter, TDN = Total Digestible Nutrients, Cr Prt = Crude Protein.

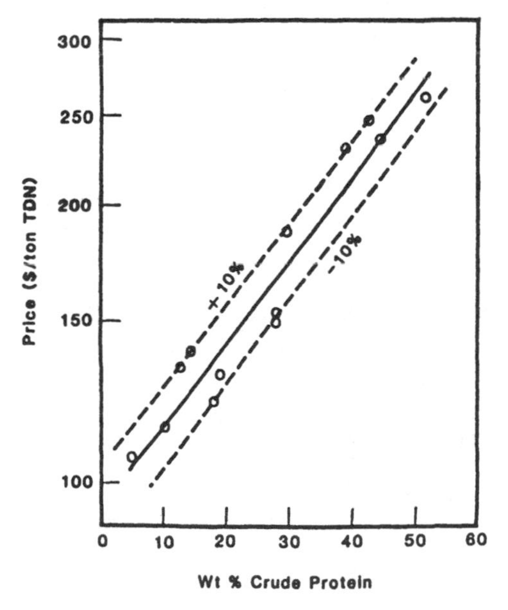

Figure 1. Relationship between feed costs and nutritional content.

TABLE 2. COMPOSITIONS AND PRICES OF VARIOUS CATTLE FEED COMPONENTS

Component	Centrifuge cake		Centrifuge liquid	
	Wt % as rec'd[a]	Wt % dry basis[b]	Wt % as rec'd	Wt % dry basis
Fiber	9.59	44.08	.82	15.44
N-free extract	8.07	37.08	.43	8.10
Total carbohydrate	17.66	81.16	1.25	23.54
Crude protein	2.32	10.66	2.91	54.80
Fat	.36	1.65	.08	1.51
Total volatile solids	20.34	93.47	4.24	79.85
Ash	1.42	6.53	1.07	20.15
Total solids	21.76	100.00	5.31	100.00
Moisture	78.24	--	94.69	--
Total sample	100.00	--	100.00	--
Total digest. nutr.	11.57	53.17	4.45	83.80
Total digest. protein	1.44	6.62	2.53	47.65
Fertilizer components				
Nitrogen			.47	
Phosphate			.23	
Potash			.26	

[a]The above "As rec'd" analyses were prepared by Harris Labs, Inc., Lincoln, Nebraska, and reported to Energy Cycle, Inc. as Lab Sample No. 31720 on Feb. 24, 1982.
[b]"Dry basis" calculated from "As rec'd."

ANAEROBIC DIGESTION READING LIST

The following list of publications treats the subject matter of anaerobic digestion and the specific topics covered in this paper. It is felt that such a list will be of more value to the audience than references to facts stated above.

Anaerobic Digestion Fundamentals and General Papers

Hashimoto, A. G., Y. R. Chen and V. H. Varel. 1981. Theoretical aspects of methane production: state-of-the-art. In: Livestock Waste: A Renewable Resource. pp 86-91. Amer. Soc. of Agr. Engrs., St. Joseph, MI.

Hobson, P. N., S. Bousfield and R. Summers. 1981. The microbiology and biochemistry of anaerobic digestion. In: Methane Production from Agricultural and Domestic Waste. pp 3-51. Applied Science Publishers, London.

Messing, R. A. 1982. Immobilized microbes and a high-rate, continuous waste processor for the production of high btu gas and the reduction of pollutants. Biotech & Bioengr. 24:1115.

Humenik, F. J., M. R. Overcash, J. C. Baker and P. W. Westerman. 1981. Lagoons: state-of-the-art. In: Livestock Waste: A Renewable Resource. pp 211-216. Amer. Soc. of Ag. Engrs., St. Joseph, MI.

McInerney, M. J. and M. P. Bryant. 1981. Review of methane fermentation fundamentals. In: Fuel Gas Production from Biomass, Vol. I. pp 19-46. CRC Press, Boca Raton, FL.

Office of Technology Assessment. 1980. Anaerobic digesters. In: Energy from Biological Processes, Vol. II. pp 181-200. OTA, Washington, D.C.

Anaerobic Digestion Processes

te Boekhorst, R. H., J. R. Ogilvie and J. Pos. 1981. An overview of current simulation models for an anaerobic digester. In: Livestock Waste: A Renewable Resource. pp 105-108. Amer. Soc. of Agr. Engrs., St. Joseph, MI.

Dahab, M. F. and J. C. Young. 1981. Energy recovery from alcohol stillage using anaerobic filters. In: Biotech & Bioengr. Sym. No. 11. pp 381-397. John Wiley & Sons, New York.

Hawkes, D. L. 1980. Factors affecting net energy production from mesophilic anaerobic digestion. In: Anaerobic Digestion. pp 131-148. Applied Science Publishers, London.

Langton, E. W. 1981. Digestion design concepts. In; Fuel Gas Production from Biomass, Vol. II. pp 63-110. CRC Press, Boca Raton, FL.

Sievers, D. M. and E. L. Iannotti. 1982. Anaerobic processes for stabilization and gas production. In: Manure Digestion, Runoff, Refeeding, Odors. pp 1-10. Iowa State Univ. and USDA Pub. MWPS-25.

Varel, V. H. and A. G. Hashimoto. 1981. Effect of dietary monensin or chlortetracycline on methane production from cattle waste. Appl. Environ. Microbiol. 41:29.

Anaerobic Digestion Economics

Hashimoto, A. J. and Y. R. Chen. 1981. Economic optimization of anaerobic fermenter designs of beef production units. In: Livestock Waste: A Renewable Resource. pp 129-132. Amer. Soc. Agr. Engrs., St. Joseph, MI.

Maramba, F. D. 1978. Biogas utilization and economics. In: Biogas and Waste Recycling: The Philippine Experience. pp 143-180. Regal Printing Co., Manila.

Martin, J. H., Jr. and R. C. Loehr. 1981. Economic analysis of biogas production and utilization. In: Livestock Waste: A Renewable Resource. pp 327-329. Amer. Soc. Agr. Engrs., St. Joseph, MI.

Scheller, W. A. 1982. Commercial experience with a plug flow anaerobic digester for the production of biogas from agricultural and food processing wastes. In: Energy from Biomass: 2nd E.C. Conference. pp 492-496. Applied Science Publishers, London.

Feeding Digester Effluent

Fontenot, J. P. 1982. Assessing animal-waste feeding. Animal Nutrition and Health. March 1982.

Lizdas, D. J., W. B. Coe and M. H. Turk. 1981. Field experiments with an anaerobic fermentation pilot plant. In: Fuel Gas Production from Biomass, Vol. I. pp 201-224. CRC Press, Boca Raton, FL.

Prior, R. L., R. A. Britton and A. G. Hashimoto. 1981. Nutritional value of anaerobically fermented beef cattle wastes in diets for cattle and sheep. In: Livestock Waste: A Renewable Resource. pp 54-60. Amer. Soc. Agr. Engrs., St. Joseph, MI.

Prior, R. L. and A. G. Hashimoto. 1981. Potential for fermented cattle residue as a feed ingredient for livestock. In: Fuel Gas Production From Biomass, Vol. II. pp 215-237. CRC Press, Boca Raton, FL.

9

UTILIZATION OF CATTLE MANURE FOR FERTILIZER

John M. Sweeten

NUTRIENT CONTENT

Cattle manure is considered a good "shotgun" fertilizer. It contains nitrogen, phosphorus, and potassium and is an excellent source of micronutrients such as iron and zinc. Nutrient concentrations of manure vary widely because of differences in diet, weathering, and handling practices. On the average, cattle feedlot manure contains 35% moisture, 1.3% nitrogen (N), 1.2% phosphorus (P_2O_5), and 1.8% potassium (K_2O) as shown in table 1 (Mathers et al., 1972). On a dry-matter basis, feedlot manure can be expected to contain an average of 2.1% N, 1.9% P_2O_5, 2.8% K_2O, and 0.32% Fe.

This chemical analysis suggests that feedlot manure containing 35% moisture may potentially be worth $13/t based on its nutrient content. However, because some soils do not need additional phosphorus or potassium, and less than half the nitrogen in manure is available to plants during the first cropping season, the first-year fertilizer value of feedlot manure may be only $1.70 to $8.80/t.

Data in table 2 show the nutrient content of manure from eight dairy corrals and four beef cattle feedlots in Arizona (Arrington and Pachek, 1981). Nutrient values for dairy corral manure were similar to those for beef feedlot manure. Stockpiled dairy manure had a 20% lower nitrogen concentration than did dairy corral manure. The data for feedlot manure revealed slightly higher values for nitrogen but lower concentrations of all other nutrient and salt parameters than did those reported in table 1 for Texas cattle. The higher nitrogen values reflected in the Arizona data are believed attributable to quick air drying in the desert climate, which retards bacterial decomposition of organic matter and conversion of organic nitrogen. Other nutrient analyses for dairy and beef cattle manure have been published by Azevedo and Stout (1974); Ward et al. (1978); Gilbertson et al. (1971); and Miner and Smith (1975).

TABLE 1. CHEMICAL ANALYSIS OF MANURE FROM 23 TEXAS FEEDLOTS

	Range, % Wet basis			Avg nutrient concentration		
				Wet basis %	Dry basis %	Lb/ton dry basis
Nitrogen (N)	1.16	to	1.96	1.34	2.05	41
Phosphorus (P_2O_5)	.74	to	1.96	1.22	1.86	37
Potassium (K_2O)	.90	to	2.82	1.80	2.75	55
Calcium (Ca)	.81	to	1.75	1.30	1.98	40
Magnesium (Mg)	.32	to	.66	.50	.76	15
Iron (Fe)	.09	to	.55	.21	.32	6
Zinc (Zn)	.005	to	.012	.009	.014	.3
Sodium (Na)	.29	to	1.43	.74	1.13	23
Moisture	20.9	to	54.5	34.5	.0	-

Source: A. C. Mathers, B. A. Stewart, J. D. Thomas, and B. J. Blair (1972).

TABLE 2. AVERAGE CHEMICAL ANALYSIS OF BEEF AND DAIRY CATTLE MANURE IN ARIZONA[a]

	Fresh cattle manure, %	Beef feedlots, %	Dairy corrals, %	Dairy manure stockpile, %
Moisture	76.50	16.20	34.00	4.30
Nitrogen	2.20	2.37	2.13	1.72
Phosphorus (P_2O_5)	1.31	.87	.82	.76
Potassium (K_2O)	.81	1.56	1.96	1.43
Calcium	1.52	1.46	1.81	2.04
Magnesium	.50	.55	.61	.67
Iron	.059	.14	.16	.23
Zinc	.0079	.0048	.0071	.0045
Sodium	.32	.72	.44	.45
Sulfur	.28	.23	.21	.12
Manganese	.011	.0098	.012	.014
Copper	.0019	.0025	.002	.0019
Ash	15.00	35.10	33.90	46.20
Soluble salts	3.70	4.96	4.74	3.78

Source: R. M. Arrington and C. E. Pachek (1981).
[a]All data except moisture are on a dry weight basis.

GENERAL CROPPING CONSIDERATIONS

Manure should be applied to crops that generally respond best to nitrogen fertilization (e.g., grain sorghum and corn) and should be used primarily as a phosphorus source, or on problem soils to correct chemical imbalances.

Recommended manure application rates for solid manure from open cattle feedlots in the southern Great Plains states are normally as follows: for corn and grain sorghum, 10 t/acre/yr; for wheat and cotton, 5 t/acre/yr; for alfalfa, 10 t/acre/yr (wet basis). These application rates were developed on the basis of many years of research and farmer experiences with manure utilization as fertilizer. Farmers should apply more manure on sandy soils because of lower soil fertility and greater leaching losses.

Extensive research in Texas, Kansas, and Nebraska generally has shown that corn and grain sorghum yields from 10 t/acre/yr are as high as those from 25 t/acre/yr to 100 t/acre/yr manure application rates. Moreover, it is more profitable to fertilize with 10 t/acre than at the higher rates.

The fertilizer value of manure can be measured as the cash value of increased crop production resulting from its use, less the application costs. For example, USDA research at Bushland, Texas, determined grain sorghum yields after applying feedlot manure for 5 yr (1969-1973), as shown in table 3 (Mathers et al., 1975). Annual application of manure at 10 t/acre/yr (wet basis) consistently produced maximum yields of grain sorghum. Incremental yield increases, in terms of dollars returned per ton of manure applied, strongly favored the 10 t/acre/yr application rate, which produced 2,150 lb/acre more grain sorghum annually than did the no-fertilizer treatment. At $6.00/cwt, the extra grain sorghum production would amount to a return of $12.90/t of manure applied. The 30 t/acre application rate resulted in $4.00/t return on manure applied. Heavier application rates would not pay the collection, hauling, and spreading cost.

Yields were decreased by manure application of 120 t/acre/yr and 240 t/acre/yr because of higher salt and ammonia concentrations in the root zone. However, yields increased dramatically the first year after such heavy applications were discontinued. As leaching of salts continued during the recovery period, peak yields were achieved on these heavily treated plots.

NITROGEN

Available nitrogen and salt are the major limiting factors in determining land application rates of livestock and poultry manures (Gilbertson et al., 1979). Application rates are usually based on nitrogen because it is the most widely used fertilizer element and the most mobile nutrient from the standpoint of surface and groundwater pollution control. Mathers et al. (1975) found that when 30 t/acre or more manure was applied, nitrate accumulated in the top 6.5 ft of soil, and some nitrate moved to a depth of 20 ft.

TABLE 3. VALUE OF FEEDLOT MANURE IN GRAIN SORGHUM PRODUCTION, 1969-1973, BUSHLAND, TEXAS

Annual treatment	Avg yield, lb/acre/yr	Yield increase,[a] lb/acre/yr	Incremental yield value	
			$/acre/yr	$/t
Check--no fertilizer	4,490	-	-	-
N (240 and 120 lb N/acre)	6,440	1,950	117	-
N-P-K (240 and 120 lb N/acre)	6,410	1,920	115	-
Manure - 10 t/acre	6,640	2,150	129	12.90
30 t/acre	6,490	2,000	120	4.00
60 t/acre	6,360	1,870	112	1.87
120 t/acre	5,120	630	38	.32
240-(3 yr trmt and 2 yr recovery)	900/6,800	-1,230	-74	-.10
240-(1 yr trmt and 4 yr recovery)	330/6,750	976	58	.24

Source: A. C. Mathers, B. A. Stewart and J. D. Thomas (1975).
[a]Yield increase relative to check plot.

TABLE 4. DRY TONS OF MANURE NEEDED TO SUPPLY 100 LB OF AVAILABLE NITROGEN OVER THE CROPPING YEAR[a]

Length of time applied, yr	Nitrogen content of manure, % dry basis					
	1.0	1.5	2.0	2.5	3.0	4.0
	Tons of dry manure/100 lb N					
1	22.2	11.6	7.0	4.6	3.1	1.4
2	15.6	9.0	5.8	3.9	2.8	1.4
3	12.7	7.7	5.1	3.6	2.6	1.4
4	11.0	6.9	4.7	3.4	2.5	1.3
5	9.8	6.3	4.4	3.2	2.4	1.3
10	6.9	4.9	3.7	2.8	2.2	1.3
15	5.6	4.2	3.3	2.6	2.0	1.2

Source: C. B. Gilbertson, F. A. Norstadt, A. C. Mathers, R. F. Holt, A. P. Barnett, T. M. McCalla, C. A. Onstad, R. A. Young, L. A. Christensen and D. L. Van Dyne (1979).
[a]Manure tonnage values are for repeated annual applications on the same acreage.

A method for determining proper manure application rates based on nitrogen content was developed at the USDA Research Center at Bushland, Texas. This technical guide takes into account the slow rate of release of organic nitrogen in manure. Recommended manure application rates are shown in table 4. For example, suppose cattle manure contains 2% nitrogen on a dry matter basis. As shown in table 4, it takes 7 t of manure (dry basis) the first year to supply 100 lb/acre of available nitrogen. In succeeding years, release of residual organic nitrogen lowers the manure requirement to 5.8 t/acre in the second year and to 4.4 t/acre in the fifth year. Because of nitrogen losses after manure is applied to soil, application rates listed in table 4 should be increased by approximately one-third if manure is to be applied rather than incorporated into the soil.

PHOSPHORUS

When adequate N is supplied by manure, P and K are usually adequate for crop production as well. Although applied phosphorus often exceeds crop requirements, it rarely causes toxicity problems (Gilbertson et al., 1979). Approximately half of the N requirement could be provided by manure and the other half by commercial fertilizer to reduce P and K application rates.

Soils with less than 10 ppm phosphorus respond exceptionally well to feedlot manure in terms of increased crop yields, but soils with more than 20 ppm P do not show a phosphorus fertilizer response (Pennington, 1979). Manure is a cheap source of phosphorus, but it is an expensive source of nitrogen relative to anhydrous ammonia. Therefore, feedlot manure could be applied on fields lowest in phosphorus, with the extra nitrogen requirement (if any) purchased as anhydrous ammonia.

To illustrate, 261 lb/acre of 18-46-0 fertilizer would be needed to supply 120 lb/acre of P_2O_5, which is the amount of available phosphorus in 10 t of feedlot manure (Mathers et al., 1972). At a price of $294/t, commercial 18-46-0 fertilizer would cost $39.90/acre, including $1.50/acre for spreading. But an additional 107 lb of anhydrous ammonia at $260/t would be needed to supply the 137 lb of available nitrogen present in 10 t of feedlot manure. Including spreading costs, the total cost for supplying the commercial N and P_2O_5 would be $54.30/acre. This is equivalent to $5.43/t of manure. Alternatively, if 10-34-0 fertilizer is used as the phosphorus source, the equivalent value of feedlot manure would be $6.65/t. Therefore, if feedlot manure can be bought and spread for less than $5.40/t, then it is a better buy than commercial phosphorus plus anhydrous ammonia with the above price scenario.

IRON DEFICIENCY

Iron deficiency in grain sorghum may cause yield losses of 30% to 40% in areas with high calcium (calcareous) soils. States west of the Mississippi River produce grain sorghum on 12 million acres of calcareous soils.

Applying feedlot manure to calcareous soils in the Texas High Plains corrected iron chlorosis and increased yields of grain sorghum (Thomas and Mathers, 1979; Mathers et al., 1980). Experiments using Arch fine sandy loam soil, found around lakes, compared the effects of manure and iron (Fe) treatments on grain sorghum. In greenhouse experiments, manure increased dry matter yields by 400% as compared to check and N-P-K treatments (Thomas and Mathers, 1979). Apparently, the manure had a chelating effect that increased Fe availability. Commercial fertilizer worsened iron chlorosis and decreased yields slightly, unless supplemental iron was added.

In field experiments, feedlot manure on Arch soil in West Texas increased grain sorghum production from nearly zero on untreated plots to 7,000 lb/acre after manure application (Mathers et al., 1980). Based on $6/cwt grain sorghum prices, this was a yield increase of $420/acre, or $28 to $42/t of applied manure. Application rates were 0 t, 5 t, and 15 t of manure dry matter/acre. Yields from the 5-t treatment were almost equal to the higher rate. Grain sorghum that had been badly affected by iron deficiency responded extremely well to 15 t of feedlot manure (dry matter basis). Application of 20 lb Fe/acre for 3 yr also seemed to solve iron deficiency problems.

LIQUID BEEF CATTLE MANURE AS FERTILIZER

Liquid beef cattle manure is produced in beef cattle confinement buildings at the rate of .6 to 1.0 cu ft/hd/day depending upon ration and live weight. Spilled drinking water or flush water, if any, is not included in this value. The nutrient content of liquid beef cattle manure varies because of different diets and nutrient losses resulting from various handling systems. A literature search yielded the nutrient content shown in table 5.

TABLE 5. NUTRIENT CONTENT OF LIQUID BEEF CATTLE MANURE FROM CONFINEMENT BUILDINGS

Nutrient	Daily nutrient production, lb/hd	Slurry concentration, % w.b.		Slurry content, lb/1000 gal	
		Range	Mean	Range	Mean
N	.27	.3 to .8	.55	25-65	45
P_2O_5	.20	.18 to .58	.3	15-48	25
K_2O	.23	.20 to .47	.3	17-39	25

The maximum nutrient value of 1,000 gal of undiluted beef manure slurry is approximately as follows: N--$5.85; P_2O_5--$7.25; K_2O--$1.75; or total nutrient value = $14.85/gal. As previously indicated, not all this nutrient value can be realized.

Nitrogen loss during storage is usually less for liquid beef manure than that for solid feedlot manure exposed to the outdoor elements, as evidenced by the higher nitrogen percentage on a dry matter basis (e.g., 3.7% in liquid vs 2.1% in solid beef cattle manure). However, most of the nitrogen in liquid manure is in the ammonium form and is readily volatilized upon agitation and land application. Beauchamp (1979) measured NH_3 volatilization losses ranging from 24% to 33% (29% average) in beef manure slurry within 6 days to 7 days following land application.

Crop yields resulting from application of liquid beef cattle manure have been studied by many researchers. Representative of this research is the work of Evans (1979) in western Minnesota. Liquid beef manure from the deep pit of a total confinement building and solid beef manure from a bedded feeding floor under roof were applied each October to corn for the crop years 1973 to 1978 (table 6). Manure treatments gave consistently higher corn grain yields than did the inorganic fertilizer treatment. The crude protein content of corn grain was also consistently higher for manure treatments (11.3% crude protein) as compared to the check and inorganic fertilizer treatments (9.8% and 10.7% crude protein, respectively).

The lowest manure application rates gave the highest returns per ton from manure application (table 6). For solid manure (21% to 41% DM), doubling and tripling the application rate (2X and 3X rates) provided only $1.12/t and $1.48/t (d.b.) yield increases, respectively, which did not pay manure-handling cost. The lowest application rate of liquid manure (4.7% to 10.8%) provided $11.00 per dry ton return through corn, which is equal to about $3.56/1,000 gal.

Subsoil injection or knifing of manure slurries containing up to 12% solids is recommended in many instances. Nitrogen losses are only 5% after soil injection as compared to 25% or more with surface spreading. Subsurface injection reduces odors and prevents water pollution, but it takes more tractor horsepower. However, nitrogen savings will probably more than pay for soil injection (Scarborough et al., 1978).

TABLE 6. CORN GRAIN YIELD AND MANURE VALUE (PRODUCTION BASIS) AS AFFECTED BY APPLICATION RATE, 1973 TO 1978, MORRIS, MINNESOTA

	Avg application rate, wt/acre/yr						Value of yield increase[a]	
Treatment	Manure DM, t	N, lb	P_2O_5 lb	K_2O lb	Avg grain yield, lb/acre/yr	Yield increase, lb/acre/yr	$/acre/yr	$/dry t
Check	-	0	0	0	4,654	-	-	-
Inorganic fertilizer	-	100	40	40	5,237	582	38.44	-
Solid beef manure								
1X	20	460	244	476	5,567	913	60.23	3.01
2X	40	920	488	952	5,531	677	44.68	1.12
3X	60	1,380	732	1,427	6,009	1,354	89.37	1.48
Liquid beef manure								
1X	3.9	203	91	95	5,304	649	42.86	11.00
2X	7.7	407	182	190	5,639	985	64.99	8.40
3X	11.6	610	272	285	5,673	1,019	67.23	5.80

Source: S. D. Evans (1979).

[a]Assumes price of corn at $6.60/cwt.

RESIDUAL NUTRIENT BENEFITS OF MANURE

Application of dairy manure for 3 yr greatly improved in the yield of coastal Bermuda grass in Alabama, both during and after the manure treatment was applied, because of the residual nutrients in manure (Lund et al., 1975; Lund and Doss, 1980). Solid dairy manure was surface-applied for 3 yr at annual rates of 20 t and 40 t/acre (dry matter). Liquid dairy manure was simultaneously applied at dry matter rates of 20 t, 40 t, and 60 t/acre. Manure composition averaged 1.95% organic N, .58% P, and 1.20% K (dry weight basis). Thus, organic nitrogen application rates were very high: 780 lb, 1,560 lb, and 2,340 lb/acre/yr. Check plots received commercial fertilizer (N-P-K) at annual rates of 420-200-420 lb/acre/yr. Field experiments were conducted on two soils: Dothan loamy sand and Lucedale fine sandy loan.

The check plots that received commercial fertilizer produced a total of 19.7 t/acre of coastal Bermuda grass during the 3-yr treatment period, as shown in table 7 (Lund et al., 1975). Most manure treatments produced less forage the first year than did the commercial fertilizer. Yields from the lowest manure application rate of 20 t/acre caught up with those of the commercial fertilizer treatment by the second or third year. The 40 t/acre application rate produced more hay in the second and third years than did the commercial fertilized plots. However, 3 yr of manure application at 40 t and 60 t/acre became detrimental to grass stand, and weed encroachment was noted. Forage nitrate levels essentially quadrupled, with the highest manure rate causing nitrate levels of 0.25% NO_3-N in forage.

TABLE 7. TOTAL YIELDS OF COASTAL BERMUDA GRASS HAY (TONS/ACRE) DURING AND AFTER 3 YR OF DAIRY MANURE TREATMENT, 1971 TO 1976

	Dothan loamy sand			Lucedale fine sandy loam		
Fertilizer treatment	3 yr during t/acre	3 yr after t/acre	Total 6 yr t/acre	3 yr during t/acre	3 yr after t/acre	Total 6 yr t/acre
Check (420-200-420)	19.7	9.1	28.8	23.6	3.8	27.4
Dairy manure dry t/acre						
20[a]	17.4	20.2	37.6	17.7	14.7	32.4
40[a]	21.1	26.2	47.3	24.4	22.0	46.4
60	23.0	30.2	53.2	26.8	24.9	51.7

Source: Z. F. Lund and B. D. Doss (1980).
[a]Combined average results for liquid and solid manure treatments.

Residual nutrients from manure greatly improved yields of coastal Bermuda grass hay for 3 yr after manure application ceased (Lund and Doss, 1980). When commercial fertilizer treatment was halted, hay yields the following year dropped to only 3.7 t/acre on the Dothan loamy sand and 1.7 t/acre on the Lucedale fine sandy loam and to only 1.8 t and 0.7 t/acre, respectively, in the third year without fertilizer (table 8). By contrast, the manured plots produced 7.5 t to 10.8 t/acre of hay the first year and 2.4 t to 7.9 t/acre in the third year after manure treatment stopped (table 8). Residual fertilizer effects were greatest for the highest manure application rates. Response to liquid and solid manure was essentially the same at the same dry manure application rates.

TABLE 8. YIELDS OF COASTAL BERMUDA GRASS HAY AFTER MANURE TREATMENTS CEASED, AUBURN, ALABAMA, 1974 TO 1976

Prior fertilizer treatment	Hay yields (t/acre) in years following last treatment					
	Dothan loamy sand			Lucedale fine sandy loam		
	1	2	3	1	2	3
Check (420-200-420)	3.7	3.6	1.8	1.7	1.4	0.7
Manure (dry basis)[a]						
20 t/acre	7.5	8.3	4.4	7.5	4.8	2.4
40 t/acre	8.2	11.7	6.3	10.1	7.4	4.5
60 t/acre	8.6	13.7	7.9	10.8	8.3	5.8

Source: Z. F. Lund and B. D. Doss (1980).
[a]Combined average results for liquid and solid manure.

For all 6 yr of this project, the 20 t/acre (dry basis) manure application rate was the best treatment, outyielding the commercially fertilized plots by 8.8 t/acre of hay (31%) on the Dothan loamy sand and 5.0 t/acre (18%) on the Lucedale fine sandy loam (table 7). All of this beneficial difference was attributable to the residual nutrients released from the manure in the last 3 yr.

IMPROVEMENTS IN SOIL PHYSICAL PROPERTIES

Animal manures added to soils tend to increase soil porosity, permeability, and water holding capacity and to decrease bulk density and modulus of rupture. Hafez (1974) reported that soil bulk density was reduced 4%, 13%, and 23% by adding manure at the rate of 2.5%, 5%, and 10% of the soil mass, respectively. Fibrous manures (dairy, beef, and horse manure) were more effective in reducing bulk density than was poultry manure.

In the Texas High Plains, feedlot manure was applied to Pullman clay loam soil at rates of 0 t, 10 t, 30 t, 60 t, and 120 t/acre (wet basis) for 4 yr (Unger and Stewart, 1974). Major effects on soil properties were as follows:

- Reduced soil bulk density by 3% at 10 t/acre and up to 18% at 120 t/acre as a result of increased

soil organic matter and improved soil aggregation.
- Increased the soil organic matter content by 50% at 10 t/acre and up to 100% at 60 t/acre.
- Increased soil porosity by 2% at 10 t/acre and up to 33% at 60 t/acre.
- Increased the percentage of larger water stable aggregates at the two higher application rates, indicating greater soil structural stability and possibly higher water infiltration rates.

According to Unger and Stewart (1974), the 10-t/acre application rate, which was adequate for crop fertilization, did not cause statistically significant improvements in most soil physical properties as compared to the check treatment. However, manure effects on soil conditions became significant as application rates increased. At high manure application rates, no detrimental effects on soil conditions were noted, even though large amounts of salts were added. Most salts were leached from the root zone by irrigation water. The lower bulk densities, large differences in soil water content between saturation and .2 bar matric potential, and higher porosity values resulting from larger manure application rates, suggesting that water infiltration rate would be higher and surface moisture retention near the soil surface would be lower. Higher irrigation water infiltration rates were observed on the field plots that received the 60 t and 120 t/acre/yr manure applications, as compared to those with lower application rates.

Feedlot manure applied to Pullman clay loam soil nine times in 11 yr significantly increased the soil organic matter content (Mathers and Stewart, 1981). Nine manure applications at 10 t/acre increased the organic matter to 3.2%, as compared to 2.0% for check plots and 1.9% to 2.1% for commercially fertilized plots that received no manure. The 30 t/acre application rate resulted in 4.5% organic matter in the soil. Increased organic matter content was maintained only by continued applications of manure rather than single high rate treatments. Soil bulk density was reduced slightly (3% to 4%) by manure applied for 9 yr at 10 and 30 t/acre/yr. The very high application rate of 120 t/acre for 5 yr lowered the soil bulk density by 10%, as compared to that of the check plots. A high correlation was found between soil bulk density and percentage organic matter.

Infiltration rate depends on the proportion of larger pores in the soil surface, stability of surface soil aggregates, soil moisture content, and surface cover conditions. Infiltration rates in fine-textured soils are often increased by application of animal wastes (Gilbertson et al., 1979; Hafez, 1974). McCalla (1942) determined that composted cattle manure mixed with soil at 4% concentration resulted in 92% to 144% increase in water infiltration rate, as compared to that in the untreated soil. Smith et al. (1937) showed that manure increased infiltration rates in

Clarion loam. Mazurak et al. (1955) showed that manure increased water infiltration rate of Tripp very fine sandy loam. Swader and Stewart (1972) showed that manure decreased bulk density and increased infiltration rate of Pullman clay loam. Lemmerman and Behrens (1935) showed that permeability was increased on plots receiving farmyard manure as compared to similar nonmanured plots.

Mathers and Stewart (1981) determined that there was significant correlation between hydraulic conductivity and percentage of organic matter in the soil, and that hydraulic conductivity quadrupled in soils that received feedlot manure for 9 yr at 10 t/acre/yr. Although sodium levels were increased by manure application at all rates and durations (6% to 29%), they were not sufficiently high to cause serious problems on these soils. The beneficial effect of organic matter on water infiltration rate was greater than the detrimental effect of increased sodium content (Mathers and Stewart, 1981).

MANURE APPLICATION EQUIPMENT

Selection of equipment and procedures for manure application will depend on manure moisture content, transportation distances, application rates, and economic factors. Types of equipment available for transportation and land application of manure are categorized in table 9 according to the manure moisture content. Manure containing less than 4% solids can be pumped readily with less than 10% increase in hydraulic friction (head) loss as compared to that in pumping irrigation water. Large diameter (big gun) sprinklers with 1 in. to 2 in. nozzles provide efficient disposal of large volumes of liquid manure and wastewater. Conventional medium-bore sprinklers with .25 in. to .5 in. nozzle diameter can be utilized only for mechanically screened liquid manure, for second-stage lagoon effluent, or for runoff stored in holding ponds. Open ditches, furrows, or borders are not recommended for distribution of liquid manure because of the settling of the solids and nonuniform distribution of organic matter.

Most vacuum-loaded tank wagons or trucks discharge liquid manure directly behind the unit, although some commercial tank wagons have side discharge nozzles. Application rates can be controlled by compressed air pressure and ground speed.

Ammonia loss can reach 25% to 70% depending upon soil pH, temperature, and other factors. To retain nitrogen and minimize odors, surface application of liquid manure should be followed by disking to a depth of 4 in. to 6 in. Soil injection attachments for knifing or chiseling liquid manure to 6 in. to 10 in. soil depth from tank wagons are widely used for liquid manure application in row crops during early spring and late fall. This will limit volatilization of ammonia to 5% or less.

TABLE 9. TYPES OF MANURE-SPREADING EQUIPMENT

Manure consistency	Total solids content, % w.b.	Types of spreading equipment
Solid manure	35-90	(a) Spreader trucks (300 to 570 cu ft) (b) Tractor-drawn box spreader (67 to 325 cu ft)
Semisolid or semiliquid manure	10-35	Tractor-drawn, side-discharge flail manure spreader (81 to 240 cu ft)
Liquid slurry	2-15	Tank wagon or tank truck (160 to 470 cu ft) (a) Surface spread vs soil incorporation (b) Vacuum vs pump loaded
Liquid manure (with fiber separation)	0-3	Irrigation (a) Big gun sprinkler (1 to 2 in. nozzle), or (b) Gated pipe
Lagoon or holding pond water	0-1	Irrigation (a) Conventional sprinkler nozzle, or (b) Big gun sprinkler, or (c) Gated pipe

Open-tank, flail-chain spreaders are used to spread semisolid or semiliquid manure. This includes fresh manure with little or no dilution water, as well as wet corral-shaped manure collected after rainy weather or from pot holes on the feedlot surface. Capacities of flail-chain spreaders range from 80 cu ft to 240 cu ft. They require tractor sizes of 55 hp to 100 hp. Short chains attached to a PTO-driven center shaft discharge the manure to either side.

Tractor-drawn box spreaders for solid manure are available in capacities of 67 cu ft to 325 cu ft. If the manure source is nearby, or if manure has been stockpiled near the fields, it can be reloaded and spread efficiently. Tractor power requirements range from 43 hp for the smallest spreader to 100 hp for the largest models.

Trucks built for spreading solid manure have 300 cu ft to 570 cu ft beds mounted on a single- or dual-axle truck chassis. Chain-driven flights on the truck bed move the manure to the rear during unloading. Serrated augers or beaters at the rear of the truck beds reduce particle-size and improve manure-spreading uniformity. These manure spreader trucks are widely used in the cattle feedlot industry where large tonnages must be moved each year to fields up to 15 miles away.

SUMMARY AND CONCLUSIONS

Cattle manure supplies nitrogen, phosphorus, potassium, and essential micronutrients to plants. In this respect, cattle manure is aptly called a "shotgun fertilizer." Depending on ration and manure handling practices, cattle manure can be expected to contain roughly 2% nitrogen, 1% P_2O_5, and 2% K_2O on a dry weight basis. Less than half the nitrogen and phosphorus are available the first year after application. Transportation costs limit the economical haul distance of cattle manure to less than 20 miles in most cases, depending upon manure quality and cropping circumstances. The highest cash returns from cattle manure can be obtained by applying manure 1) on crops requiring nitrogen and phosphorus; 2) at limited application rates of 5 to 15 dry t/acre in most circumstances; or 3) on soils with certain types of chemical imbalance.

Substantial yield benefits are obtained from residual nutrients in manure up to 3 yr after application is halted. Grain and hay yields following application of beef and dairy cattle manure (liquid or solid) at normal agronomic rates usually will equal or exceed yields from comparable levels of commercial fertilizers. However, higher application rates are not economical in most instances, and excessive application rates can contribute to soil salinity and nitrate problems.

Changes in soil physical properties--including increased infiltration rate, organic matter content, and water holding capacity--often occur following application of manure at high rates, such as 50 t to 100 t/acre/yr of manure dry matter. But these rates require careful soil management to prevent soil salinity problems. When manure is applied annually at normal agronomic rates to supply plant nutrient needs, changes in soil physical properties may be difficult to measure initially but are significant over many years of continued manure application. Water infiltration rate may be increased due to manure application because of improved soil structure, greater porosity, and reduced bulk density. On most soils, the beneficial effects of manure organic matter on water infiltration outweigh any detrimental effects of increased sodium content.

REFERENCES

Arrington, R. M. and C. E. Pachek. 1981. Soil nutrient content of manures in an arid climate. In: Livestock Waste: A Renewable Resource. Proc. 4th Int. Symp. on Livestock Waste - 1980. pp 150-152. American Society of Agriculture Engineers. St. Joseph, MI.

Azevedo, J. and P. R. Stout. 1974. Farm animal manures: An overview of their role in the agricultural environment. California Agr. Exp. Stat. and Ext. Serv. Manual 44. Univ. of California.

Beauchamp, E. G. 1979. Liquid cattle manure--corn yield and ammonia loss. Presented at 1979 Summer Mtg., Amer. Soc. of Agr. Eng. and Canadian Soc. of Agr. Eng., Winnipeg, Manitoba, Canada.

Evans, S. D. 1979. Manure application studies in west-central Minnesota. ASAE Paper No. 79-2119. Amer. Soc. of Agr. Eng., St. Joseph, MI.

Gilbertson, C. B., F. A. Norstadt, A. C. Mathers, R. F. Holt, A. P. Barnett, T. M. McCalla, C. A. Onstad, R. A. Young, L. A. Christensen and D. L. Van Dyne. 1979. Animal Waste Utilization on Cropland and Pastureland: A Manual for Evaluating Agronomic and Environmental Effects. URR 6. U.S. Department of Agriculture, Agricultural Research Service. Washington, D.C.

Gilbertson, C. B., T. M. McCalla, J. R. Ellis and W. R. Woods. 1971. Characteristics of manure accumulation removed from outdoor, unpaved beef cattle feedlots. In: Livestock Waste Management and Pollution Abatement. Proc. 2nd Int. Symp. on Livestock Waste. American Society of Agricultural Engineers, St. Joseph, MI.

Hafez, A. A. R. 1974. Comparative changes in soil physical properties induced by admixtures on manures from various domestic animals. Soil Sci. 118:53.

Lemmerman, O. and W. U. Behrens. 1935. On the influence of manuring on the permeability of soils. Z. Pflanzenernahr, Dung. Bodenk 37:174.

Lund, Z. F. and B. D. Doss. 1980. Coastal Bermuda grass yield and soil properties as affected by surface-applied dairy manure and its residue. J. of Environ. Quality. 9:157.

Lund, Z. F., B. D. Doss and F. E. Lowry. 1975. Dairy cattle manure--its effect on yield and quality of coastal and Bermuda grass. J. of Environ. Quality 4(3):358.

Mathers, A. C. and B. A. Stewart. 1981. The effect of feedlot manure on soil physical and chemical properties. In: Livestock Waste: A Renewable Resource. Proc. 4th Int. Symp. on Livestock Wastes - 1980. pp 159-162. American Society of Agricultural Engineers, St. Joseph, MI.

Mathers, A. C., B. A. Stewart and J. D. Thomas. 1975. Residual and annual rate effects of manure on grain sorghum yields. In: Managing Livestock Wastes. Proc. 3rd Int. Symp. on Livestock Wastes - 1975. American Society of Agricultural Engineers, St. Joseph, MI.

Mathers, A. C., B. A. Stewart, J. D. Thomas and B. J. Blair. 1972. Effects of cattle feedlot manure on crop yields and soil conditions. Tech. Rep. No. 11. Texas Agr. Exp. Sta. and USDA Southwestern Great Plains Res. Center, Bushland, TX.

Mathers, A. C., J. D. Thomas, B. A. Stewart and J. E. Herring. 1980. Manure and inorganic fertilizer effects on sorghum and sunflower growth on iron - deficient soil. Agron. J. 72:1025.

Mazurak, A. P., H. R. Cosper and H. F. Rhoades. 1955. Rate of water entry into an irrigated chestnut soil as affected by 39 years of cropping and manurial practices. Agron. J. 47:490.

McCalla, T. M. 1942. Influence of biological products on soil structure and infiltration. Soil Sci. Soc. of Amer. Proc. 7:209.

Miner, J. R. and R. J. Smith. 1975. Livestock waste management with pollution control. Midwest Plan Serv. Handbook MWPS-19, North Central Res. Publ. No. 222. Iowa State Univ., Ames.

Pennington, H. D. 1979. Personal communication. Texas Agr. Ext. Serv., The Texas A&M Univ. System, Lubbock.

Scarborough, J. N., E. C. Dickey and D. H. Vanderholm. 1978. Sizing of liquid manure tank wagons and the economic evaluation of liquid manure injection. Transactions of the ASAE 21(6):1181.

Smith, F. B., P. E. Brown and J. A. Russell. 1937. The effect of organic matter on the infiltration capacity of Clarion loam. J. of Amer. Soc. of Agron. 29:521.

Swader, F. N. and B. A. Stewart. 1972. The Effect of Feedlot Wastes on Water Relations of Pullman Clay Loam. ASAE Paper No. 72-959. Amer. Soc. of Agr. Eng., St. Joseph, MI.

Thomas, J. D. and A. C. Mathers. 1979. Manure and iron effects on sorghum growth on iron deficient soil. Agron. J. 71:792.

Unger, P. W. and B. A. Stewart. 1974. Feedlot waste effects on soil conditions and water evaporation. Soil Sci. Soc. of Amer. Proc. 38:954.

Ward, G. M., T. V. Muscato, D. A. Hill and R. W. Hansen. 1978. Chemical composition of feedlot manure. J. of Environ. Quality 7:159.

Part 3

INFORMATION CHANNELS AND INSTITUTIONAL STRUCTURES

10

FINDING AND USING PROBLEM-SOLVING TECHNOLOGY AND INFORMATION IN LIVESTOCK PRODUCTION

Dixon D. Hubbard

THE INFORMATION SOCIETY

Although many continue to think we live in an industrial society, we have in fact changed to an information society. Today's information technology--from computers to cable television--did not bring about this new information society. It was underway by the late 1950s. Today's sophisticated technology only serves to hasten the development of the information society. The problem is that our thinking, attitudes, and, consequently, our decision making have not caught up with reality.

In 1950, only about 17% of the people in this country worked in information jobs. Now more than 60% work in information. It is the number one occupation in the U.S. We now mass-produce knowledge, and this knowledge is the driving force of our economy. We are getting out of the work-hard-and-you-will-succeed-complex and into the thinking business. Change is occurring rapidly and the future success of agriculture will be governed by our ability to adapt to this change.

The pace of change will accelerate even more as communications technology "collapses information float". Communication requires a sender, a receiver, and a communication channel. Sophisticated information technology has revolutionized this process by vastly reducing the amount of time information spends in the communication channel. If I mail a letter to someone, it takes three or four days for them to receive it. If I send them a letter electronically, it takes a couple of seconds; that is "collapsing the information float". If they respond to my electronic letter in an hour, we have communicated in an hour rather than in a week.

Change is occurring much faster because of reduced information float. The speed with which information can presently be transmitted is awesome. However, we probably have not seen anything yet.

We are literally being inundated with various types of information--it is all around us. It is coming at us from every direction in ever-increasing quantities. The problem is to sort out the information that is applicable to the decisions we are having to make on a daily, hourly, or minute-by-minute basis. We are having to spend less time working and more time thinking. We are having a life and death struggle with our worship of the activity trap--equating activity with results--and reorienting our lives to planning and management by objective. We no longer have the luxury of postponing decisions and basing them predominantly on what we learned from our past. We are not having to learn from the present how to anticipate the future. Ultimately, we are probably going to need to be able to learn from the future the way we formerly learned from the past and make decisions accordingly.

BACKGROUND

From whence did we come in agriculture information and technology? The land grant colleges were created by the passage of the Morrill Act in 1862. This granted land to each state for establishing and supporting an institution to teach agriculture in addition to other areas of higher learning. These institutions, presently known as land-grant universities, were established especially for working people and were originally known as "the people's universities."

At the time the land grant universities were established, the U.S. was predominantly rural, and agriculture was the principal occupation. Farmers only had knowledge derived from experience, observation, or handed down from one generation to another. Traditional ways and empirical knowledge were valuable, but they were inadequate to meet the needs of a developing agricultural industry and of a developing nation.

Soon after the establishment of the early agricultural colleges, it was realized that they lacked a body of scientific knowledge and relevant subject matter to teach. Consequently, the Hatch Act, creating the agricultural experiment stations, was passed by the U.S. Congress in 1887. The scientific research conducted by these experiment stations, which were established as an integral part of the land grant universities and the U.S. Department of Agriculture, provided purposeful, effective, and dependable information for teaching agriculture. However, this information was available only to the few people who attended the land grant colleges. Thus, the Smith-Lever Act creating the Agricultural Extension Service was passed by Congress on May 8, 1914.

The passage of the Smith-Lever Act provided for cooperative extension work in agriculture, home economics, and related subjects between the land grant college of states and the U.S. Department of Agriculture. The cooperating states were required to furnish supporting funds that at lest equaled in amount those appropriated by the Congress.

Congressman Lever, who introduced the Smith-Lever Act and was Chairman of the House Agricultural Committee, said that the purpose of cooperative extension was to set up a system of general demonstration teaching throughout the country. The agent in the field, representing the college and the department, was to be the mouthpiece through which this information reached the people.

The major responsibility of the Agricultural Extension Service, as stated in the Smith-Lever Act, is "To aid in diffusing among the people of the United States useful and practical information on subjects relating to agriculture and home economics, and to encourage the application of the same." It further states in the Act that cooperative agricultural extension work "shall consist of giving instruction and practical demonstrations in agriculture and home economics and subjects related thereto to persons not attending or resident in said colleges (land grant universities) in the communities, and imparting information on said subjects through demonstrations, publications, and otherwise." Little did the Congressman realize the magnitude of all the subjects that would ultimately be related to agriculture and all the methods that would ultimately be available for imparting information on this subject. If he had, cooperative agricultural extension might have never gotten off the ground.

Seventy years after its birth, extension's mission is essentially the same. Today the Cooperative Extension Service interprets, disseminates, and encourages practical use of knowledge. It transmits information from researchers to the people and the people's problems to the researchers. But it also is an agency of change. It functions as a dynamic educational system oriented to the development of educational programs designed to meet the changing need of diverse publics. A major strength of extension is the involvement of people in the program-development process in determining, planning, and operating programs that meet their needs.

The Morrill Act establishing the land grant universities, the Hatch Act creating the Agricultural Experiment Stations, and the Smith-Lever Act creating the Cooperative Agricultural Extension Service were (and still are) key factors in the development and delivery of information and technology for livestock producers and agriculture in general. The land-grant system in cooperation with the U.S. Department of Agriculture still produces and extends a major portion of the information and technology available to agriculture producers in this country and many other countries. It also develops most of the agriculture scientists and educators for government, universities, and industry who continue to perpetuate the flow of information and technology to producers.

The wisdom, and possibly luck, of the leaders of this nation who fostered the land grant university system boggles the mind. This system tied to free enterprise is the backbone of the most dynamic and efficient food-producing machine in the world--American agriculture. Directly or indirectly, this accomplishment is responsible for most of the factors that contribute to the U.S. having the highest quality of life of any nation in the world.

I have talked with a lot of people who have visited and compared American agriculture with agriculture in other countries. I have yet to find one who has not developed a deeper appreciation for the land-grant system and free enterprise. They all will tell you that this is what makes American agriculture great. We not only have a system for generating needed information and technology but we also have a system for getting it to the people, and the people have the economic incentive to utilize it.

USING THE INFORMATION AVAILABLE

The point in all this discussion is that the land-grant system, in cooperation with USDA, provides livestock producers with the best resource in the world for finding and utilizing problem-solving information and technology. However, finding information and technology is like laying a water line. Nobody likes digging the ditch. However, it is the only way to get the water line laid. Since we can not survive without water, we either dig the ditch or we have to haul water.

Many livestock producers do not fully benefit from all the information and technology available to them through the land-grant system. The reasons for this vary: some people have a strong enough economic base that they do not have to improve efficiency, whereas others feel that they do not have adequate cash flow to implement new information and technology. I recall making recommendations to both types. Producers who are comfortable with the way they are doing things feel no need to change and will only accept new information and technology if it does not alter their management system significantly. This is fine as long as they can afford to be this independent, and some I have known can afford it a long time. On the other hand, I have known some that have gone down the tube before they ever knew they were in trouble.

It is not wise to ignore information and technological developments. The best example I can give of the plight of producers with inadequate cash flow relates to a fellow who really wanted to make a change I was recommending because it would make him money. However, after a little thought he said to me, "That's like recommending acupuncture for hemophilia". At least he had not lost his sense of humor. He also knew that had he sought help earlier he probably would not have been in the condition he was in.

Livestock producers who do not take advantage of the information and technology available through the land-grant system are wasting their tax dollars. They are helping to pay for information and technology they are not using. Granted, there are some weak spots in this system, but it is still the best in the world. If it is not responsive to your needs, make it responsive. When you contact your local county agent and you do not get what you need, then keep going up the ladder until you reach the state extension director, if necessary. But make it work--it is in your best interest as well as those who will come after you.

In addition to the USDA and the land-grant system, numerous other delivery systems have developed through which livestock producers can receive information and technology. These include commodity organizations, general farm organizations, agribusiness, financial institutions, private foundations, and others. Also, there is some networking between and among these systems. This is why we frequently receive the same information from a multiplicity of sources.

The information and technology delivered through these systems comes in the form of personal contact, telephone, letters, newsletters, bulletins, magazines, newspapers, books, meetings, seminars, symposia, workshops, radio, teletype, television, movies, cassette tapes, slide sets, computers, and various combinations of these methods. Thus, livestock producers, like everyone else in this country, are exposed to a constant flow of information and technology in verbal, written, and visual form. In fact, we are exposed to so much information and technology that many of us are becoming insensitive or overloaded to the degree that we are probably letting some good things go by. Also, the quantity of information and technology is still increasing along with improved methodology for delivering this information and technology. This is why planning is so important. We must set goals and establish objectives that will accomplish those goals; then we can search out the information and technology that will help us accomplish our objectives. If the information and technology are not available for us to accomplish our goals and objectives, we have reason to be active in getting research initiated that will provide us with what we need.

Networking

There are livestock producers who basically follow most of the recommendations I have made regarding such planning. They have a well-structured plan for their operation, set goals, establish objectives to reach these goals, manage by objective, and measure results. They are using basically all the information and technology available to them that is applicable to their operations. They are also actively involved in helping set the research priorities for their industries.

Something that I have noted about these people is that they do a lot of networking. Simply stated, networks are people talking to each other, sharing ideas, information, and resources. They are structured to transmit information in a way that is quicker, more high-tech, and more energy-efficient than any other process we know. They are a very appropriate form of communication and interaction that is suitable for the energy-scarce, information-rich future of the 1980s and beyond.

The type of networking I have seen among livestock producers is done by phone calls, conferences, grapevines, mutual friends, coalitions, tapes, newsletters, photocopying, parties, etc. There are probably millions of networks of a similar nature, to one or more of which most of us belong--the informal networks among friends, colleagues, community organizations--that never grow into the organizational stage.

One of networking's great attractions is that it is an easy way to get information--much easier, for example, than going to a library, university, or government. Experienced networkers claim they can reach anyone in the world with only six interactions. It has been my experience that I can reach nearly anyone I want in the U.S. with two or three exchanges.

Although sharing information and contacts is their main purpose, networks can go beyond the mere transfer of data to the creation and exchange of knowledge. As each person in a network takes in new information, he or she synthesizes it and comes up with other new ideas. Networks share these newly-forged thoughts and ideas.

I would encourage any serious livestock producer who is not part of a good informal network to give it serious consideration. Sharing ideas, information, and resources in this way can be very fruitful and save a lot of time and money.

Holistic Thinking

Another thing I have noted about livestock producers who take their businesses seriously is that they think holistically. They are always concerned about how altering one component of their management scheme might affect another. For example, the research data on performance testing of livestock are solid. Therefore, any serious livestock producer should be utilizing this technology. However, if you, as a cattle producer, add a performance-tested bull with a high growth rate to your herd, there are several other things you should consider in your management scheme. If you plan to stock at the same rate, then increased feed will have to be produced. Management of replacement heifers will need to be improved or they may not breed back as wet two's. Growth rate is highly correlated with birth weight, so there may be an increased calving difficulty if your cow herd can not accommodate larger birth weight. In other words, the use of performance-testing technology is good management; but, good management must know the limits to the use of this or any other technology.

When I was an extension specialist in Texas, there were over 100 recommendations with a sound research base that could be made relative to some aspect of the production of cotton. The quickest way for any cotton producer in the state of Texas to go broke was to try and implement all of these unsystemized recommendations simultaneously.

The point is that all information and technology (independent of how well-founded it is) must be tailored to fit the operation and management scheme of a producer. Some of it will not fit at all and, therefore, should not be applied. Good information and technology improperly applied can be an economic disaster. Sorting out the information and technology that will provide the highest economic returns and getting it effectively applied are what management is all about. This is why it is said that there is no substitute for good management.

Future Considerations

The vast quantity of information and technology available for producing livestock in this country is even beginning to stymie the best managers. Thus, there is a concerted effort on the part of the major livestock commodity organizations in this country to get the USDA and the land-grant system to be more responsive to this problem. They are insisting on the integration of disciplines and the functions of research and extension (in concert with industry) in both the identification and solving of problems. They are saying they cannot handle all the information and technology they receive in component parts anymore. It appears that the Congress of the U.S. will ultimately pass legislation to ensure that the USDA and the land-grant system become more responsive in this area. In the meantime, both the USDA and several land grant universities are making adjustments to accommodate this need.

I have not gone into detail in this paper on the specific methods of finding and utilizing information and technology. Basically, what I have said is that there is a lot of it around. All that determines the amount that livestock producers receive is the degree to which they wire themselves into the various sources that are available. However, the primary source is still the land-grant-university system.

Being inundated with information and technology, on the other hand, will not solve many problems. The name of the game is to have the ability to sort out the information and technology that will result in the greatest economic returns on a particular operation and to apply it. This requires good management that sets goals and manages by objective. This also requires getting good advice. Select some people in whom you have confidence and who can give you a knowledgeable and unbiased answer to your questions. This may be the least expensive and most effective source of help a producer can obtain in evaluating information and technology. Set up an informal network if you can. Share information and ideas, solicit help, listen to what knowledgeable and unbiased people tell you, show your appreciation for their assistance, and then make your own decisions.

The livestock commodity organizations, the USDA, and the land grant universities are aware of the need to improve information and technology by reducing the number of component parts a producer must integrate into the decision-making process. Seemingly there is help on the way in this area.

In the meantime, remember: no decision is any better than the information on which it is based. There is ample information and technology available to livestock producers in this country to make good decisions.

11

INTEGRATED MANAGEMENT: THE DELIVERY METHOD FOR THE FUTURE

L. S. Bull

INTRODUCTION

The dairy or livestock farmer, rancher, or manager of today, and certainly of tomorrow, must integrate a wide variety of subjects and information in his problem solving. As the productivity levels of the farm or ranch have increased and as operating margins have shrunk, sound management has become even more crucial.

The growth of technology in all areas of animal agriculture, and the increasingly complex factors influencing production and marketing problems suggest that inputs from a problem-oriented team would be most effective for extension work, rather than one individual specialist. Extension has the responsibility to deliver new information to producers. Usually, each extension specialist has developed his/her own program and responded to a management question within the narrow scope of a specialty. It is not uncommon to hear of cases where several specialists have visited a farm or ranch with a specific problem with each specialist attributing the problem to a different cause. It is little wonder that some farms have been heard to say that "Colleges of agriculture have departments and farmers have problems."

A new approach is now being used by extension to address farm problems--integrated management using the expertise of a team. The idea is not new or revolutionary, but the potential benefits are great.

INTEGRATED MANAGEMENT

The application of integrated management to animal systems is an offshoot of Integrated Pest Management (IPM), which has been well-defined and successful (CAST, 1982). The first application to animal systems evolved from the National Extension-Industry Beef Resource Committee, made up of members of state and federal research and extension groups and various beef industry organizations (Absher and McPeake, 1981).

A prioritization of needs resulted in an identification of reproduction as the most important problem. With that start, further events demonstrated that reproductive efficiency can be improved through an interdisciplinary, integrated approach. With IPM taking the lead and encouraging legislation, the concept of Integrated Reproductive Management (IRM) was born (Absher and McPeake, 1981).

Why is reproduction the ideal candidate for an integrated approach to problem solution? First, reproduction is an endpoint function of key importance to any animal system. It involves many disciplines of wide and diverse nature including physiology, genetics, pathology (veterinary), microbiology, nutrition, behavior, endocrinology, and husbandry. In the context of management, one can add engineering, economics, and marketing--all of which determine both financial and biological success. Therefore, considering the complexity of animal production, a team using the integrated approach can and should address the problems and evaluate the causes.

The Benefits from an Integrated Approach to Management

Research. The basis of new knowledge is research. However, the successful application of research depends on the molding of results into practical recommendations. A major feedback from the team approach to management and problem solving is the identification of new and practical research problems.

Demonstration. The key to success of extension programs is the demonstration on a cooperating farm or farms. Since early extension programs in Terrell, Texas, these practical applications of research have become the foundation of the benefit-cost ratio in agriculture. Similarly, integrated management demonstrations serve the same purpose by demonstrating the benefits of a complete program (Absher and McPeake, 1981).

Education. It is not possible to teach management. It is, however, possible to introduce the "students" to methods of identifying the components of a problem and of assembling the information needed for finding an objective solution. The successful classroom curriculum and extension demonstration will integrate specific disciplines into the solution of problems. It should be emphasized that the success of both the classroom and extension experiences is dependent upon the people who manage the information.

THE IRM EXAMPLE

The livestock industry has identified reproductive performance in herds and flocks as a top priority problem, and has demanded a mechanism that goes beyond research, teaching, and extension that will put everything we know into a package for implementation. Two IRM programs serve as examples: 1) the PEGRAM Project of Idaho (Card, 1982) and 2) the Vermont-Pennsylvania Dairy IRM Project (Gibson, 1982; O'Connor, 1982).

The PEGRAM Project stemmed from an analysis of tremendous calf losses which were thought to be caused by a number of kinds of problems. An integrated program was organized to focus on the management of the cow herd specifically to reduce calf losses. The expertise of researchers, teachers, extension specialists, the veterinary profession, and the various support industries (feed, supply, etc.) was used. Calf mortality was reduced from over 20% to 3% as a result of the program (Card, 1982; Card and Duren, 1978).

The Vermont-Pennsylvania IRM Project deals with reproductive efficiency in dairy cattle. The economic impact of improved production (market availability assumed!) would be **$135 billion** if the average calving interval were reduced by 15 days nationally (Sechrist, 1981)! Although the dairy IRM project is not completed, several indicators have surfaced that suggest a high degree of success: 1) nutritional problems (especially minerals and vitamins) have been identified that have resulted in demonstration research projects and altered management practices; 2) vaccination programs, as well as diseases have contributed problems, and an education program with producers and veterinarians has resulted for improving those situations; and 3) an economic analysis model for determining how much an IRM program contributes to the net profit of the operation, as well as future field programs, is being developed.

As a result of the present IRM projects, there is an impetus to develop programs within and between states that involve producers, industry, veterinarians, academic units, and extension. The results will be seen in new research, new recommendations, and new and rapid ways to deliver information to the field in useful form.

SUMMARY

Integrated management in livestock production combines the expertise of all specialty areas dealing with livestock and focuses on the solution to problems. It represents groups of people who work together with confidence in the ability of each member of the team without having interdisciplinary gaps and squabbles.

To be successful, integrated management should follow the "heterosis" concept in genetics, i.e., the result is greater than the additive inputs of the sources. Therefore, the responsibility of the livestock industry, through legislative action, advisory group input, research, teaching, and extension is to maintain this vigorous concept.

REFERENCES

Absher, C. W. and C. A. McPeake. 1981. Integrated reproductive management - funded or not. Annu. Mtg., Southern Section, Amer. Soc. Anim. Sci., Atlanta, GA, Feb. 2, 1981.

Card, C. S. 1982. Pegram project - a model for IRM. In: Proc. Northeast Integrated Reprod. Management Conf., Beltsville, MD, May 25-27. Univ. of Vermont Agr. Exp. Sta., Burlington.

Card, C. S. and E. Duren. 1978. Pegram project beef herd health program. CIS No. 430. Univ. of Idaho, Moscow.

Council for Agricultural Science and Technology. 1982. Integrated pest management. Rep. No. 93. Ames, IA.

Gibson, K. S. 1982. IRM: the Vermont-Pennsylvania project. In: Proc. Northeast Integrated Reprod. Management Conf., Beltsville, MD, May 25-27. Univ. of Vermont Agr. Exp. Sta., Burlington.

O'Connor, M. L. 1982. Pennsylvania-Vermont IRM project--organization and procedures. In: Proc. Northeast Integrated Reprod. Management Conf., Beltsville, MD, May 25-27. Univ. of Vermont Agr. Exp. Sta., Burlington.

Sechrist, R. S. 1981. Position of the dairy industry: concerning integrated reproduction management. Presented at mtg. of Integrated Reprod. Management Developmental Committees, St. Louis, MO, Dec. 8. National Dairy Herd Improvement Assoc., Columbus, OH.

12

THE UNIVERSITY FARM: WHAT IS ITS ROLE IN ANIMAL AGRICULTURE AND HOW DO WE SUPPORT IT?

L. S. Bull

INTRODUCTION

The model efficiency in American agriculture stems from a combination of research, development, production, marketing, and incentive. The average American farmer now produces enough food and fiber for nearly 80 people, a remarkable jump from the 1:30 ratio of only 20 yr ago and a striking success story compared to other U.S. industries.

This paper focuses on the role of U.S. university farms, their role in animal agriculture, and the financial support for them.

MISSIONS AND GOALS OF UNIVERSITY FARMS

The farms first associated with colleges or universities were part of the mechanism for providing food (and fiber?) for the student population. The Morrill Act in 1862 (extended in 1890 by the second Morrill Act.) formally established a land base for public institutions and became the catalyst for the first agricultural experiment stations in each state. This legislation provided guidelines for using the proceeds from the products of "land grants" to support agricultural and mechanical arts education. The Hatch Act of 1887 granted federal funds to support agricultural research in each state, later including marketing and forestry. The extension service was established by the Smith-Lever Act of 1914. Thus, the land-grant complex, as originally defined, has three fundamental obligations: education, research, and extension or public service. Too often individuals and interest groups fail to recognize that the university farm role includes this triad of obligations.

Because agricultural-systems economics are generally governed by supply and demand, the use of tax dollars to support university farms that create competition for private farmers in the marketplace has been questioned. Within the university system, this view appears to be short-sighted, but it reflects the need for an explanation of the advantages of the university (land-grant) agricultural complex and the benefit-to-cost ratio obtained through teaching, research, and extension at university farms. State agricultural experiment stations are the backbone of agricultural research in the U.S. and internationally. The mission of the agricultural experiment station and the associated farms is to conduct research into the mechanisms of agricultural systems (plant, animal, physical) and through formal teaching and extension, to deliver that new knowledge for the betterment of farmers and all consumers. The goal has been to improve production efficiency, product quality, and supply to the consumer thus providing a high standard of living at a minimal cost (relative to disposable income). Both mission and goal have been a model of success--so successful in fact that maintenance of short-term support is increasingly difficult! With a return on investment of 35% to 50%, far above that for other public expenditures, there is evidence that there is insufficient investment on the national level. Private firms consider a return of 10% to 15% adequate to attract investment. A partial list of returns for various agricultural research inputs is shown in table 1.

TABLE 1. A SELECTED LIST OF AGRICULTURAL RESEARCH PRODUCTIVITY

Commodity	Time period	Annual rate of return (%)
Poultry	1915-1960	21-25
	1969	37
Livestock	1969	47
Dairy	1969	43
Aggregate	1937-1942	50
	1947-1952	51
	1949-1959	47
	1957-1962	49
Research and extension	1949-1958	39-47
	1959-1968	32-39
Technology--Southern U.S.	1948-1971	130
Northern U.S.	1948-1971	93
Western U.S.	1948-1971	95
Farm management research and extension	1948-1971	110

Source: V. W. Ruttan (1982).

UNIVERSITY FARMS IN ANIMAL AGRICULTURE

In the discussion that follows, I use a university dairy farm as the model because it is the one with which I am most familiar. However, the concepts apply to most animal species.

How Good Should the Animals Be?

In many states, animals (genotype and phenotype) have been considerably below-average quality, particularly those assigned to research. The attitude among those making assignment decisions in these stations has been that research is harmful and that the better animals should be protected.

If research is to be useful to the user, given the time that it takes to turn results of research around and a modest amount of genetic progress, the animals upon which the work is based must be representative of the target population. The metabolism, behavior, health characteristics, and management needs of a cow producing 20 times her body weight in milk per lactation are radically different from those of a cow producing at 10 times her weight, but more important, the effects of these factors are not predictable from one cow to another. The dramatic increases in milk production that we have seen in recent years are a result of the conduct of critical research on animals of high genetic merit and the application of that research in the field.

It is the absolute obligation of the chief administrator in any university farm unit to see that only appropriate animals are selected for research, teaching, and demonstration. In short, the most up-to-date practices in selection, genetic progress, and management must be used to ensure the quality of animals needed.

How Should the Farms and Animals Be Managed?

The commercial producer or farmer who visits a university farm often comments negatively about the management practices used. The same holds for the labor force--which is usually larger than that found on "operating" farms. The management practices for animals on farms associated with universities should be **superior** to those found in the field. Without this advantage in management, the credibility of the recommendations made to students or by extension personnel is limited. For example, if the milking procedures promoted by extension and written in texts are not followed, it is hard to counter the statement, "If it is so critical that we follow this, why don't you do it at the University?" Routine and normal management of animals must be of top quality. However, there are some circumstances associated with research protocols that dictate a departure from "recommended" management. The casual observer may not be aware of these research needs, however.

The labor question has many facets. It is common for university farm employees to work shorter "weeks" than their private counterparts. A 6-days-on/2-days-off schedule is often used, with a 40-hr to 54-hr workweek. (As an example, the difference between 40 hr and 54 hr at the University of Vermont represents work in lieu of rent on homes provided for employees.) On the other hand, when animals must be individually fed and records kept of all feed consumed, often with many different diets used, the labor requirement is substantial. Labor costs are a major part of the cost of research. When groups of animals must be provided for classwork and demonstration, the labor requirement increases. Also, since university farms deal with public funds, the accounting procedures and accountability needs are high. This adds labor costs to the operation. A very important function for university farms is public relations to enhance the image of agriculture. At many institutions, thousands of children and adults from nonfarm locations visit each year. Upkeep and cleanliness of facilities and animals should be exceptional to present the most favorable image. The labor cost associated with this function is significant.

The level of dedication among university farm employees can vary widely, and in most instances attitude and dedication are suboptimal; the blame can at least be partially attributed to lack of interest or concern by researchers, teachers, extension specialists, and(or) administrators.

FINANCIAL SUPPORT FOR UNIVERSITY FARMS

Figure 1 lists the sources of support available for university farms. The ratio of federal to state support ranges from 1:1 to 25:1, and the contract, grant, and other nonfederal or state funds range from less than 5% to more than 70% of the total support among the states. A meaningful average for the inputs cannot be calculated since some states do not have direct access to income and others use no

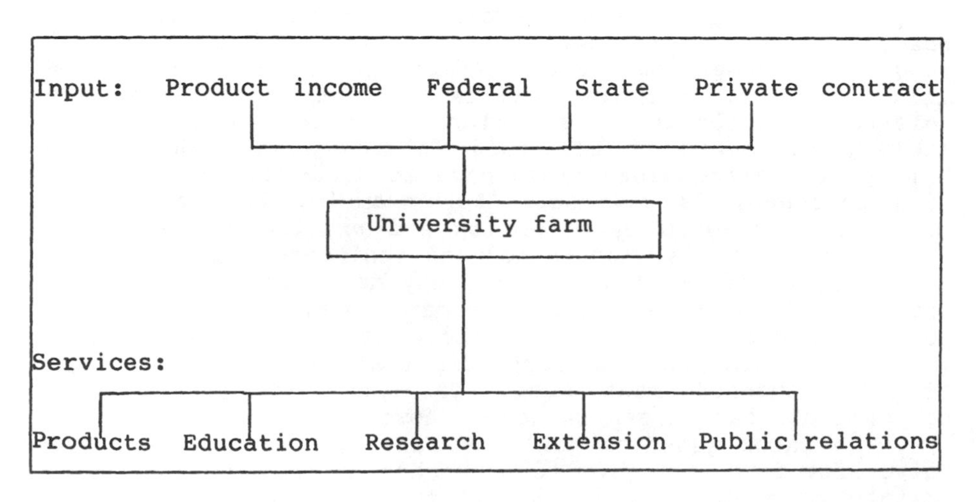

Figure 1. Sources of input support for university farms

federal funds for salary or operating support. The allocation, as a percentage, from grants and contracts varies considerably from year to year. On the income side, some states have marketing agreements that are similar to those of commercial farms, whereas others must sell products at a price disadvantage. For example, some universities sell milk at the market price for the area while others are forced to sell at a lower price. This difference can amount to 20% of sales.

Regardless of the inputs, the overwhelming success of agricultural research and education has created support problems. The high return rate, coupled with the production increases and the political aspects of world trade, have created large and expensive surpluses in many areas. The short-term response to this has been a tendency to shift funding priorities to nonagricultural areas, especially away from production. The underinvestment situation, characteristic of the past, is becoming more critical, and there is concern about a decline in the number of trained people to fill critical needs in agriculture. The current shortage amounts to 13% for agricultural scientists at all levels, with some areas short by 30%. This shortage seems likely to generate a further shortage of support funds.

How Can the University Farms Be Supported?

All states are finding it more difficult to support and(or) maintain university farms with the funding reductions that are imposed at various levels. Some serious questions are being asked, usually prompted by an escalating labor cost in a labor-intensive enterprise. A few of the proposed solutions are as follows:

- Option A. Concentrate the burden of funding more directly upon the user of the results.
- Option B. Operate the farms at an activity level that is directly related to short-time needs for specific research, teaching, and demonstration funded by each project on a real-cost basis.
- Option C. Operate the farms in direct competition with the purebred breeding establishments and depend upon the specialty sale arena to underwrite the operation.
- Option D. Enter into regional agreements and consolidate efforts based on the priority commodity industries within each state in the region.

Some Examples and Possible Solutions

- Option A. In the dairy area, some successful programs have been based on an assessment of a small fee on each hundredweight of milk produced, with those funds directed to the support of research, teaching, and extension related to the dairy industry. The key to the success of such a program is that the use of the funds is under the direct sanction of an advisory group made up of dairy farmers and leaders. This creates a short "loop" of identified need, direction of funds to meet that need, and return of information. The impact of such an assessment is potentially enormous. For example, in Vermont, which produces 2.5 billion lb milk (ranking 12th nationally), an assessment of $.02 per hundredweight would raise $480,000 annually with 100% participation. This amount is greater than the combined faculty salaries plus operating money for teaching, research, and extension in dairy science at the university. The size of the contribution by the average farm would be less than $.50 per day.

- Option B. Certain animal operations can function on a purchase-as-needed basis; for example, poultry, feeder cattle, hogs, and some dairy research units are operating in this way. There are risks, but the maintenance cost for the animal herd and flocks is reduced, providing the labor cost can be handled on the same basis. Where student/temporary labor makes up the bulk of the workforce, this is possible. Such a system enables the true cost of the activity to be measured. A variation on this procedure is to charge a per diem rate per animal, billed to the research project, from the beginning of the project until that animal is used to start the next project.
- Option C. Many university farms combine some degree of purebred breeding with their other functions. The involvement varies as do the success and the attitude of the private operators who are in competition. In some states there is very strong and proud support for this activity. In others the reverse is true. There is a very strong tendency for a split to develop between research and purebred livestock interests, which results in a serious productivity problem for such programs. I feel, however, that such friction is unnecessary and cite the following example as support for my opinion. In 1974, every dairy cow that freshened at a certain university research unit was assigned to nutrition research projects. During that year, the herd received an award for having the greatest increase in milk production in the testing association, an increase

of 1443 lb. The split between cattle interests and reseachers disappeared.

A classic example of the success of such an effort is the University of Vermont Morgan Horse Farm. This farm operates as a separate entity, financially supported by sales, gifts, and contributions. It pays taxes, bills, and the salaries of employees like any private farm. It has a strong endowment based on gifts and support from the industry, guided by an advisory board. In addition, it serves as the resource for the teaching and extension programs in light horse management.

- Option D. A major problem is that of local politics; however, this option may be the most financially attractive from an operating basis, as long as the balance of expertise needed for teaching and extension is available in the region. If such a program is to work, there must be an integration of efforts in developing research, teaching, and extension programs. Since a large percentage of the return on investment in agricultural growth has been due to "spillover" from one state to another, there is a sound base for this activity. The dairy situation in New England is a good example. Approximately half of the dairy production in this small six-state area is in Vermont where 90% of agricultural income is derived from the dairy industry. It seems reasonable to pursue the concept of a concentration of expertise in dairy research, extension, and teaching in Vermont. Poultry and forestry have strong concentrations in Maine, and fruits and vegetables in Massachusetts. A beginning has been made for such a mechanism to exchange students for concentrated study without tuition penalty. Extension is beginning to exchange expertise on a regional basis. Whereas regional research mechanisms have been in place for many years, special regional facilities have not been developed--but there are encouraging signs that these are being planned.

SUMMARY

Farms should be retained as a part of the animal agriculture research, teaching, and extension complex. When funding is stretched and(or) restricted, there is a tendency to divert funds to other areas. If producers (users) are to continue to benefit from the developments that have made the return on investment so great, new and(or) innovative mechanisms for support must be developed. This paper has attempted to identify some areas of concerns and of possible solutions, and to stimulate further innovation and development of ideas.

REFERENCES

RICOP, National Association of State Universities and Land-Grant Colleges. 1983. Human Capital Shortages: A Threat to American Agriculture.

Rockefeller Foundation. 1982. Science for Agriculture. Rep. of a Workshop on Issues in Amer. Agr. Res. Rockefeller Foundation, New York, NY.

V. W. Ruttan. 1982. Agricultural Research Policy. Univ. of Minnesota Press, Minneapolis.

G. H. Schmidt. 1981. Personal communication. Ohio State Univ., Columbus.

J. M. White. 1981. Personal communication. Virginia Polytechnic Institute and State Univ., Blacksburg, VA.

Part 4

ACTIVITIES AND SAFETY FOR HORSE OWNERS

13

ACTIVITIES FOR THE PLEASURE-HORSE OWNERS

Cecile K. Hetzel

You know the old saying that "the outside of a horse is good for the inside of a man." Well, that's true! Riding is a wonderful exercise that is good for people of all ages. Just think, no one is too old or too young to ride; this sport can be done in your golden years and without a lot of effort. It is a sport that is followed year-around.

Activities available for the pleasure-horse owner form a long list. If we discuss a few activities in detail, perhaps you can obtain some new ideas about how to enjoy your pleasure horse.

TRAIL RIDING

Trail riding is probably one of the most enjoyable things a person can do with a pleasure horse. Many horses are marketed with the stamp of "good trail-horse prospect" attached to his list of desirable qualities. Trail riding can be divided into three types: endurance, competitive, and social.

Social trail riding finds a group of friends riding together for relaxation and probably enjoying the beauties of nature. Horseback trips are made into many national parks and wilderness sites. The U.S. Forest Service has a service where one can rent guides, horses, and all necessary gear. Trail Riders of the Wilderness is a division of the American Forestry Association, and they hold lots of pleasure rides annually around the country (figure 1).

Pleasure cross-country trail rides offer a wonderful source of entertainment for the pleasure-horse owner or rider. An example of such an activity is held in Eminence, MO where cross-country trail riders have two 1-wk summer rides. There is an unlimited amount of camping area, and for those who prefer sleeping in a bed there are motels and cabins close by. The camp is located on the banks of the Jack's Fork River where 400 outside tie stalls are available and hay and grain may be purchased, if you do not wish to bring your own. Meals in camp are served in a screened dining tent and entertainment consisting of horse shows,

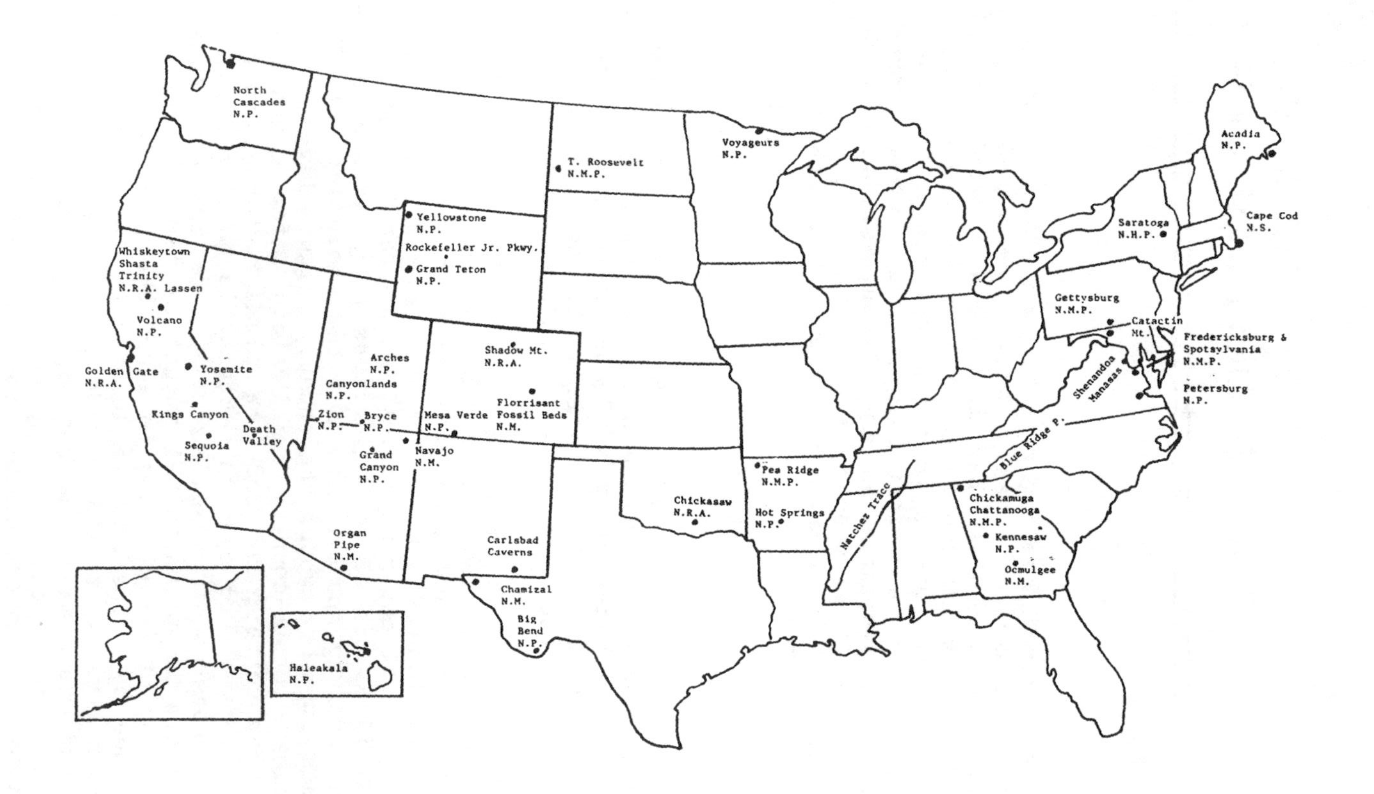

Figure 1. National parks that allow horseback riding.

movies, and outdoor dances, plus an old swimming hole, helps make everyone have an enjoyable time. The trails are arranged so that those who wish to move at a faster pace go together, while those with slower mounts or less experience do not have to keep up. This is a privately operated camp.

Competitive and endurance trail riding are a bit more challenging. The NATRC (North American Trail Ride Conference) was organized for the purpose of sanctioning competitive trail rides. A uniform judging system is used, which is supported by a judges manual, a rule book with specifics to follow in competition, plus a management manual. These publications are available from NATRC, 1995 Day Road, Gilroy, CA 95020.

Objectives of the NATRC are to:

- Stimulate greater interest in the breeding and use of good horses possessed of stamina and hardiness as mounts for trail use.
- Demonstrate the value of type and soundness in the proper selection of a horse.
- Learn and demonstrate the proper methods of training and conditioning horses for competitive riding.
- Encourage good horsemanship as related to trail riding.
- Demonstrate the best methods of caring for horses during and after long rides without aid of artificial methods or stimulants.

Competitive riding asks riders to cover a specific span of overland country within a designated time. You are penalized if you arrive ahead of that time, or after that time.

Endurance riding might be considered a race. The object of the rider and horse is to cover a designated distance under a minimum time allowance. The head of the group is the most important thing in endurance riding where the least amount of time to get to the end of the ride is the objective. The competitive rider is seeking the optimum time, not the fastest.

The most famous endurance ride is the Tevis Cup 100 Mile Ride, which is a 1-day ride from Tahal, New Mexico, to Auburn, California. Horses are capable of covering 100 mi in less than 24 hr. Winners have finished well under 13 hr.

In an endurance ride, horses are examined by a vet at mandatory rest stops to determine that they are sound and ready to continue. Awards are made to horses that complete the ride, to the first one to finish, and to the horse that finishes in the best condition. The main governing body is the American Endurance Ride Conference, Inc., P.O. Box 1605, Auburn, CA 95603.

Let us suppose that you have decided to ride on a competitive ride. If you choose a Class A ride, you will ride for 2 consecutive days, for 30 mi to 40 mi on fairly level terrain. If the terrain is steep and rugged, the distance will be shortened to make it possible to complete the course in 6 1/2 hr to 7 hr riding time. If you choose a Class B ride, you will ride for 1 day with the same requirements. You must then decide which division is best for your horse to enter.

I. Open Division--requires that the horse be 5 yr old and older. It has three classes: a) Lightweight Class--rider and tack must weigh a minimum of 140 lb and less than 190 lb. b) Heavyweight Class--rider and tack 190 lb or more. c) Junior Class--rider 10 yr to 17 yr of age. No weight restrictions.

II. Novice Division--which is designed to prevent young horses from being overworked. It is also to introduce new riders to competitive riding. The minimum age of the novice horse is 4 yr or 48 mo. Riders must be 10 yr old and older. The length of a novice course does not exceed two-thirds of the distance of the Open Division (or 20 miles). The Novice Division has the same three classes.

In the Open and Novice Divisions both lightweight and heavyweight contestants must be weighed in, with their tack, prior to the start of the ride. These weight limitations must be properly maintained for the duration of the ride.

After you have decided which division and class to enter, now what happens? Riders and their horses are checked in on the evening before the start of the ride for preliminary examination by the judges and the farrier. A riders' meeting is held each evening prior to the start of the next day's ride to brief the riders on the course, trail markings, available water, etc. The judges discuss their judging methods and procedures. The rules of trail etiquette and trail safety are explained. All of the riders are given maps and a schedule of the estimated times and(or) distances and elevations for various points on the ride.

All horses are kept under uniform conditions from the preliminary check-in until the awards are presented. On the first day, horses may not be saddled until 1 hr before starting time. After the first day's ride, riders are not allowed to groom or handle their horse after 10 p.m., except in case of an emergency. On the second day, the horses may be fed and watered, while remaining tied, as early as 5 a.m. However, they cannot be moved, groomed, handled, or untied until the judge gives instructions. During the competition, all riders are required to care for their own mounts.

In competitive trail riding, the riders must remain in the saddle when the horse is in forward motion over the course, but they may dismount and rest themselves and their horse at any time, provided they do not advance. Each day from an identified point approximately 2 mi from the finish line, the riders must maintain forward motion and may not stop or dismount from this point to the finish line. Forward motion must be via the most direct route following the marked trail.

Timekeepers keep an accurate record of each entrant's time. Riders are started at 30 sec intervals, which provides the judges an opportunity to observe horse and rider individually at the start of the day's ride. A minimum and maximum time are allotted for riders to complete the ride, with a 30 min range between the minimum and maximum time. Any horse completing the day's ride in less than the minimum time and more than the maximum time is penalized by 1 point/min.

JUDGING

The judging is done by a team of qualified personnel of two to three judges, one of whom must be a qualified veterinarian. The preliminary judging takes place on the evening before the beginning of the ride. At this time, the judges determine that the score card accurately describes the horse being judged. The judges also identify all marks and blemishes prior to the start of the ride and evaluate the horse's way of going. When the horses have completed the ride for the day, the rider is allowed to cool the horse and groom and feed him until 10 p.m. The judges then monitor the horses during the evening, observing symptoms that might reflect colic, depression, exhaustion, etc.

Horsemanship is an integral part of how well a horse will perform on the trail. In competitive trail riding, the horse's performance and the rider's horsemanship are evaluated separately. Horses are evaluated on four criteria:

1. Soundness, 40%--
   - Lameness and stiffness--the judges evaluate the horse's way of going during the ride to determine any change in gait that is the result of an inability to adapt to the stresses placed on his tendons, muscles, ligaments, bones, and joints.
   - Heat and swelling--in the area of joints and tendons, this is of particular importance, indicating inflammation. Stocking is not considered detrimental when found after resuming activity.
   - Muscle soreness--particularly in the back muscles.
   - Saddle and cinch sores.
2. Condition, 40%--
   - Fatigue.
   - Pulse and respiration--recovery after extended exercise is a definite indication of the condition and performance of the horse. There is a minimum of 2 P & Rs taken during each day's ride. If the P & R is 72/72, the horse and rider are held over for 10 min for additional recovery time and are deducted 5 points. (The P & R scoring is located at the end of this report.)

3. Manners, 15%--
   - Consideration must be given to the manners of the horse on a ride including throwing the head, fighting the bit, kicking, standing quietly during mounting, response to command, and the general type of ride the horse offers his rider.
4. Way of going, 5%--
   - This includes any abnormal gait or action that may be detrimental to the horse or rider, i.e., interfering, forging, scalping, etc.

GAMES

Gymkhana events or games on horseback are very popular with young riders, but they also are enjoyed by many adults who like to be competitive and are challenged by the harmony required between horse and rider. Some of the standard gymkhana events are keyhole races, figure eight stake race, quadrangle stake competition, figure eight relay race, pole bending, cloverleaf barrel race, scurry competition, rescue competition, and a speed barrel race. In gymkhana competition, the horse is shown on the rail after completing a timed event. A horse must be able to work in the ring on the rail, as well as have the ability to perform in the events.

DRESSAGE

Dressage is making a surprising upswing in popularity with pleasure-horse owners. The principles of riding this seat dates back to a famous Greek horseman, Xenophon, in 400 B.C. Dressage teaches a horse to be obedient, willing, supple, and responsive. It is a system of skilled horsemanship that links a rider to his horse. All light horse breeds are capable of dressage and the riding is considered an art.

The rules state: "The object of dressage is the harmonious development of the physique and ability of the horse. As a result it makes the horse calm, supple, loose and flexible, but also confident, attentive, and keen, thus achieving the perfect understanding with his rider." Basic dressage training is the best preparation a horse can receive for any number of tasks, notably jumping and hunting, but also pleasure riding, driving, cutting, and reining. Any horse and any rider can benefit from basic dressage.

Dressage is increasing in popularity as a competitive sport. Many shows are being held all over the country. Local dressage organizations are sponsoring shows, clinics, and educational meetings. The U.S. Equestrian Team has a dressage team which has shown in international competitions.

The goal of a dressage rider is to produce a horse that is calm and obedient. The horse is to move freely and easily at the walk, trot, and canter. Training begins on an egg-butt snaffle bit and the position of the horse's head is to show flexion at the poll. Smooth transitions between the gaits, halting on command, lengthening and shortening of the strides at the trot and canter with curves, simple turns and circles showing the horse's ability to bend its neck and body, are major requirements of dressage training.

COMBINED TRAINING

Combined training stresses both the physical and mental fitness of the horse and rider by asking them to perform in three successive events: 1) dressage, 2) a cross-country course with fences that take the rider and horse over a natural terrain with obstacles to jump, and 3) stadium jumping, which requires the horse to prove that he has retained the suppleness, energy, and obedience necessary to continue service after a severe test of endurance.

Combined training offers the opportunity for the participants to compete at the level in which they feel most comfortable: pretraining, training, and preliminary levels, each of which is frequently divided into novice, junior, and adult sections. A rider and his mount can move up or return to any level for schooling and development, depending on the ability that his team feels is most suited.

Among the things that seem to make combined training unique and fast-growing in popularity is the great variety of the courses. Unlike the fences in ordinary hunt and jump competition, where you find essentially the same type of fences in similar arrangements at every show, each fence on a cross-country course is different. A variety of obstacles is used, such as poles, stone walls, natural hedges, earth banks, streams, ditches, barrels, tractor tires, hay bales, logs, and farm wagons, just to mention some I've seen. The positioning of these fences creates the interest.

Obstacles may be placed at the top or bottom of steep hills, before or after a sharp turn, at the edge of a tree line, in woods, or on slopes. The simplest type of jump can become a real challenge because of its placement on the terrain. One very popular type of obstacle is a fence where the rider has a choice of several ways of going over the obstacle. Usually the fence is arranged so that the shortest route over requires the greatest jumping ability, while the safer or easier way involves a loss of time.

Cross-Country

Phase A. Roads and tracks. The distance is usually 2 mi to 4 mi taken at a speed of 240 m/min, which is about 9 mph. This is accomplished over a marked trail that includes all types of terrain but no jumps.

Phase B. Steeplechase. The course is 2 mi long; which is taken at 550 m/min, or 20 mph. Bonus points are given for faster time. The course has 10 jumps at heights of up to 3 ft 3 in. in the solid parts and up to 4 ft 7 in. at the top of the brushy part.

Phase C. Roads and tracks. A route 8 mi to 10 mi long is taken at 9 mph over varied terrain, including up and down slopes. There are no scored obstacles, thus no jumping-fault penalties are counted. (A 10-min compulsory break occurs and the veterinarian check follows Phase C. The rider dismounts and sponges off the horse. The vet check eliminates any over-tired or unfit horses.)

Phase D. Cross-country. The course is 5 mi long and is taken at 45 m/min or 17 mph over 30 obstacles. The maximum height of the jumps is 3 ft 11 in.; the jumps can be 6 ft 7 in. wide at the top and 9 ft 10 in. wide at the base.

The only jumping faults are refusals or run outs and falls. A horse may refuse twice at each fence, but he gets a huge penalty score. The third refusal requires elimination from the event. (The second fall in the steeplechase requires elimination from the event.)

All four phases are timed with stopwatches. Penalties are given for going too slow in the steeplechase and cross-country, bonus points are given for faster time, up to a stated maximum.

In stadium jumping, penalties are given for refusals, falls, knockdowns, and exceeding the time allowed.

I want you to know that all combined-training competition is not limited to grueling advanced events as I have described above. There are smaller events held where the phases are all ridden on the same day. The levels of difficulty are greatly reduced from those described above. "Pre-training Level" offers a dressage test of training level, plus a 1-mi or 2-mi cross-country phase taken at slow speeds over obstacles no higher than 3 ft 3 in. and low fences in the stadium-jumping round.

Combined training tests more facets of a horse's training and abilities than does any other single type of competition. Many good event horses can go on to excell as dressage horses or competitive trail horses, or as excellent all-around pleasure-horses.

The trend for pleasure riders to work their horses in three-day events is growing by leaps and bounds.

DRIVING

For thousands of years, horses have served as driving animals. They were used in agriculture, transportation, battle, and sports. Most of the first uses for driving a horse have died away, but the sport of driving is flourishing. The interest in pleasure driving has gained such popularity that in 1975 an organization called the American Driving Society was formed to bring together, inform, and aid people interested in this fascinating aspect of equine sports. The American Driving Society members are horsemen who are dedicated to the promotion of this style of horsemanship.

Driving calls for the horse to move quietly and under complete control of the driver. The driver under proper title is called the "whip."

The early training of a driving horse is essentially the same as that for any horse. He must first be handled and made to feel at ease--leading without resistance and relaxed and supple. Use of quiet voice aids is instituted followed by lunging. Lunging is done first with a caveson or halter; then with a roller pad around the girth; then with a crupper under the tail. An open bridle with a light bit is put on in the stall and is worn while the horse is on the lunge. The lunge line goes to the caveson and not to the bit. Next, long lines are attached to each side of the caveson and the horse is driven from behind (the rider walking). If this work is done in a ring, the horse will progress faster since the ring imposes confidence in an animal. After the horse moves freely and quietly with a good degree of relaxation, the long lines can then be attached to the bit and the driver moves behind the animal at the walk and trot. Light contact is maintained with the mouth through the lines.

A weight (approximately 10 lb to 15 lb) is attached by means of two lines that go through the roller pad to a breast-plate. The horse thus pulls with something dragging behind. A blinker bridle replaces the open bridle. The horse must stand quietly, respond to the voice, and move forward with ease. We want a picture of ease and unhurried activity. Our next step is to put the horse to a two-wheel breaking cart.

The cart is hitched to the horse, the driver (not in the cart) asks the horse to move forward. It is good to have assistants on each side with lead shanks attached to a halter that is placed over the bridle. These assistants help the horse around corners since the shafts, and the restrictions caused by them, often frighten the horse. After a few applications of this procedure, the driver gets into the cart. At this point you should have the makings of a good driving horse.

The good whip (driver) uses voice aids and uses the lines and whip with smoothness and in the most subtle way possible. The driver's right hand carries the whip, which should be ready for use to enforce a voice aid. Even control on the bit is maintained through the reins. The whip (driver) is to handle the reins in such a way as to convey the appearance of complete ease. Any adjustments of the reins should be made without looking down.

Driving of one horse is referred to as a single, of two horses as a tandem, of three horses as a unicorn (a pair on the wheel and one out front), and two pairs of horses (one pair in front of the other) as a four-in-hand.

Aside from pleasure-driving classes offered in many shows, classes are offered that measure the whip's competence by asking the driver to execute a number of difficult maneuvers.

Timed classes can be extremely exciting. Obstacles are placed only 8 in. to 12 in. wider than the widest part of the vehicles. Usually a course consists of 20 or so sets of obstacles to drive through. A time allowance is given for the course.

Gambler's Choice classes are great fun; spectators love this class as much as do the participants. Eight or more obstacles are each given a point value of 20 to 100, depending on the degree of difficulty. Obstacles can be bridges, farm yards, water ponds, serpentines, a T, and a circle. In the T, the whip drives into the top portion of the T from its side, then backs up into the vertical portion of the letter and drives out the same way he came in. Time is the factor! Normally 2 min are given to complete as many obstacles as the whip chooses. The harder obstacles give more points. The winner is the whip who accumulates the most points during this time period.

For many persons, the long hours of preparation that are demanded of competitors and the rigors of the showring take away the pleasure. However, the pleasure of driving behind a horse is reward enough for some people. In addition to the fun of driving, many enthusiasts enjoy restoring carriages and harnesses. Sleighing in winter is something which is still enjoyed by those who do not mind the cold.

Anyone attracted to the sport of driving should select a suitably trained horse or pony if they have not had a lot of experience in driving. Safety is the most important factor in driving. The vehicle and the harness should be in good repair and should fit well. An old harness found in an old barn or antique shop may look all right, but it could be rotting in crucial points. Avoid accidents at all costs!

Every whip, regardless of the level of skill he or she aspires to, will share equally in the pleasure of driving a horse or pony to a vehicle.

14

DISTANCE RIDING FOR YOU AND YOUR HORSE

Matthew Mackay-Smith

Until fairly recently, trail riding was considered an entertainment for riders and horses incapable of other equestrian endeavors. Recently distance riding has begun to develop as a distinct discipline with advantages for riders of every level and horses of every performance category.

Two events of recent times have given rise to greater interest in this area. The Green Mountain Horse Association revived long distance riding in 1936 in commemoration of the old calvary tests held through World War I. Their stated purpose was to recognize and encourage the development and training of better mounts for riding long distances over natural country. In the late 1950s, a casual challenge between two California lumbermen brought about the birth of modern endurance riding. Wendell Robie established an annual event to negotiate 100 mi of rugged Sierras within 24 hr. From that impetus, modern endurance riding has grown. In the last 15 yr to 20 yr, the advantages to horses and riders alike of this kind of athletic activity has become increasingly apparent. There are at least nine benefits conferred by a serious and systematic distance riding program. This requires a commitment on the part of the horseman--we're not talking about 1 day of weekend joy riding and 6 days of neglect--one of the images of distance riding in the past.

The first advantage of long-distance riding is that it uses the horse's most natural activity. Evolution on the short grass plains developed a mechanical and physiologic entity which is perfectly adapted to the long, slow-to-moderate travel the horse lived by in grazing and fleeing from predators. This natural ability opens up distance riding to the widest possible variety of horses with the greatest assortment of conformation and athletic tendencies.

The second advantage is that long-distance and slow-paced work is the proper background for all well-conceived athletic conditioning. Because the supporting tissues of hoof, bone, tendon, ligament, and joint are the ones most vulnerable to overexertion and athletic injury, and because they are the slowest to respond to the demands of increased exercise, starting a young horse at a slow pace for long distances and slowly, progressively increasing the level of demand builds the greatest possible durability for his eventual performance, whatever that may be. Slow long-distance travel builds a foundation of fitness upon which strength and speed can be built as well as the skills required of the agile or versatile horse.

A third compelling attraction of long-distance riding is that it is a natural teacher of the horse's balance, agility, and ability to handle the rider's weight. The leverages applied to the horse's motion by the rider sitting above the center of gravity create complex demands on coordination as well as the horse's structure. The constant variety of gaits and body maneuvers required by long-distance riding over natural terrain gives the greatest possible flexibility to the horse's developing balance and agility.

Fourthly, long-distance riding is a natural reenforcement to all the rider's aids. The variations in terrain will suggest to the rider whether he needs more or less leg aid, more or less rein, or a shift in balance which would guide and cue the horse to properly negotiate various turns and obstacles. This welding of the requirements of the terrain to the rider's aids simultaneously improves the rider's effectiveness and the horse's responsiveness. It's amazing how much dressage you can teach a horse riding him through the woods or across the range.

A fifth inducement to long-distance riding is that it develops a deep, relaxed, and comfortable seat in the rider's equitation. Thirty or 40 min in the saddle is barely enough to develop an appreciation of a natural seat. Six or 7 hr in the saddle, and the faint suggestions from early in the ride become screaming imperatives to relax, sit deep, follow the motion, and avoid chafe.

A sixth benefit of distance riding is that it can be a sedative and a psychological comforter to both a fretful horse and a tense rider. Also, the horse who by reason of youth or high spirits is inattentive to the rider becomes more responsive after 2 hr or 3 hr on the trail.

A seventh advantage of distance riding is its endless variety, its fun, and its being a great way to see the countryside. Groups of riders of very different ability and persuasion can enjoy trail riding together without the unnecessary comparisons that more formal riding or competition impose.

An eighth attraction of distance riding (which perhaps should be elevated to first place) is that it's good for your body. Long-distance riding will take inches off your waist and add years to your life. As Wendell Robie has often said, "Ride for your life. Really ride!"

Ninth, distance riding is a great and growing competitive field. It offers various objectives in relative and absolute performance to horsemen of various levels and directions of ambition and expectation. In order of increasing athletic demand, they could be listed as judged pleasure rides, competitive trail rides, and endurance rides. Judged pleasure rides cover from 5 to 20 or 25 mi. Horsemen and veterinary judges are asked to select placements reflecting the suitability of the horse for distance work and its apparent tolerance of the work level imposed. These rides generally operate at 3 1/2 mi to 5 mi per hr and are considered introductory to the second and third levels of competition.

The competitive trail ride is conducted at a standard speed which is between 5 mi and 7 mi per hr, depending on the terrain, over distances of 25 to 40 or 45 mi for a 1-day ride and of up to 100 mi divided into segments for a 3-day competitive ride. The 3-day-100-competitive rides offer a level of challenge to conditioning and performance equivalent to endurance participation-for-completion. Competitive trail rides are judged on objective and subjective evidence of fatigue and wear and tear on the horse resulting from the effort of the ride. Pulse and respiration recoveries are a principle objective measure. Skin pinch resilience and mucous membrane refill time are also quantitatively assessed. Subjective criteria include evidences of fatigue such as loss of impulsion, decrease in the elasticity of the gait, visible evidence of fatigue in the expression of the face and the carriage of the ears, and evaluation of the gait for lameness. A cumulative score on objective and subjective criteria enables the judges to place the horses in decreasing numerical rank.

Endurance riding is conducted without direct control of the pace but with close regulation of the competing horses' conditions through periodic veterinary examinations and the reaching of predetermined objective and subjective criteria of fitness-to-continue. These fitness checks are conducted every 12 mi to 20 mi along the route. Endurance rides vary in length from 50 mi up to 100 mi in 1 day, or in 2 or 3 days up to 150 mi; each day's segment must be at least 50 mi. The concept of endurance riding places the absolute responsibility for the horse's condition on the rider. The rider who is most sensitive to the horse's condition from minute to minute and from mile to mile along the trail will have consistently the best record of completion and the highest average position of finishing. The awards in endurance riding are based on the time of finish, all control criteria having been met, and on the horse's recovered condition at the end of the ride as judged by a veterinary estimate of the horse's fitness to continue, factored in with the speed at which the horse covered the course and the weight he carried. This factoring of the condition score with speed and weight attempts to recognize the horse that has given the best performance overall on the course for the day.

If you are interested in distance riding as a general adjunct to your present program or as an end in itself, there are several general rules for safety and effectiveness.

First of all, do it regularly. A long ride once a week or every second or third week will contribute little to any of the goals of distance riding.

Secondly, start out gradually and increase first the distance and then the speed of your trail riding. Never increase speed and distance for the first time at the same time. If you are starting out with a very young horse, remember that while you can gain quicker and greater benefits, it requires a less intensive program to do so and the penalties for overexertion are more immediately seen. After each increase in either distance or speed, stay at the new level for a week or more to permit the animal's body to catch up with the increased demands. Most importantly, don't just slop along as a passenger, but use this contact time with the horse to develop his skills and your relationship in every way. Perhaps that feature can be illustrated by the experience of one of our Olympic riders who had occasion somewhat unexpectedly to ride in the Tevis Cup. An acquaintance asked what the experience had been like. The rider said she had expected to be bored and had taken measures to remedy that. But to her great surprise the only three things she had done for the preceding 20 hr was to ride, ride, and ride.

If you are interested in distance riding as an end in it itself or as a competitive sport, there are some additional considerations. The primary one is selection of the horse or evaluation of the one you have and intend to use. Horses best suited to distance riding tend to be of moderate size and relatively slight build. Horses which are very large and(or) beefy have difficulty in transporting their own extra weight and in dispersing the unavoidable accumulation of body heat. The horse should have an elastic and efficient but not spectacular gait. Horses with extremely long extended gaits either acquire shorter ones in serious distance training or pay a great deal more in energy per mile and in fatigue for the imposing extended gait. Horses with short choppy gaits incapable of elastic adaptation to the conditions will likewise tire excessively after many miles. Good distance horses have excellent feet, resistant to the effects of hard use. These horses are also of cheerful and willing temperament. Both the runaway and the lazy horse will prove difficult to get to the end of the course one way or the other. If the luxury of a wide selection of horses is available, try to choose a horse which has a low resting pulse rate, 32 or below. Horses with resting pulse rates in the 20s have larger hearts which can pump more blood per beat and recover more blood per beat and recover more rapidly to the designated values in competitive and endurance ride control. Be sure that the horse you take distance riding is one whose company you enjoy. You're going to have a lot of it.

THE SERIOUS DISTANCE RIDER

In training your horse for serious distance competition, if you work in average countryside having terrain which offers less than 200 ft of rise and fall per mile, your goal in the first 2 mo to 4 mo of preparation should be to work up from approximately 5 mi in 1 hr to 10 mi in 1 hr, alternately picking up the distance and then the speed in 10-day to 2-wk increments. A horse that can do 10 mi in 1 hr for 4 days or 5 days a week, retain a good attitude and a good appetite, and stay sound, is ready to go on most competitive trail rides of the 1-day or 2-day variety. If a more ambitious program is required, toward the end of this initial phase, longer rides at reduced speeds can be substituted 1 day and then 2 days a week, moving on into the latter part of the first year of training.

Horses expected to move into endurance work can generally handle 25 mi in 2 1/2 hr to 2 3/4 hr, once or occasionally twice a week on a regular basis. Once a horse has reached this level, additional training is obtained through competition and through progressive addition of demand in easy stages as the horse's development and achievements will dictate. Horses at the highest level of endurance performance will complete 50 mi in easy terrain in approximately 4 riding hr with an additional hour or less of rest interspersed during the ride. Even the most difficult 100-mile endurance rides are being completed in 11 hr to 12 hr of riding time by the champion horses. Completion in good order requires an additional 3 hr to 4 hr in the most severe tests such as the Tevis Cup in California and the Old Dominion in Virginia. Whether your goal in distance riding is to improve your horse and yourself for general riding or other sports, or to begin a progressive development in distance riding competition, remember that the best teacher you have is the horse itself. It will respond to your demands in ways that are specific to its nature and its biology and guide you in its further development.

15

PROTECTING THE HORSEMAN

Betty M. Bennett

Our wonderful horses are justifiably protected from harm through the use of bell boots, pads, shoes, good nutrition, immunizations, safe stabling, and head bumpers. The list of comfort and safety equipment for our favorite equines is endless. Consult any tack catalog for the latest necessities in tack, training equipment, or supplements. All of this tender care is fine for the horse, but what has the horse world done to protect the horseman?

Protection from injury is based on knowledge and experience--confidence and safety, may or may not result. An experienced horseman is one who has come to know the hazards and threats around horses, and knows how to cope with them. He is not devoid of fear and caution but has sublimated those fears and cautions into constructive habits and attitudes of handling and approaching horses.

AWARENESS OF POTENTIAL DANGER

Fear of horses should not be eliminated entirely, rather it should be transferred into attitudes of respect and recognition. Horses are large, unpredictable, and dangerous to ride. We should "fear" them, just as we fear the potential of a speeding automobile.

Recall the trepidations you first experienced when you placed the bit in the horse's mouth and felt his teeth. Recall your early attempts to mount the horse when you underestimated the energy it took to hoist yourself aboard. Recall your first canter. Did you feel that you had complete control? Recall the first time you drove a car. Do you remember the excitement and apprehension as you practiced starting, stopping, and steering? Both riding and driving involve education prior to experience. Continued education in both disciplines adds up to minimizing the risks in each.

As the horseperson becomes more familiar with equines, he is more likely to recognize and respect the complexities of the horse. He is more keenly aware of the great variability in kinds of horses, the individuality of horses. He is cognizant of the unreliability of all rules and principles of behavior in the equine.

On the basis of their own experience and knowledge, most adults would not allow a youngster to drive a car without having passed a driver education course. And yet many of the same adults would put a youngster on a horse, hand him the reins, expect him to ride through woods and fields, along roads, and to return unharmed. Having watched horsemen on TV and in movies, they expect riding to be an easy, natural sport. Such people feel there is no need for instruction to learn how to ride a horse--nor to learn to understand the mind of the horse.

Stopping a speeding auto or a runaway horse presents a similar risk, yet they are quite different tasks. The ignition key turns off the motor of the car, but the only key to turn off the mind and motor of the runaway horse is an educated experienced rider. In 1980, in Wisconsin, three riders were killed when their runaway horses refused to listen to the riders. One was a 24-yr-old dressage rider who was killed in full sight of her family. Her well-trained horse took off for no apparent reason, flinging the young woman to the ground. Later her parents felt that a hard hat might have saved her life, or might have at least minimized her injuries. A trainer was found unconscious on the ground, his mount nearby. He was 57 yr old. . .experienced. The other, a 9-yr-old girl was dragged to her death by her runaway horse. Spooked by an automobile, the horse panicked and kicked the child mercilessly. The child's foot was wedged in the stirrup, her tennis shoe still on her foot.

Other problems may arise that the uneducated rider or driver may find frustrating. How about a stalled motor? Whether it is a stalled car or a stalled horse, invectives and kicks may accomplish little forward motion. And there is the problem of making smooth transitions. Without instruction, shifting from low gear to high or from trot to canter can be most uncomfortable. Once these transition skills are learned, it takes many hours of supervised practice to ensure the safety of all.

UNDERSTANDING THE MECHANISM

The first step toward efficient driving is to take a look at what is under the hood. Learning basic car control includes a knowledge of the instrument cluster, the control used in driving, the starter, accelerator, steering wheel, and braking system.

Efficient, safe horsemanship also requires the rider to look under the hood--into the mind of the horse. What is under that hood? What makes him start, shy, stop, and kick? Why does he behave like a placid plug one minute only to rush snorting down the trail like a runaway locomotive the next?

The horse (a complex, powerful, and flexible piece of machinery like the car) has limitations beyond which it will function inefficiently or not at all. But the horse, unlike the car, has a mind of its own. Although the car is almost completely predictable, the horse is not. The rider, therefore, must learn to cope with the mechanics of directing and controlling his/her mount and must anticipate the unexpected reactions of a very large animal. The results of the rider's actions are sometimes compelling, frightening, and often disastrous.

The Mind of the Horse

The horse expresses emotions such as jealousy, fear, suspicion, and loneliness. He should be approached with respect and consideration for the power and destruction of which he is capable. Despite this power (of which many horses are unaware), they submit readily to human domination--without reward. A unique animal, the horse will transfer his loyalty from herd to human. He will permit his trainer to teach him unnatural feats such as jumping a 6-ft wall or high-diving into a pool of water. His confidence in man can later save both from injury in a difficult situation.

Understanding the mind of the horse brings one to the realization that the horse actually prefers the leadership and decision-making of man. If the novice rider is not skilled in the use of his aids in a compromising situation, the horse must make his own choice.

In moments of stress, horses (like humans) may revert to patterns of behavior that are entirely natural to facilitate their survival. Although these primitive responses have become dulled through his contact with man, they are never lost. The urges to flee from danger and to fight with teeth and hooves when threatened are inherent tendencies. Quick to react to threats or stimuli, the horse may consider some of the actions of the novice rider a threat to his survival. Jabbing the horse's mouth by jerking on the reins is a common thoughtless punishment. Any bit, whether snaffle or curb, can punish the horse when used indiscreetly. If the punishment is excessive, the horse may react by rearing, kicking, or refusing to move. Beginning riders and drivers generally overestimate their ability. This leads them into trouble. A speeding car filled with teenagers too often makes the news as a tragic story. Eager cowboys anxious to get maximum speed from their steeds often get tossed and bruised--and (sometimes) killed. In both instances, education, coupled with supervision and good judgment, might have prevented such drastic results.

Overloading the Stimuli

Safety seems to lack real significance to the average person. People enjoy the pleasurable risk of dangerous sports such as hang-gliding, car racing, and riding without predicting the consequences of their actions. Most accidents are the result of the emotional fluctuations of the individual. Antisocial, aggressive, nonconforming behavior, plus irresponsible lack of communication contribute to the accident data. This statement does not suggest that high-risk sports should be eliminated from the activities of people, rather, the aim is to motivate parents and leaders to educate themselves and others in safety practices.

Horseback riding ranks third on the list of the most dangerous sports by an insurance company. Could riding outrank car racing and motor cycling in risk potential? Why is such a pleasurable sport so dangerous? What can be done to alert the public to safer riding habits? Experts agree that the following contribute to this high rating: lack of education or supervision; improper attire or lack of protective headgear; overestimation of the rider's ability; and ignorance of equine behavior.

A. Seaber (in "How Hardheaded Are We?") insists that the rider must wear protective headgear that will stay on the head. Seaber says that during acceleration, an increase of intracranial pressure occurs and the entire brain smashes up against the side of a rigid skull and causes tearing and bleeding into the cranial vault. Permanent brain damage is often caused by bruising. High velocity impacts are most common in riding accidents.

MANIPULATING HUMAN AND EQUINE BEHAVIOR

It would seem that in light of collected data, appalling reports from insurance companies and unsuccessful efforts to educate riders, more emphasis must be placed on methodology. Perhaps behaviorial changes and attitudes may be altered toward safety through positive and negative reinforcement.

Research studies reveal incredible patterns of behavior in humans and horses. Applied behavioral analysis departs from the traditional concept of horse psychology by rejecting the needs, impulses, and desires of the animal. More emphasis is being placed on the situational, external environmental, and social stimuli that influence equine behavior. In the past, animal intelligence was measured by food response, using the maze as the measuring device. We know now that the strong, inherent desire for survival would preclude the horse's desire for food and also inhibit him from entering the restrictive maze. J. Fiske, equine psychologist, has shown that even the survival instinct can be altered. Experiments have indicated that horses respond to varying signals and stimuli that can be changed frequently.

What this all amounts to is that the tone of your voice, the determination in it (or lack of it), and your odor may have more to do with the behavior of your horse than his need for food or survival. The layman can hardly be expected to understand much more than what he sees on the screen or hears from a buddy who got bucked off. The problem remains as to the method of educating the public in safe horsemanship practices without taking all the fun out of riding.

<u>Games As Learning Tools</u>

Horses play games just as people play. Sometimes called "horsing around," these games can often be useful as a learning tool and a method of changing behavior. Behavioral change is explicit in any activity where hazards abound simply because it is a critical factor in the creation of accident potential. It is also a factor necessary in the reduction of accidents.

Simple competitive games lessen the rider's fear of falling, divert attention from the rider's self-image, and relax the horse from the usual boring "round the ring" routine. Games offer a goal for horse and rider, but more important, they create a unity that requires teamwork. Having a good time heightens the learning process and develops cooperation between man and beast (figures 1 and 2).

Figure 1. Games reinforce learning skills.

Suitable for all ages, the simple game of "Red-Light, Green-Light," for example, increases the sensitive communication. Played by two or more mounted riders who line up abreast at one end of the riding ring, they move their horses forward on the oral command, "Green-Light." The command "Red-Light" indicates a complete halt (much like the motorist signals). Using hands, legs, and weight in a tactful manner accomplishes the desired result. Rough handling or kicking creates anxiety in the horse and inhibits him from coming to the desired halt...or even encourages rushing forward. Riders whose horses are not calmly halted, or that are observed jigging, backing, or rushing, must return to the starting line. Obviously, the first to reach the finish line is the winner. We might compare this activity to drivers who rush the stop lights, race the motor, kill the engine, and perform other senseless acts. They often lose the race.

Figure 2. Games improve communication with horses.

Behavior Reinforcement

Behavior can be influenced through negative and positive reinforcement. A pat on the shoulder and a kind word encourage repetitive action in man and beast. The instructor carries the responsibility of instilling in his students correct attitudes that are best taught by his example or an appeal to common sense. Most of us recall a particular teacher who was as effective in "attitude" education as in the subject matter.

Behavior of animals and humans is generally improved through calm, reassuring instruction. We know that harsh, abusive language produces defensive action. A good instructor will always be conscious of the unpredictable actions of the horse and develop in his students an alertness to warning signs. Defensive strategy, controlling, and maneuvering are then much the same as that taught by the driver instructor. All instruction to be retained must be presented in a calm, interesting manner. These techniques apply to horse training as well as to humans.

Ground School

Science classes, sometimes called ground-school classes, are an effective method of teaching safety. Subjects dealing with equine attitudes and behavior, such as we have discussed in this article, breeds of horses and their characteristics, conformation, anatomy, and general care of the horse heighten the appreciation and respect of students. Most children love horses and cannot be convinced that they would ever be harmed by a horse. In their love, they seldom anticipate that the toss of the horse's head or the kick at a fly could cause permanent injury. They are just as likely to refuse to believe that a horse could run away with them or that protective apparel is important.

SAFETY EDUCATION

I mentioned the 9-yr-old girl who was dragged to her death by her frightened horse. Her tennis shoe had caught in the stirrup when she fell. Proper footwear with a raised heel might have allowed her foot to slide out of the stirrup. A secure helmet may have prevented the concussion that contributed to her death.

On the bright side of the coin, effective training of youngsters involved under expert supervision in 4-H, Pony Clubs, and some camps and stables have held down the accident rate. Unfortunately, there are several factors hindering this progress. There are not enough trained instructors, there are no national standards of efficiency or competency, and there is no uniform curriculum in schools for riding instructors.

Drivers' licenses are issued only upon completion of written tests and road tests. We have rules and regulations governing most hazardous activities that deal with life and death situations. By strict observance to these rules, careless accidents are avoided. We are often unaware that riding is the most dangerous of driving, riding, or swimming. A camp director would not hire a good swimmer to supervise his waterfront but would insist that a water-safety instructor or lifeguard be employed. By the same token, one often wonders about the rationale of a camp director who hires a "good rider" to manage his high-risk riding program. Too often, uninformed adults equate horseback riding with other activities such as tennis, archery, and golf. They forget that 1,100 lb of horseflesh that cannot read the rules of the stable is a lethal weapon. It is time that adults realize that a "lifeguard" (an educated, trained instructor) is necessary in the riding program.

Critical thinking, the ability to quickly analyze the situation, and to make a correct decision are skills developed through education and experience. "Common sense" is generally not inherent in individuals and is much less so in children. The novice rider knows little about controlling a rearing horse, just as the novice driver is incapable of controlling a car skidding on ice-covered roads.

Learning and safety are progressive. Education in a logical sequence improves the competency of rider and driver. Competency and comprehension ensure the safety of the individual and ultimately contribute to the enjoyment of the sport. The key to happy horsemanship is safety education first.

16

RISK MANAGEMENT IN A RIDING INSTRUCTION PROGRAM

Betty M. Bennett

OUR SOCIETY DEMANDS PROTECTION

All providers of goods and services must protect the consumers and clients from harm. Our society has become more punitive in the penalties it exacts when this duty has been breached. The direction and insistence of these demands has forced the issues of safety and risk management to a central position in the long-range planning of riding schools, 4-H, and Pony Club riding programs. The professional trainer preparing a horse for a client or teaching an individual is not exempt from this pressure. To ignore these demands jeopardizes the very existence of some riding instruction programs.

The Litigation Climate

The litigation climate has changed dramatically. Our society reflects a harsher attitude today than it did only 5 yr ago. Clients are more prone to sue when injured or displeased with products or services.

People sue for a variety of reasons, but primarily because of the attitude and the economics of the times. Philosophically, we should realize that we are apt to be sued since today's society is impersonal. People feel that they should get as much as they can out of an incident. When people are squeezed for money they also seek all sources of revenue, especially from those who they think have "got plenty!" Since more people participate in riding activities than ever before (an estimated 80,000,000 rode in 1980), there is considerably more exposure to accident and more possibilities for lawsuits.

We cannot eliminate the risk activities from society to avoid litigation. People will seek dangerous, challenging activities to test their strength and endurance. It is our duty as competent horsemen to provide these activities but with a minimum of danger to the participants.

There is a psychological and physical need to pursue high-risk activities. Riding, hang gliding, and car racing all carry the element of obvious excitement and exhilaration. Eustress, defined as "pleasurable risk," is inherent in most people, regardless of age, and increasingly more apparent in older people! The urge to compete, conquer, excel, and achieve compels riders to participate in speed activities, such as racing, gymkhanas, rodeo, jumping, and endurance riding. All of these sports are good in themselves, but they place much more responsibility on the management and organizers.

Providing a safe environment for these sports is not sufficient. Today we must also educate participants to take more of the responsibility from the shoulders of management. People are inclined to blame others for their mishaps. It is important to educate riders to assume responsibility for their actions.

In Chicago, a suit was filed against a tack shop for the injury to a woman whose saddle slipped under the horse, resulting in a back injury to the woman. Ridiculous? Certainly the woman was at fault for not securely tightening the girth, and yet the case was settled in her favor. The claim was that the tack shop and manufacturer should have given better instructions in the methods of securing the saddle on the horse.

In New York, a ski operator lost a suit to a young man who broke his leg when he departed from the main ski slope and caught his ski under a tree root. These examples testify to the tolerance of a jury, and the cleverness of the legal profession. It is well known that most juries will favor a child in case of injury regardless of the circumstances surrounding the case. It is particularly significant that increased care must be taken when instructing children.

PREVENTING LEGAL LIABILITY

Types of liability that are legally enforceable:

- Contractual--two parties for performance of services or provision of product. Failure to perform usually results in monetary damages.
- Violation of laws--environmental laws, civil rights laws, health and safety regulations, and personal laws.
- Criminal acts--crime against public, murder, rape, and burglary.
- Tort--wrongdoing against a person who suffers damages.

We are concerned with torts, which may be classified in various ways; however, they generally fall into the following categories:

- Intentional
- Unintentional (negligence)
- Nuisance
- Strict liability

Many accidents involving horse and rider result from unintentional wrongs based on negligence. For example:

- A 4-H leader neglects to insist on protective footwear when working around horses. Horse rears at sound of electric clipper, comes down on child's foot, breaking metatarsal bones.
- Riding academy ties handicapped rider in saddle. Horse spooks. Saddle rolls under horse's belly, dragging rider upside down.
- Riding school overmounts aggressive demanding rider only to have rider severely injured when horse bucks rider off. No protective headgear worn, which may have minimized concussion. In all three cases negligence was determined, resulting in monetary damages.

Who is Liable?

For acts based on negligence, the corporate entity, i.e., the school, the camp, the organization, the named owner or sponsoring agent is responsible for the action or lack of action on the part of the employees when they act within the scope of duty. Not exempt are the administrative staff or the supervisory personnel. It is their duty to employ competent personnel, to devise rules and regulations, and maintain advisory communication with them. Volunteers and student trainees are responsible to a lesser degree, and generally their negligence is imputed to the corporate entity if they were acting within the scope of duty.

Elements of Negligence

An instructor must recognize that he has a duty to provide a safe environment for the students, not only in location but in selection of horses, equipment, and program of instruction. That duty is breached when the standard of care is insufficient and results in injury. An instructor leaves his students to answer the phone. Unattended, two horses get into a kicking battle, breaking the leg of the rider who did not know how to avoid the encounter.

By accepting the task of instructing, we hold ourselves to be competent, aware of the best professional practices, and up to date in methodology--whether or not we are paid for the job.

The nature of supervision demands that we have a plan of supervision. Who is going to take the class if the instructor should become ill? The location of the class--is it safe from natural hazards? Do we observe a reasonable standard of approximately four students mounted per instructor? Do we have a written first aid plan? An emergency plan? Are the phone numbers posted in an easily accessible place? Is there additional help nearby if needed? Is there first aid knowledge or emergency vehicles nearby, or emergency medical help on call?

More specifically, good supervision demands that we inform the participant of the risks involved. Be alert to the rider's physical condition, endurance, and attitude toward horses and others. Sometimes an aggressive rider bothers a horse to the extent that the animal retaliates by harmful action toward another horse. Be sure that riders adhere to rules that have been clearly stated. Keep control of the class at all times. Be alert to changing conditions that may influence equine behavior, such as an impending storm or heavy traffic. Check the condition of the horse before the rider mounts. Is he alert, healthy? Are his feet in good condition? Check the tack. Is it safe, free from stress cracks and weak areas that may break? The rider needs particular attention. Is his apparel suitable? Long pants, shoes or boots with a raised heel so foot can quickly slide out of the stirrup in case of a fall? Is he wearing protective headgear? Western headgear with a protective crown is now available. Many styles of protective headgear have met the medical standards of penetration resistance, distribution and absorption of shock, and strap retention. In 1977, a court case was won in favor of a child who suffered a concussion when thrown from a horse. The defendant did not provide protective headgear. It is interesting to note that as far back as 1977, a case could be won based on the fact that headgear was available and the instructor should have known to supply it for the students. Think about what a jury might award today in a similar case! There is just such a case pending with a $5 million suit involved for lack of protective headgear.

Theories of failure to adequately supervise, failure to adequately instruct, failure to warn of known dangers, and failure to provide competent personnel have surfaced in recent years as the most popular methods of determining a school's or instructor's liability. The general rule is: what would a reasonable instructor have known or reasonably foreseen under the circumstances.

The school or instructor is neither an insurer or a guarantor of total safety, nor are they required to foresee every hazard. They must, however, exercise reasonable care.

WHEN INJURY OCCURS

Follow established procedures which should be in written form and posted. Stay calm, reassure the other riders that all will be well, that most falls are not serious. Apply first aid to the injured while keeping an eye on other potential secondary problems. Sometimes hysteria and emotion precipitate a second accident. Control of the group is essential through a calm insistent tone of voice and actions that reinforce your position. Often it is best to have the other riders dismount and lead their horses well apart from each other. An assistant instructor, if available, should supervise the other riders or go for medical aid, if needed. The person who goes for aid should be reminded to stay at the road or entrance to the accident site to lead the medical people to the victim. Too often, delays occur because the assistant has returned to the accident. A first aid course and CPR training are essential if one is to conduct classes where there is little or no immediate medical service. A vehicle should also be on hand for emergency transportation. Be careful of remarks that may incriminate you such as, "We are sorry, it was our fault."

As soon as possible after the accident, notify the proper authorities. File an accident memo, but stay within the facts, avoiding any opinions on why and how it happened or how it might have been avoided. Be human, courteous, and concerned, but refer all and any questions to your attorney or insurance carrier. Statements such as, "That horse always kicks," or "I should have given you an easier horse to ride," can be used against you as admissions of legal fault.

Take the names of all witnesses to include in your report to be retained in case of suit. Consider pictures of the accident scene or equipment and horse. Have a vet examine the horse and determine the probable cause of accident. Be careful of remedial action after an accident because it may be used against you. Consult your attorney before selling the horse, or rerouting the trail away from that steep hill.

Notify parents or relatives of injuries to indicate condition or to obtain permission for medical treatment. Most riding schools will have a permission form that lists family doctor, hospital preferred, and allergies. Don't assume payment of medical bills since that also may indicate fault. Your attorney will advise if that is the desired action. Follow up with concern and interest in the victim, again being careful not to implicate yourself through careless statements.

POSITIVE RISK MANAGEMENT

The law will force us to monitor our actions if we don't take the initiative to develop a basic systematic risk management plan. To minimize the likelihood of injury and legal liability, we must implement a plan and develop procedures that should be followed by instructors and volunteers. Many resources are available through lawyers, insurance brokers, riding schools, AHSA and AQHA standards, and risk management professionals. Increased expectations in terms of safety in horse sports and excellence in program demand professional deligence. What may have been acceptable 5 yr ago is not the criterion for today's riding programs.

Who Should Teach?

Our academic programs insist on qualified teachers who have earned their education degrees. Our ski instructors, driver education teachers, and water safety instructors all have earned certification through a process of education requiring evaluation and testing. In the U.S. there is no single standard for certifying riding instructors. They may attend a variety of schools for 3 mo to 4 yr and earn a degree that may or may not include practice teaching.

Unlike the European and British schools, where many years of apprenticeship and teaching are required, our teachers may hang up a shingle after having had a minimum of instruction themselves.

Because of the diversity of our teaching methods and the hard-headedness of breed and show organizations, it appears unlikely that the U.S. will achieve the unanimity of methodology that is prevalent in the older countries. A few states such as Massachusetts and Minnesota require testing of instructors and(or) licensing.

The question remains, "Who should teach?" Those who have reached a degree of proficiency and recognition in the show ring may be likely teachers, that is, if they have the skills needed to teach. Perhaps through an examination of lesson plans, progressive instruction, and many hours of practice teaching under the scrutiny of independent examiners, a license to teach could be provided.

Instructor Certification

A certification process should be developed for teachers of children. Children are often directed by an adult who knows less about horses and teaching than do the students themselves. Whatever system is used, it is wise to have documentation from a school or certifying body as to the ability of the instructor. There are several 40-hr instructor courses now offered to horsemen who have advanced skills in riding. These courses are recognized by the American Camping Association. Many college equine programs now offer more practice teaching time than before on the insistence of camp directors and riding school administrators. Any instructor needs to exhibit patience, empathy, and fairness, in addition to voice projection, group control, and personal skill at the level he is teaching.

Selection of Horses

Beginners need horses that are quiet enough to be handled and ridden but not so lethargic that they move with reluctance. A good school horse is said to be worth his weight in gold since he must tolerate all the errors a beginner makes. He must overlook the conflicting signals of the rider and yet be alert enough to carry the better rider on a fulfilling ride. It is necessary to eliminate all kickers, biters, rollers, and other creatures of obnoxious habits from the school string. Good conformation, willingness, size, and temperament are far more important than breed and color.

Although raising horses for a riding school is the better way to assure safe, well-mannered mounts, it is also very expensive. The cost averages out to $1,000/yr/horse when facilities, feed, labor, vet, and farrier costs are included. The benefits outweigh the costs in most cases because the horse has not developed any bad habits by growing up in the riding school, thus ensuring the safety of the student.

A School Program

Our riding school and camp at Hoofbeat Ridge in Mazomanie, Wisconsin, has 50 horses, 80% of them raised on the ranch. Our record over the 22 yr of existence is impeccable. We instruct as many as 400 students per week during the busy seasons. In addition to the well-mannered horses, our staff is trained to advance the students according to individual ability. Ground school (horse science) classes are mandatory with every mounted lesson.

A program such as this is feasible when there is a city within 25 miles to 30 miles with at least 200,000 population. A system of instruction affiliated with the public schools enhances the enrollment at Hoofbeat Ridge Riding School. Students are bussed to the Riding School directly from their schools in Madison, Wisconsin.

Organization and good management require that all riding programs be viewed within the perspective of preventing legal liability. Risk management, therefore, is an ongoing responsibility. Considering the costs, loss of time and concern brought on by major litigation, the time spent in developing a positive risk management program is well worthwhile.

Part 5

NEW FRONTIERS IN BIOLOGY

17

GENETIC ENGINEERING AND COMMERCIAL LIVESTOCK PRODUCTION

H. A. Fitzhugh

Christmas came early for commercial livestock producers in December 1981 and 1982. The presents were major breakthroughs from genetic engineering. First, the development of a safe vaccine against foot-and-mouth disease (FMD) was reported December 4, 1981, in Science magazine; then a year later the successful interspecies transfer of a growth hormone gene from rats to mice was reported in the December 16, 1982, edition of Nature magazine. Probably, few producers realized the significance of these breakthroughs at the time. However, they were both major steps toward substantial improvements in livestock productivity.

Although, these breakthroughs are the most dramatic, they are only two of a rapidly expanding series of major advances in genetic and reproductive biology that fit the definition of genetic engineering.

Before describing these developments in more detail, two points should be made. First, these advances are the culmination of decades of publicly funded, basic research. However, the relevance of this research to livestock production often has not been immediately obvious. The lesson is that investment in good, basic research generally pays great dividends but often in ways that cannot be anticipated. Second, the success of genetic engineering depends on the synergistic application of many technologies--some old, some new. A commonly cited example of this synergism is the impact obtained when artificial insemination (AI), which allows the same sire to produce progeny in multiple herds, is combined with the recording and analysis of performance data (milk yield, growth rate, etc.). Without AI, it would not be feasible to measure progeny performance in different environments against different sets of contemporaries. As one example, most of the impressive genetic gains for milk yield of the national dairy herd over the past 2 decades--currently worth $70 million according to Foote (1981)--have been achieved by combining use of AI and performance testing.

REPRODUCTIVE TECHNOLOGIES

Genetic changes in livestock populations are associated with the reproductive process. The benefits from selection and crossbreeding are obtained by controlled mating of selected males and females. Reproductive technologies, such as artificial insemination and (more recently) embryo transfer, provide additional opportunities for control, e.g., through partitioning X- and Y-bearing sperm (sexed semen) and by changing the genotype of the embryo through micromanipulation.

Artificial Insemination (AI)

Foote (1981) reviewed the current status of AI for the major commercial species (table 1). Good success can be achieved for all species with fresh semen, but only bovine semen is routinely frozen and used successfully. The lack of success with freezing sperm cells from most species probably stems more from lack of resources and research efforts devoted to those species than from any other factor. As Leibo (1981) pointed out, the specifics for successfully freezing vary widely with different types of cells and across species so that protocols must often be worked out by trial and error--a costly process.

TABLE 1. DIFFERENCES AMONG SPECIES IN TECHNICAL FEASIBILITY OF AI

Species	Potential progeny/ sire/yr	Semen fertility	
		Fresh	Frozen
Cattle	50,000	Good	Good
Sheep	5,000	Good	Fair
Goats	5,000	Good	Fair
Swine	5,000	Good	Fair
Horses	750	Good	Fair

Source: R. H. Foote (1981).

Because the necessary time and resources have been devoted to cryopreservation of semen, frozen bovine semen supports a major and growing industry in the U.S., especially for dairy cattle (table 2). This industry promotes genetic improvement as a principal justification for choosing AI over natural service. AI has been a major factor responsible for substantial genetic improvements in the U.S. dairy population--improvements that have created major export sales for semen from superior U.S. dairy sires. On the other side of the coin, AI made possible the rapid spread of genes from continental breeds through the U.S. beef herd starting in the 1960s and continuing today.

TABLE 2. SALES OF FROZEN BOVINE SEMEN IN THE U.S., 1979

	Dairy		Beef	
	No. doses (1,000)	% change from 1978	No. doses (1,000)	% change from 1978
Domestic sales	12,467	5	1,086	6
Export sales	1,836	13	240	61
Custom frozen	682	16	1,125	10

Source: R. H. Foote (1981).

Embryo Transfer

The process that has come to be known as embryo transfer is actually the successful combination of a number of separate, research-derived technologies: estrus synchronization, superovulation, surgical and(or) nonsurgical harvesting, and transfer of embryos (Seidel, 1981). In vitro culture and cryopreservation of embryos are additional technologies that greatly simplify the logistics, reduce costs, and increase the applications of embryo transfer. For example, freezing embryos makes possible long distance--even intercontinental--transport of embryos. This can open the way to increased sales of U.S. seedstock or, perhaps, to the importation of exotic stocks such as the many *Bos indicus* breeds from Asia. Such introductions will depend on the successful resolution of concerns about disease introduction; however, there is reason for optimism that there are safe procedures for movement of livestock embryos (Waters, 1981).

MICROMANIPULATION OF GAMETES AND EMBRYOS

Collection of sperm, eggs, and embryros has developed to the point of being routine reproductive technologies, largely because of commercial demand for these services. A spin-off has been that the experience gained in harvesting and culturing the gametes and embryos has facilitated development of cloning, nuclear transplants, in vitro fertilization, sex control, and other technologies classified as micromanipulation (Seidel, 1982). These technologies remain at the research stage, but several have commercial potential.

Cloning

The excitement (and fears) accompanying the possibility of cloning mature individuals, whether they be 50,000-lb-yield milk cows or charismatic politicians, appears unjustified because cloning of mature body cells now seems unlikely (Markert and Seidel, 1981). Current evidence strongly indicates that mature body cells have lost their "totipoteny," i.e., their ability to develop from a single cell to a complete organism. Perhaps, as the embryo develops, cell differentiation to specialized tissues and organs is accompanied by irreversible changes in the cellular genetic messages. A possible exception is spermatogonium, the diploid cell of the testes that divides to produce sperm (Markert and Seidel, 1981).

Another type of cloning has been successfully achieved. Embryos at the 2-cell, 4-cell, or 8-cell stage have been split to produce identical cattle twins (Willadsen et al., 1981; Ozil et al., 1982). Further splitting to produce identical quadruplets also is feasible. This technique can increase the number of viable embryos available for transfer from valuable matings. Future commercial value may lie in the opportunity to evaluate the genotype of one of the identical "sibs" before making the investment in gestating and raising the other identical sibs. For example, one individual could be developed to the point of determining sex or other genetically determined traits such as polledness. In the long-run, however, the major value of availability of identical sibs may well be to increase efficiency of research in health, nutrition, physiology, and genetics to benefit commercial production.

Nuclear Transplants

Several different types of genetic engineering involve the transplant of nuclear material. The early basic work in this area was done with frogs and other amphibia (McKinnell, 1981). This is a good example of how seemingly irrelevant research can impact on commercial production. Nuclear transplants offer a method of extreme linebreeding to superior individuals. Through nuclear transplant, it will be possible to breed a bull to himself or a cow to herself. Offspring will be 100% related to their single parent instead of the usual 50%; sibs will be 100% related to each other instead of the usual 25% for half sibs or 50% for full sibs. And this can be accomplished in one generation--not in the decades invested in achieving a fraction of the degree of linebreeding to Duchess, Anxiety 4th, and other individuals thought to be special.

The technique is quite simple in concept and not all that difficult in practice for those experienced in micromanipulation of cells. The concept is that the haploid nuclear material from two gametes from the same individual are placed in the same cell in such a way that fertilization initiates embryonic development (Seidel, 1982).

Fertilization resulting from union of two male gametes is called androgenesis. One process is to induce polyspermy in which two sperm from the same or different males enter the same ovum with the consequence of three pronuclei ($_1$, $_2$,) in the cell. The female pronucleus ( ) is extracted from the cell; the male pronuclei ($_1$, $_2$) unite and embryonic development starts. The ratio of progeny will be approximately two-thirds males and one-third females because if both male pronuclei are carrying the Y-male chromosome, the resulting YY union is lethal. Gynogenesis, the union of two female pronuclei ($_1$, $_2$), is also possible by fusing two ova or by microinjection of female pronuclei into an egg. All resulting progeny would be female (XX).

Successful gestation of embryos produced by either androgenesis or gynogenesis has not yet been reported for mammals. One problem is that "selfing" exposes homozygous lethals and decreases fitness (inbreeding depression). Better success may be expected from mating different males or different females rather than by selfing. In this manner, progeny could be directly produced from two highly selected sires without the genetic dilution of the female.

In Vitro Fertilization

In vitro fertilization has been in the news primarily because of the success of "test tube" babies born to parents who had been unable to effect conception for reasons such as oviduct blockage.

Applications to commercial livestock production (Brackett, 1981) include directed multisire fertilization of oocytes collected after superovulation and(or) frozen for long-term storage. In vitro microinjection of sperm could spread the influence of individual sires for which only small amounts of semen were available. Gene banks could be more practically maintained because only a few doses of semen would have to be kept per sire. One of the more useful applications of in vitro fertilization will be an assessment of the ability of semen to fertilize. Current assessment is indirect depending on observed motility and structural soundness of sperm. Direct measurement of fertilizing ability of sperm before and after freezing could be made by exposing oocytes to sperm in vitro.

Normal development of embryos from in vitro fertilization has been obtained with rabbits, mice, rats, cows, and humans. However, percentages of normal development have been low for all except the laboratory species (Brackett, 1981).

Sex Ratio Control

Considerable attention has been devoted to modifying the sex ratio of domestic species. Obvious applications are to increase the proportion of males with the objective of rapid growth to produce lean meat or to increase the proportion of females with the objective of milk production.

Efforts to separate sperm-carrying female (X) and male (Y) sex chromosomes have focused on discovering and utilizing differences in density (centrifugation), electrical charge (electrophoresis) or haploid expression of either X- or Y-linked genes (Amann and Seidel, 1982). None of these procedures have shown significant promise for domestic livestock.

Technologies involving determination of the sex of the embryo show more promise. Amniocentesis works well, but analysis of chromosomes from cells in fetal fluid is delayed until 70 days to 90 days after conception. Abortion of calves of the unwanted sex can then be safely done but at the cost of extending the interval between parturitions. Alternatives include chromosomal analysis of cells obtained by biopsy of embryos at 12 days to 15 days of development (Betteridge et al., 1981), for which a 68% success rate has been achieved. Pregnancy rates of sexed bovine embryos averaged only 33% (table 3). Such low rates are not likely to be acceptable for commercial application.

TABLE 3. PREGNANCY RATE FOR 12 to 15 DAY SEXED BOVINE EMBRYOS

Sample	No. embryos	No. sexed	%	No. sexed embryos transferred	No. pregnancies	%
1	26	15	58	7	1	14
2	117	87	74	4	2	50
3	40	26	65	6	2	33
4	31	23	74	25	10	40
5	69	41	59	29	12	41
6	21	20	95	20	3	15
7	43	25	58	19	6	32
Total	347	237	68	110	36	33

Source: K. J. Betteridge, W. C. D. Hare, and E. L. Singh (1981).

The sexing of one of a set of identical sibs produced from split embryos was previously described. Sexing could be accomplished by biopsy of the embryo, by amniocentesis, or even by waiting to birth. Sex of the remaining embryos would be known and could then be transferred or discarded.

GENE TRANSFER

The two examples of genetic engineering given in the introduction fall in the category of gene transfer or recombination DNA. This technology first came to widespread public attention with the announcement of potential commercial production of insulin by bacterial "factories." The process consists of inserting the DNA code for a desired gene product (e.g., insulin, interferon, growth hormone) into the chromosome of a suitable bacteria, often *E. coli*, which then proceeds to produce the gene product in substantial quantity. With the enthusiasm for the potential of gene transfer came widely publicized concerns that some "recombinants" might have unanticipated dangerous effects, such as new diseases. Fortunately, experience suggests that most of these concerns have little substance (Motulsky, 1983).

Growth Hormone

As important and exciting as is the splicing of mammalian genes into single-celled bacteria, the possibility of interspecies transfer of genes into large complex animals is even more so. Prospects seemed dim until a few years ago. First came the reports of introducing rabbit β-globin into mice (Wagner et al., 1981). This report did not generate the public interest of the announcement a year later that the rat gene for growth hormone had been successfully transferred to mice (Palmiter et al., 1982). A major reason for the excitement greeting this latter report was the dramatic increase in growth rate of the recipient mice and the evidence that this advantage was passed to their progeny. Applications to commercial livestock production were obvious.

The procedure fused the rat GH gene to the mouse metallothionein gene. Multiple copies of this fusion gene were then injected in vitro into newly fertilized mouse eggs that were then transferred to foster mothers. Twenty-one mice developed from the injected embryos; seven carried the fusion gene.

Results from this experiment are summarized in table 4. Comparison of growth rates to 74 days shows that treated females weighed an average of 56% more than did the untreated female littermates; the advantage for treated males was 39%. Reports of favorable response in milk production to prolactin, an analogue of growth hormone, further extend the potential impact of this experiment. The metallothionein gene offers additional potential advantage. This gene that "promotes" activity of the growth hormone gene is sensitive to the presence of zinc. The addition of zinc to the diet could turn on the production of the growth hormone; deletion of zinc could turn off production. While results from the experiment with regard to the effect of dietary zinc were not clear, hormonal stimulation of growth might be regulated to fit available feed supply.

Foot-and-Mouth Disease Vaccine (FMD)

The U.S. livestock industry has not had serious problems with this viral disease (FMD) for several decades; however, recent epidemics in Great Britain have had major economic impact. The usual approach where FMD is not endemic is to slaughter all animals that may have been exposed to carriers. Where the disease is endemic, vaccines made of inactivated viruses are used. Outbreaks of disease have been tracked to use of the vaccine or to the escape of

TABLE 4. EFFECTS OF TRANSFER OF FUSED MOUSE METALLOTIONEIN-RAT GROWTH HORMONE GENES ON GROWTH OF MICE

Mouse	Sex	74-day wt, g	Ratio[a]
2	Female	41.2	1.87
3	Female	22.5	1.02
21	Female	39.3	1.78
Avg for treated females		34.3	1.56
Avg for untreated female littermates		22.0	1.00
10	Male	34.4	1.32
14	Male	30.6	1.17
16	Male	36.4	1.40
19	Male	44.0	1.69
Avg for treated males		36.3	1.39
Avg for untreated male littermates		26.0	1.00

Source: R. D. Palmiter, R. L. Brinster, R. E. Hammer, M. E. Trumbauer, M. G. Rosenfield, N. C. Birnberg and R. M. Evans (1982).
[a]Ratio of treated individuals' 74-day weight to average weight of untreated littermates of same sex.

the live virus from vaccine production facilities. Exports of U.S. livestock are hindered by the usual necessity of vaccination and, as a consequence, the vaccinated animals sometimes contract the disease.

Thus, the biosynthesis of a "safe, stable, and effective polypeptide vaccine for FMD" by a team of USDA and private laboratory scientists was a major breakthrough (Kleid et al., 1981). The basis for this advance was the discovery that one of the four polypeptides that make up the coat of the virus serves as an antigen to stimulate resistance to the virus itself. Because this coat protein has no virulent effect by itself, a vaccine based on this protein is both effective and harmless.

The next step was to produce the desired polypeptide in large quantities. This required identification of the DNA code for approximately 211 sequential amino acids and then the introduction of this code into a special type of *E. coli* for mass production. Vaccine based on this polypeptide produced immunity in both cattle and swine.

The battle is not yet won because there are many strains of FMD virus. Although an all-effective vaccine is yet to be produced, prospects seem good within the decade (Abelson, 1982). Once again, basic research has yielded results with major potential for application to commercial production. Perhaps other research will lead to safe, effective vaccines for diseases such as brucellosis, infectious bovine rhinotracheitis (IBR), or even the common cold (not much help to livestock but a definite boon to producers who have to feed livestock on cold, wet days).

Other Possibilities

The most important performance traits of domestic livestock appear to be conditioned by many genes. Thus, potential impact for interspecies transfer of single mammalian genes is somewhat limited. However, there are a few commercially important traits primarily conditioned by one or a few major genes. Examples include polledness, twinning in sheep breeds (such as the Booroola Merino) and double muscling. In the case of polledness, mating of polled to horned individuals and selecting for polledness will probably be the simplest course. A similar course would likely be most appropriate for double muscling. The sheep gene for twinning might be transferable to cattle. Another candidate for gene transfer--either between or within species--is the gene for cryptorchidism, especially in combination with sex control. The resulting testosterone-producing, infertile males would retain the growth advantage of intact males (10% to 20%) without the problems.

Before the potential of gene transfer can be realized, however, much basic work is needed to determine which genes have what effects and where genes are located on chromosomes. Our state of knowledge on gene location and action in domestic livestock is extremely limited.

REFERENCES

Abelson, P. H. 1982. Foot-and-mouth disease vaccine. Science 218:4578.

Amann, R. P. and G. E. Seidel, Jr. 1982. Prospects for Sexing Mammalian Sperm. Colorado Associated University Press, Boulder, CO.

Betteridge, K. J., W. C. D. Hare and E. L. Singh. 1981. Approaches to sex selection in farm animals. In: B. G. Brackett, G. E. Seidel, Jr. and S. M. Seidel (Ed.) New Technologies in Animal Breeding. pp 109-126. Academic Press, New York.

Brackett, B. G. 1981. Applications of in vitro fertilization. In: B. G. Brackett, G. E. Seidel, Jr. and S. M. Seidel (Ed.) New Technologies in Animal Breeding. pp 141-162. Academic Press, New York.

Foote, R. H. 1981. The artificial insemination industry. In: B. G. Brackett, G. E. Seidel, Jr. and S. M. Seidel (Ed.) New Technologies in Animal Breeding. pp 14-40. Academic Press, New York.

Kleid, D. G., D. Yansura, B. Small, D. Dowbenko, D. M. Moore, M. J. Grubman, P. D. McKercher, D. O. Morgan, B. H. Robertson and H. L. Bachrach. 1981. Cloned viral protein vaccine for foot-and-mouth disease: Responses in cattle and swine. Science 214:1125.

Leibo, S. P. 1981. Preservation of ova and embryos by freezing. In: B. G. Brackett, G. E. Seidel, Jr. and S. M. Seidel (Ed.) New Technologies in Animal Breeding. pp 127-140. Academic Press, New York.

Markert, C. L. and G. E. Seidel, Jr. 1981. Parthogenesis, identical twins, and cloning in mammals. In: B. G. Brackett, G. E. Seidel, Jr. and S. M. Seidel (Ed.) New Technologies in Animal Breeding. pp 181-200. Academic Press, New York.

McKinnell, R. G. 1981. Amphibian nuclear transplantation: State of the art. In: B. G. Brackett, G. E. Seidel, Jr. and S. M. Seidel (Ed.) New Technologies in Animal Breeding. pp 163-180. Academic Press, New York.

Motulsky, A. G. 1983. Impact of genetic manipulation on society and medicine. Science 219:135.

Ozil, J. P., Y. Heyman and J. P. Renard. 1982. Production of monozygotic twins by micromanipulation and cervical transfer in the cow. Vet. Rec. 110:126.

Palmiter, R. D., R. L. Brinster, R. E. Hammer, M. E. Trumbauer, M. G. Rosenfeld, N. C. Birnberg and R. M. Evans. 1982. Dramatic growth of mice that develop from eggs microinjected with metallothionein-growth hormone fusion genes. Nature 300:611.

Seidel, G. E. Jr. 1981. Superovulation and embryo transfer in cattle. Science 211:351.

Seidel, G. E. Jr. 1982. Applications of microsurgery to mammalian embryos. Theriogenology 17(1):23.

Wagner, E., T. Stewart and B. Mintz. 1981. The human $_h$-globin gene and a functional viral thymidine kinase gene in developing mice. Proc. Nat'l Acad. Sci. USA 78:5016.

Waters, H. A. 1981. Health certification for livestock embryo transfer. Theriogenology 15(1):57.

Willadsen, S. M., H. Lehn-Jensen, C. B. Fehilly and R. Newcomb. 1981. The production of monozygotic twins of preselected parentage by micromanipulation of nonsurgically collected cow embryos. Theriogenology 15(1):23.

18

THE POTENTIAL OF IN VITRO FERTILIZATION TO THE LIVESTOCK INDUSTRY

R. L. Ax

INTRODUCTION

Artificial insemination (AI) has greatly improved genetics in livestock and progeny testing permits comparisons among males so that the most superior are utilized in breeding programs. If the female is to make significant genetic contributions, she too, must be evaluated in progeny-testing programs. Current embryo transfer techniques, however, will not result in females being statistically evaluated in AI as we do for males because the offspring from a single mating are sired by the same male.

In vitro fertilization (which implies union of an egg and a sperm in laboratory glassware) offers the potential for females to provide many unfertilized eggs that can be fertilized with semen from various males. The purpose of this chapter is to discuss the potential of in vitro fertilization in relation to embryo transfer and to procedures yet on the horizon for animal breeding programs.

SOURCES OF EGGS (OOCYTES)

A viable oocyte is essential for successful fertilization. Oocytes can be obtained in the oviduct shortly after ovulation or in the follicle on the ovary just prior to ovulation. To do this, a surgical instrument called a laparascope is used to locate and inspect the oviduct or ovary. Superovulation can be induced with hormonal treatments to provide the opportunity for harvesting many oocytes.

The advantage of collecting ovulated oocytes is that they are in the proper meiotic configuration for fertilization. The other alternative would be to aspirate oocytes from follicles and mature the oocytes in vitro prior to in vitro fertilization. We have previously reported that oocytes collected from ovaries of slaughtered cows could be successfully matured and fertilized in vitro, so the next step is to harvest oocytes from live cows with a laparascope.

What is the advantage of maturing oocytes in vitro rather than setting up a superovulation program? The main advantage is that there is still a tremendous amount of uncertainty with superovulation procedures--it is hard to predict how many follicles will develop in a particular animal injected with the hormones. By maturing the oocytes in vitro and examining them under a light microscope, only those oocytes that appear normal are used for subsequent in vitro fertilization. Another advantage is that about 20 follicles can be found on an ovary at any time--even during pregnancy. Therefore, a female could lead a normal reproductive life and provide a continuous supply of oocytes for in vitro fertilization.

Oocytes obtained from small follicles on the ovary are surrounded by tight layers of cells termed cumulus cells. Before fertilization can occur, the cumuli push apart in a process termed "expansion." This ordinarily happens in a follicle coincident with ovulation. Gonadotropins and cyclic AMP derivatives have been found effective at inducing expansion in vitro, but steroids are without effect. In cattle, exposure of cumulus-enclosed oocytes to cyclic AMP for 6 hr, and then culture without cyclic AMP, leads to expansion and maturation of the oocyte within 24 hr after the start of the culture (Ball et al., 1984). Continuous exposure to FSH induces the same effects. The quality of in vitro-fertilized oocytes is reported to be better if sperm are added to cultures of oocytes that have demonstrated expansion of the cumulus cells (Ball et al., 1983).

SOURCES OF SPERM

Three potential sources of sperm are: 1) epididymis, 2) fresh ejaculate, and 3) frozen extended semen. Epidiymal collections provide concentrated samples of sperm that have not been exposed to seminal plasma and the decapacitating effects of seminal components. Fresh ejaculates offer concentrated specimens that are in seminal plasma, so sperm need to be removed from seminal plasma. Frozen extended samples are diluted and contain the cryoprotectants in the extender.

Sperm must undergo two processes prior to being able to fertilize an oocyte; capacitation and the acrosome reaction. Capacitation involves a time that sperm must reside in female reproductive tract secretions and become diluted from the decapacitating effects of seminal plasma. After capacitation has occurred, the acrosome reaction occurs within .5 to 1.0 hr, in the presence of calcium. The acrosome reaction is a morphological change in the sperm head and is accompanied by activation of proteolytic enzymes to aid in digestion of vestments surrounding the ovum.

We are gaining a clearer understanding of induction of capacitation and acrosome reactions in vitro. High molecular weight polysaccharides termed glycosaminoglycans are effective at promoting in vitro acrosome reactions using bull epididymal or ejaculated samples (Handrow et al., 1982; Lenz et al., 1983). Glycosaminoglycans are found in secretions of the bovine female reproductive tract (Lee and Ax, 1983), in follicular fluid (Ax et al., 1983), as well as in the intercellular spaces surrounding cumulus cells that exhibit expansion (Ball et al., 1982). It appears that Mother Nature has built in an "overkill" to guarantee that sperm are exposed to materials to prepare them for fertilization prior to contact with the oocyte.

IN VITRO FERTILIZATION IN PERSPECTIVE WITH EMBRYO TRANSFER

Embryo transfer is a valuable tool needed to ensure that oocytes fertilized in vitro are placed into recipient females. As a genetic tool, embryo transfer cannot make a significant contribution to animal breeding. In contrast, in vitro fertilization provides a method where individual harvested oocytes are fertilized by semen from specific individual males. This would enable progeny testing of a female as is done with males in artificial insemination.

CURRRENT LIMITATION OF IN VITRO FERTILIZATION

After fertilization is completed, an embryo transfer cannot be performed until the embryo can be placed into the uterus by surgical or nonsurgical procedures. An embryo could be surgically deposited into the oviduct, but the chances of a successful pregnancy in the recipient would be markedly reduced. If an embryo is to be placed into the uterus, it should be cultured for several days, which would be the time ordinarily spent traveling through the oviduct. Before in vitro fertilization can be commercially widespread at a reasonable cost, major research is needed to find repeatable, reliable ways to culture embryos for up to 1 wk.

FUTURE APPLICATIONS OF IN VITRO FERTILIZATION

This section is entirely speculative in nature, but application of current technology holds promise for most of the following topics.

Predicting Fertility of Sires

Most laboratory tests to evaluate semen quality are not highly correlated with fertility. In vitro fertilization may be a means of predicting fertility of sires prior to putting them in heavy service several years later. Oocytes could be collected from ovaries at slaughterhouses and randomly distributed to culture chambers. Sperm from bulls could be added, and relative rates of fertilization could be compared among bulls. Subsequent fertility data of bulls after AI could be compared with the in vitro fertilization rates and analyzed for statistically significant correlations.

Extending the Reproductive Life of a Male

When semen samples are extended and frozen for AI, they contain ejaculates diluted to volumes ideally suited for insemination of females. It only takes one sperm to fertilize an egg, but millions are extended to account for losses in the female reproductive tract.

With in vitro fertilization, fewer sperm are necessary to ensure fertilization. Minimum numbers of sperm required have not been established for farm animals. However, a conservative estimate for the potential of oocytes fertilized from a single ejaculate would be n^2, where n is the number of females ordinarily inseminated from a single ejaculate extended and frozen. As an example, a bull who generally provides semen for 500 inseminations would be able to fertilize 500 x 500 or 25,000 oocytes. Even with conservative figures the reproductive potential of a male could be expanded greatly through in vitro fertilization. Superior males, at older ages, could provide semen samples to be processed specifically for subsequent in vitro fertilization.

Nuclear Transfer

Nuclear transfer involves transferring the nucleus of one cell into another cell having a vacant nucleus. A newly fertilized oocyte is an ideal incubator for performing nuclear transfers. The cell is programmed to undergo mitosis, thus the nuclear material would be replicated. Microsurgery is needed, and the nucleus must be inspected under a microscope so that the nucleus of the one-cell fertilized oocyte can be removed at the same time as another nucleus is inserted.

The best source of nuclei is the inner cell mass of a developing embryo. Approximately 60 cells can be harvested, and they are all exact copies. Thus, a clonal line can be established from a single embryo that is divided so that those identical cells can be transferred into newly fertilized eggs. The fertilized eggs function only as incubators, so oocytes could be harvested from ovaries obtained at a slaughterhouse, and any semen sample could be used for in vitro fertilization of the oocyte, which would serve as the incubator.

If nuclear transfers are used for our livestock species, the success rate would not have to be very high. The reason for this is that once an embryo developed from the nuclear transfer method, it could subsequently be split again, thus yielding additional exact copies for additional nuclear transfers. With frozen storage of embryos, a clonal line could be regenerated at any time deemed necessary.

Gene Transfer

Gene transfers have been successfully reported in mice--a rabbit hemoglobin gene and a rat growth hormone gene were incubated into mouse embryos in separate laboratories. As with nuclear transfers, one-celled fertilized oocytes are the ideal vehicle for introducing a gene. If a gene could be incorporated successfully into the chromatin of the one-celled oocyte, it would be replicated every time mitosis occurred. Gene transfers would open a whole new frontier for animal breeding and selection because genes for productive traits, disease resistance, and conformation could be introduced.

With recombinant DNA technology, mapping of DNA to identify genes is progressing rapidly. Optimistic estimates place odds at 1/1000 for successfully introducing genes, yet there is no guarantee that the gene would be activated when needed. In spite of these seemingly insurmountable odds, the few successes would be capitalized upon instantly and selected for intensely. If a sensitive screening procedure confirmed incorporation of the gene in an early embryo, cloning by nuclear transfer would offer a way to introduce many copies of the gene into the population for selection.

CONCLUSION

In vitro fertilization is a new tool for animal breeding with many potential applications. When commercially available, in vitro fertilization will enable a superior female to be progeny tested because harvested oocytes could be fertilized with semen from several males. Development of procedures for nuclear transfers and gene transfers will rely heavily on in vitro fertilization to supply oocytes in vast numbers. For the male, in vitro fertilization offers a potential for predicting fertility and extending the reproductive life by packaging semen in smaller units of sperm per insemination.

REFERENCES

Ax, R. L., G. D. Ball, N. L. First and R. W. Lenz. 1982. Preparation of ova and sperm for in vitro fertilization in the bovine. In: Proc. 9th Tech. Conf. on AI and Reprod. pp 40-44. Natl. Assoc. of Anim. Breeders.

Ax, R. L., R. W. Lenz, G. D. Ball and N. L. First. 1983. Embryo manipulations, test-tube fertilization and gene transfer - Looking into the crystal ball. In: F. H. Baker (Ed.) Dairy Science Handbook, Vol. 15. pp 191-197. Westview Press, Boulder, CO.

Ball, G. D., R. L. Ax and N. L. First, 1980. Mucopolysaccharide synthesis accompanies expansion of bovine cumulus-oocyte complexes in vitro. In: V. B. Mahesh, T. G. Muldoon, B. B. Saxena and W. A. Sadler (Ed.) Functional Correlates of Hormone Receptors in Reproduction. pp 561-565. Elsevier-North Holland, NY.

Ball, G. D., M. E. Bellin, R. L. Ax and N. L. First. 1982. Glycosaminoglycans in bovine cumulus-oocyte complexes: Morphology and Chemistry. Molec. Cellul. Endocr. 28:113.

Ball, G. D., M. L. Leibfried, R. L. Ax and N. L. First. 1984. Oocyte maturation and in vitro maturation. J. Dairy Sci. (In Press).

Ball, G. D., M. L. Leibfried, R. W. Lenz, R. L. Ax, B. D. Bavister and N. L. First. 1983. Factors affecting successful in vitro fertilization of matured bovine follicular oocytes. Biol. Reprod. 28:717.

Brackett, B. G., D. Bousquet, M. L. Boice, W. J. Donawick, J. F. Evans and M. A. Dressel. 1982. Normal development in vitro fertilization in the cow. Biol. Reprod. 27:147.

Brackett, B. G., Y. K. Oh, J. F. Evans and W. J. Donawick. 1980. Fertilization and early development of cow ova. Reprod. 23:189.

Fulka, J. Jr., A. Pavlok and J. Fulka. 1982. In vitro fertilization of zona-free bovine oocytes matured in culture. J. Reprod. Fert. 64:495.

Handrow, R. R., R. W. Lenz and R. L. Ax. 1982. Structural comparisons among glycosaminoglycans to promote an acrosome reaction in bovine spermatozoa. Biochem. Biophys. Res. Comm. 107:1326.

Iritani, A. and K. Niwa. 1977. Capacitation of bull spermatozoa and fertilization in vitro of cattle follicular oocytes matured in culture. J. Reprod. Fert. 50:119.

Leibfried, M. L. and N. L. First. 1979. Characterization of bovine follicular oocytes and their ability to mature in vitro. J. Anim. Sci. 48:76.

Lenz, R. W., R. L. Ax, H. J. Grimek and N. L. First. 1982. Proteoglycan from bovine follicular fluid enhances an acrosome reaction in bovine spermatozoa. Biochem. Biophys. Res. Comm. 106:1092.

Lenz, R. W., G. D. Ball, M. L. Leibfried, R. L. Ax and N. L. First. 1983. In vitro maturation and fertilization of bovine oocytes are temperature dependent processes. Biol. Reprod. (July).

Lenz, R. W., G. D. Ball, J. K. Lohse, N. L. First and R. L. Ax. 1983. Chondroitin sulfate facilitates an acrosome reaction in bovine spermatozoa as evidenced by light microscopy, electron microscopy and in vitro fertilization. Biol. Reprod. 28:683.

Newcomb, R., W. B. Christie and L. E. A. Rowson. 1978. Birth of calves after in vitro fertilization of oocytes removed from follicles and matured in vitro. Vet. Res. 102:461.

Shea, B. F., J. P. A. Latour, K. N. Bedireau and R. D. Baker. 1976. Maturation in vitro and subsequent penetrability of bovine follicular oocytes. J. Anim. Sci. 43:809.

Trounson, A. O., S. M. Willadsen and L. E. A. Rowson. 1977. Fertilization and developmental capacities of bovine follicular oocytes matured in vitro and in vivo and transferred to the oviducts of rabbits and cows. J. Reprod. Fert. 51:321.

19

EMBRYO TRANSFER, MICROSURGERY, AND FROZEN EMBRYO BANKS IN THE CATTLE INDUSTRY

J. W. Turner

Nearly everyone is now aware of "ET," the science fiction character, but to cattlemen these initials mean far more than a promotion for movie entertainment; for Embryo Transfer has become a commercial reality in cattle breeding. This technology holds some exciting opportunities for the cattle industry and promises other applications that may dramatically affect the future beef industry (Koch and Algeo, 1983; Rutledge and Seidel, 1983).

With the development of nonsurgical methods and more effective treatment control of estrous cycles of recipient cows, embryo transfer has become widely accepted in breeding purebred beef cattle. This genetic tool enhances the reproductive capacity of the cow; when superior bulls are used, the progeny obtained by embryo transfer are expected to be genetically superior and of more economic value. And because several more progeny are possible from one cow, these cows can have a greater impact on genetic change. Koch and Algeo (1983), citing Seidel (1979) relative to embryo transfer applications, listed the following cases: production of extra progeny from valuable females, transporting embryos, rescuing breeds facing extinction, testing males for recessive genes, and studies of maternal effects. Perry (1983) outlined similar applications of embryo transfer.

Most genetic change has been from bull selection because of the differential reproductive rate favoring males. Bull selection and AI have been credited as the major factors in genetic improvement of milk yields (Niedermeier et al., 1983). This is due to AI use of truly superior bulls of known breeding value (indicated by accurate DHIA testing). When embryo transfer is considered, it should be viewed similarly as a genetic tool for creating change through females of known breeding value. A common mistake is to assume that "litters" of embryo calves should be uniquely superior and identical. However, each calf is a sample of the parental genes and will differ as do full siblings produced in separate years. Embryo transfer also allows for nongenetic factors to affect the calves due to the intrauterine and maternal effects of the recipient cows. These effects have often been ignored, but their significance can be found in records of individual calves. In fact, performance testing of embryo transfer calves does not relate to accurate selection for maternal traits (Willham, 1983). For this reason, only superior cows of known breeding value should ever be considered for a donor role, which suggests accurate performance testing prior to any consideration of embryo transfer. Heifers have been used in embryo transfer to "prove" them with a large number of progeny, but such procedures may ignore performance record accuracy and the nongenetic influences of recipient cows on the embryo-transfer calves.

Currently, embryo transfer is greatly influencing the purebred beef industry, as evidenced by higher individual prices paid for donor cows. In theory, a few donors could generate a "herd" of superior calves from poorer quality recipient cows. With increased genetic value, such progeny would demand higher per head prices. The fact remains one cow gestates and lactates to produce one calf. A cattleman, however, must still manage and maintain a base cow herd. The greatest advantage to embryo transfer beyond its genetic implications may well be the increased management necessary to accomplish embryo transfer. More attention to nutrition, bull selection from AI studs, and records for estrus control should make for better cattle managers. Based upon reported costs, it seems safe to assume that embryo transfer will be restricted largely to the purebred industry.

Other technology that will impact on embryo transfer includes frozen embryo banks, microsurgery and the production of cleaved twins, and accurate sexing of embryos or semen. Should frozen embryo technology achieve a success rate similar to frozen semen (AI), cattlemen would order embryos and manage the cow herd much like they use herd AI. Estrus management and control represent a major labor constraint; improvements will be required before greater application of embryo transfer can be made to the commercial industry. A relatively small percentage of commercial beef cows are bred by AI (Koch and Algeo, 1983).

Cleaving of embryos to produce twins is exciting. There are obvious experimental advantages to this technique, but most scientists have yet to fully conceptualize its application to the industry. This process is not the same as that of asexual reproduction that we see in plants whereby a variety can come from one parent plant.

One of the more exciting aspects of immediate application in embryo transfer would be sexed semen and(or) accurate sexing of embryos. This could quickly alter production by simply providing for single-sex calf crops. Heifers could be selectively generated from the truly superior maternal parents. Should successful frozen embryo technology and sexing occur, a major effect will be seen in commercial beef herds. Commercial cattlemen could easily see the economic benefits of all male calves of controlled breeding. The technology would have to be cost effective on a commercial market basis.

Another scientific consideration affecting embryo transfer is the genetic engineering methodology whereby extra "genes" could be introduced into embryos. This has been done in some mammals and work with microorganisms has shown this to be an important concept. In higher animals, we know very little of major gene effects and location of important genes on the chromosomes. Research into embryo transfer and microsurgery may yield many more unique applications for genetic improvement (Rutledge and Seidel, 1983).

Regulation by individual breed associations will probably have some control or influence on embryo transfer. Some may allow unrestricted use while others may closely regulate its use. Logic may not always prevail concerning regulation by a board of directors. If fewer purebreds are required using AI and embryo transfer, will breed associations want to operate with fewer members and cattle numbers? Will associations require extensive testing to approve cattle for AI and embryo transfer to verify breeding value for important traits? Will the industry identify specialized breeds with closely controlled breeding programs that reflect a breed policy rather than an individual breeder's decision.

An important economic consideration in the rapid adoption of embryo transfer has been the IRS and some apparent tax advantages. Needless to say, commercial beef producers are not as interested in tax advantages when evaluating embryo transfer because they sell on the basis of production efficiency.

Technology developed in reproductive physiology to use embryo transfer must ultimately be useful as a breeding tool for more effective selection and genetic change in cattle. It is the selection aspects for genetic improvement that justify the technology and its use.

REFERENCES

Koch, R. M. and J. W. Algeo. 1983. The beef cattle industry: Changes and challenges. J. Anim. Sci. 57:28.

Niedermeier, R. P., J. W. Crowley and E. C. Meyer. 1983. United States dairying: Changes and challenges. J. Anim. Sci. 57:44.

Perry, B. 1983. New advances in bovine embryo transfer technology. In: F. H. Baker (Ed.) Beef Cattle Science Handbook, Vol. 19. pp 414-418. A Winrock International Project published by Westview Press, Boulder, CO.

Rutledge, J. J. and G. F. Seidel, Jr. 1983. Genetic engineering and animal production. J. Anim. Sci. 57:265.

Seidel, G. E., Jr. 1979. Applications of embryo preservation and transfer. In: H. W. Hawk (Ed.) Beltsville Symposia in Agricultural Research (3). Animal Reproduction. pp 195-212. Allanheld, Osmun, and Co., Montclair, New Jersey.

Willham, R. L. 1983. Fitting cattle to systems: An action plan. Red Poll News 40:3:12.

20
GENETIC ENGINEERING OF ANIMAL VACCINES

Jerry J. Callis

INTRODUCTION

The livestock population of the U.S. is among the healthiest in the world. And, although the ratio of livestock to the human population provides a self-sufficiency in animals and animal products, the U.S. imports some types of animal products and exports others. Our livestock census approximates 122 million cattle, 8 million horses and mules, 14 million sheep and goats, 55 million swine (over 100 million are slaughtered each year), and 4.2 billion poultry. The populations vary slightly from year-to-year depending upon price, need, and other circumstances. (There are also 30 million to 50 million cats and dogs.)

In international commerce, animal diseases influence trade practices, the products that are available, and the price of the commodities. Many countries that are free of certain animal diseases, such as foot-and-mouth disease (FMD), embargo animals and animal products from infected countries. Through dialogue with livestock owners, consumers, other interests, and analysis of research data, animal health authorities of the respective countries establish policies relative to animal diseases. That is, they determine those that will be eradicated, how, and which vaccines will be used.

Animal diseases can have a profound effect on animal populations. This is especially true when the so-called "epidemic diseases" gain a foothold. An example is African swine fever (ASF), which was introduced into the Dominican Republic and Haiti in 1978 and caused the destruction of the entire swine populations of those countries during the disease eradication process. Destruction of the swine was necessary to eradicate ASF because of a lack of vaccines. Another example is rinderpest, a deadly viral disease of cattle, which is inflicting heavy losses in many African countries. The eradication methods there are entirely different, however, because effective and inexpensive vaccines are available, and their effectiveness has been demonstrated in an Africa-wide control program financed by national and international sources. When the control programs reverted to national resources, however, many were not effective because of local economic, technical, or political reasons; thus the disease is on the increase in many parts of Africa.

Fortunately, in recent years none of the so-called epidemic diseases of livestock have entered the U.S. One possible exception would be the velogenic strain of Newcastle disease of poultry that entered the U.S. in 1971. This disease caused large numbers of laying hens to be destroyed in California at a cost of $72 million. Animal health authorities, practicing veterinarians, and livestock owners must be alert to the introduction of such diseases and must have the information and technology to quickly recognize and eradicate newly introduced animal plagues.

Recent advances in the development of vaccines using genetic engineering technology have the potential to provide additional and safer products for controlling certain diseases. This technology, which is being applied to some animal diseases that occur in the U.S., promises more effective, less-expensive products through genetic engineering. In the following sections, this presentation provides a brief description of the technology, its application, and a review of some of the products that may be developed through gene splicing and related technology.

DESCRIPTION OF GENETIC ENGINEERING

Genetic engineering is one of our newest technologies. As a result of this advance, some scientists feel that we are on the verge of a medical revolution based on developments of recombinant DNA technology--often referred to as gene splicing, or genetic engineering. Progress is being made in this field at a rapid rate. The production of human, animal, and viral proteins, hormones, enzymes, and interferon in microorganisms or tissue cultures has moved from theory to reality, and the technology is being applied at an ever-increasing rate. Genetic engineering is the technology of the 1980s. In laboratories all over the world, scientists are taking genes from one organism and are putting them into another. Gene splicing has been used to develop certain laboratory strains of bacteria that can produce several products for use in man and animals.

Man has been changing the genetic makeup of plants and animals in a limited way for thousands of years. This began with the planting of the seed from the wild grasses that produced the most grain and were the easiest to harvest, and the selecting of the fattest cattle, and the breeding of the sheep with the best wool. As we learn more and more about the genetic code, and how to decipher it, we can speculate about the role it plays in the evolutionary process. Genes recombine in nature every day. Viral recombination experiments have been done in the laboratory for several years, but now one can recombine genes of unrelated organisms and make artificial, but useful, molecules.

Before engaging in most types of research, one should have a plan of action or an approach; this strategy is especially applicable in genetic engineering. The phrase "before engaging in genetic engineering, one must know what one is looking for" has become commonplace. Not all microbes can or should be engineered, and the technology should not be attempted with organisms that have not been carefully studied at the molecular level. One should know as much as possible about the immunogen of the microbe so that this may be related to a particular stretch of the microbe's nucleic acid. This is because it is the gene, or a piece of the gene's nucleic acid specifying immunogenicity, that must be separated from other genes and then be inserted into the bacteria, yeast, or a tissue culture system where the desired product will be expressed and(or) produced.

The basic knowledge or technology necessary for recombinant DNA procedures has been developing steadily for decades as we have learned more and more about the molecules that make up microbes and the genetics that govern their reproduction. Three specific events, steps, or kinds of knowledge were necessary before production of products by biosynthesis could be attempted. The first occurred in 1953, when Watson and Crick proposed the DNA structure of molecules. Since then, the progress in molecular genetics has been rapid. Their description of the famed double helix "ladder," or structure, enabled scientists to fully understand the genetic blueprints for genes from bacteria to man. Their description of the model for DNA structure provided a basis for further exploration and understanding of biology at the molecular level.

These developments were quickly followed by the second prerequisite to biosynthesis, which was an improvement in the methods and knowledge about chemical and enzymatic manipulations of DNA. The description and understanding of restriction endonucleases provided a basis for separating large genomes into small segments, and the development of chemical sequencing methods provided for precise determination of thousands of base pairs on such segments.

With this knowledge about the molecular structure of genes and enzymes that could cut the gene at predetermined sites on the nucleic acid chain, scientists were ready for the third development--recombining genes and cloning them into "factories" for production. The bacterium, *E. coli*, one of the most studied microorganisms known to microbiologists, is the most commonly used factory for production of genetically engineered products. Because of the background of knowledge about *E. coli*, the technology is available to remove plasmids (extra chromosomal bacterial DNA) from the bacteria, to cut them with special enzymes, and to splice pieces of genetic material from another organism into them.

When the newly reconstructed plasmid is reinserted into the bacterium (during a process called transformation), among its products it yields the protein coded by the DNA piece from the other organism (if, of course, gene splicing occurred with the use of proper promoters, linkers, and enzymes). The yield of the engineered "factory" will depend upon the cell's ability to transcribe the foreign gene into messenger RNA and translate the messenger RNA into protein, which is not degraded in the cell. Obviously, it is also important that the growth requirements are favorable for production. In addition to *E. coli*, other organisms used as host organisms include *Bacillus subtilis*, *Streptomyces species*, and animal-tissue-cultured cells, and even certain viruses such as vaccinia. The technology is shown in figure 1.

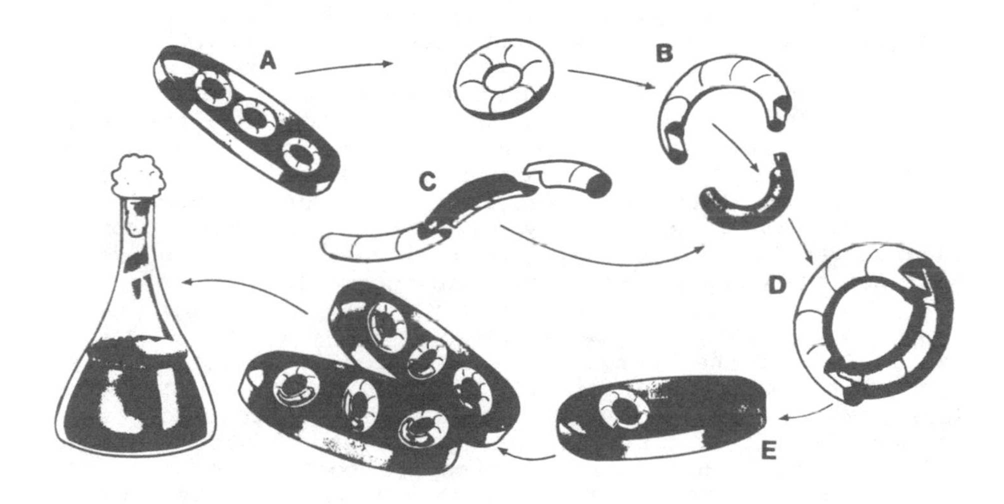

Figure 1. Small circular gene elements, called plasmids, are isolated from the common bacterium, *E. coli* (A). The circle is opened with a specific enzyme (B) and the copy of the gene for the foot-and-mouth disease virus (FMDV) vaccine protein is cut out and inserted into the plasmid (C). The plasmid is closed with another enzyme. The circular gene element is gain functional and now contains a new gene--the FMDV vaccine protein gene (D). This recombinant plasmid is inserted into *E. coli* (E), which, when grown in culture media, produces large quantities of the vaccine antigen without the infectious virus itself being present.

Source: USDA/ARS, Plum Island Animal Disease Center. 1982.

SOME GENETICALLY ENGINEERED ANIMAL VACCINES BEING DEVELOPED

The potential for application of genetic engineering is highest for use in controlling animal diseases caused by viruses. This is possibly because more viruses have been studied at the molecular level; these viruses also cause some of the more important infectious diseases. They have the ability to survive and to cross international boundaries, thus complicating international trade.

Biosynthesized Subunit Vaccines

It has been demonstrated that individual proteins that can be isolated from the surfaces of several viruses and bacteria can stimulate production of neutralizing antibodies that protect against challenge with the infectious agent. These small viral pieces are referred to as subunits or immunizing proteins. Some subunits are commercially produced, as, for example, in vaccines for influenza. These results with the natural subunit vaccines have caused scientists working on this problem to attempt to place the gene-specifying immunizing proteins into bacterial expression systems so that sufficient immunizing protein(s) can be produced and formulated into vaccines. One such product being researched is a genetically engineered monomeric protein subunit vaccine for foot-and-mouth disease (FMD), one of the most serious diseases of animals in the world.

FMD is caused by a picornavirus. There are seven immunologic types of the virus and many subtypes within each. Some of the subtypes are sufficiently different so as to require separate vaccines. In some areas of the world where FMD vaccines are used, it is necessary to incorporate as many as five different viral subtypes into a single dose. In addition, there also may be continuing antigenic shifts in the virus that require further changes in the vaccine formulation.

To recognize the antigenic shifts, there must be continuing surveillance of field strains of the virus so that new strains do not develop that are totally different antigenically from those in the vaccine. This problem is apt to continue, irrespective of the type of vaccine in use, and further illustrates the necessity of having products that provide broad antigenic coverage. In the case of whole-virus vaccines, new strains must be brought in from the field and adapted to tissue-culture systems for production of virus. In the case of biosynthesized products, the nucleic acid of the new isolate must be separated, the desired immunogenic gene located, and spliced into an expression vector for propagation. The molecular properties of FMD virus have been well-established and described. A natural subunit vaccine was produced from the virus several years ago. The FMD virion contains a single-stranded RNA molecule of about 8000 nucleotides. The nucleus is surrounded by four major proteins designated VP_1, VP_2, VP_3, and VP_4, each occurring in 60 copies. VP_1 has been identified as the protein primarily responsible for immunity, and when this protein is separated from the other three and made into a vaccine, it can protect cattle and swine from infection when they are exposed to the FMD virus.

In studies to date, the amount of VP_1 required to immunize is greater than that required for inactivated whole virus vaccines. Biosynthesis (i.e., production through gene splicing in another organism) provides an attractive potential source of FMD antigen, because in order to make the natural VP_1 subunit vaccine, it is necessary to produce large quantities of virus and to isolate and purify VP_1. To accomplish this bisynthesis, the gene for VP_1 of FMDV, type A_{12}, was isolated, cloned into an expressed plasmid of *E. coli*, reinserted into *E. coli*, and the VP_1 protein was expressed when the bacteria were propagated. The replication time of this strain of bacteria is 25 min, thus quickly producing a high concentration of bacteria. Each bacterial cell produces up to 1,000 molecules of the VP_1 protein, which is enough to formulate 4,000 doses of vaccine per liter of culture.

Worldwide, at least 17 different FMD viruses are used for vaccine production, and, as expected, early observations indicate that the biosynthesized peptides will not have any broader antigenic coverage than do whole virus vaccines. Currently, genes coding for the VP_1 of several vaccine strains of FMDV have been cloned and expressed, and the biosynthetic proteins are being evaluated. The type of immune response obtained in cattle from the use of one such vaccine is illustrated in table 1. These vaccines will not be available commercially for several years.

TABLE 1. NEUTRALIZING ANTIBODY AND IMMUNITY IN CATTLE VACCINATED WITH BIOSYNTHETIC A_{12} VP_1 VACCINE

Micrograms of antigen	Wk after vaccination											
	2	8	12	15	17	21	30	32	34	38	42	45
10	.9	.9	.9	1.7[b]	2.0[a]	1.7	1.7[c]					
50	1.0	1.2	1.0	1.8[b]	2.1[a]	1.9	2.0[c]					
250	1.1	1.1	1.0	2.0	2.0	1.8	1.8	1.9[b]	2.3[a]	2.6	2.5	2.4[c]
1250	1.2	1.3	1.3	2.0	2.6	2.0	1.9	2.3[b]	1.9[a]	2.9	2.4	2.7[c]

Source: P. D. McKercher, D. M. Moore, D. O. Morgan, B. H. Robertson, J. J. Callis, D. G. Kleid, S. Shire, D. Yansura, B. Small (1983).

[a] Titer 2 wk after revaccination.

[b] Revaccination with 10, 50, 250, or 1250 microgram dose, respectively.

[c] Challenge of immunity:

10 UG 30 wk 5/9 Immune
50 UG 30 wk 7/9 Immune
250 UG 45 wk 8/9 Immune
1250 UG 45 wk 9/9 Immune

Other animal viruses on which research is underway to produce immunogenic polypeptides by cloning genes include rabies, infectious bovine rhinotracheitis, transmissible gastroenteritis of swine, Rift Valley fever, vesicular stomatitis, pseudorabies, parvovirus of dogs, and bluetongue. Success has been reported in several instances of cloning with ensuing expression of gene products, but this technology has not yet produced commercially available vaccines against viral diseases.

Organically Synthesized Products

In the case of FMD, gene cloning has provided a means of determining the nucleotide gene sequence and amino acid-protein sequence for the VP_1 immunogen. The sequences of the immunogen of several of the 17 or more vaccine strains of virus have been published, and from this information, it is possible to predict the structure of the antigenic sites. Short 20 amino acid polypeptide sequences determining immunogenicity have been chemically synthesized, attached to carriers, and have been shown to have potential as vaccines. The peptides have been used to vaccinate guinea pigs that subsequently were protected against infection by live virus. These results are exciting and indicate that short, organically synthesized antigens also may be useful as vaccines. Polypeptides of Simian virus 40, influenza, feline leukemia virus, and hepatitis B surface antigen have been organically synthesized and have shown promise as vaccines.

Genetic Engineering of Bacteria

Genetic engineering also is being applied to the preparation of protein vaccines against bacterial diseases. Enterotoxigenic *E. coli*, (the cause of diarrheal diseases in young livestock) contain pili (proteinaceous appendages) on their surface. Distinctive immunogenic strains have been isolated from swine and calves, and the genes for these pili proteins have been cloned and expressed in other bacteria. Vaccines have been made from these products and have been licensed for use in some European countries and the U.S.

Viruses as Vectors for Immunogens

Vaccinia virus has been bioengineered to act as an expression vector for cloning foreign genes in tissue cultures and animals. Tissue-culture cells infected with vaccinia virus that carry the gene for hepatitis B surface antigen, permit the cells or the rabbits vaccinated with such a virus to express the immunizing protein for hepatitis B virus. Other viruses also are being used as vectors that may have wider acceptance than vaccinia, which currently is not used routinely anywhere in the world. Its use as a vector for another viral gene might not receive enthusiastic endorsement from public health and animal disease officials.

Interferons for Animals

Interferons are a heterogenous group of proteins divided into three classes--alpha, beta, and gamma. They have been shown to modulate several immunological reactions including antibody production. They are produced in a variety of cells and can be induced by chemicals, viruses, bacterial products, antigens, antigen-antibody complexes, etc. Recently, large-scale production methods in tissue culture systems have become available, thus, a sufficient product is available for study. More recently, interferon has been produced by recombinant technology in *E. coli* in amounts sufficient for study against some neoplasms, immune disorders, and infectious diseases. Much remains to be learned about their mode of action and therapeutic effectiveness. This work should benefit veterinary medicine in the treatment of valuable breeding stock and pets.

Monoclonal Antibody

Tissue cells that will grow in perpetuity (these so-called lines of cells are usually cancerous) can be fused with other cells that have been primed to produce an antibody of a predetermined specificity. The fused cells, called hybridomas, produce antibodies that are referred to as monoclonal because they are a homogeneous population of identical molecules. Uses of such antibodies are not yet fully explored but include purification of antigens, analysis of antigenic sites on microbes, diagnosis, and treatment of diseases. They are especially useful in mapping the antigens of a microbe. Monoclonal antibodies are now proving useful in analyzing the antigenic sites of vaccine strains of influenza, rabies, polio, foot-and-mouth disease, bluetongue, and herpes infections. Perhaps one of their most promising uses is in the development of anti-idiotype antibody vaccines. Such antibodies have sites that mimic antigenic sites of the original antigen, thus they have potential as vaccines--especially after amplification in hybridomas, or cloning and expressing in single-cell hosts. One monoclonal antibody preparation has been licensed for use in the U.S. for treatment of calf scours. It is administered to the calf as a drench soon after birth, thus, it potentiates or supplements the natural antibody in colostrum.

Animal Growth Hormones

The genes for growth hormones from cattle and chickens have been cloned in *E. coli* and sufficient quantities produced for evaluation in the respective species. Studies are currently underway in beef and dairy cattle and in poultry, but no information concerning their usefulness has been released. However, human growth hormone also has been produced by genetic engineering, and clinical trials in man indicate it is useful for treating dwarfism in some children. It also is expected to have other uses such as promotion of healing in burn victims.

ADVANTAGES OF GENETICALLY ENGINEERED VACCINES

It is generally accepted that the antigens in a vaccine represent 20% of the cost of the product; the other costs include those for vaccine formulation, bottling, labeling, sterility, potency control, shipping, storage, and marketing. For these reasons, there is a feeling that the advantages of the synthetic vaccines may be overstated. This debate will continue until several of the products have been commercialized and there is more basis for comparison.

There are some predictable and distinct advantages of genetically engineered vaccine not necessarily related to the cost. Perhaps the most important one relates to safety. Since the etiologic agent--viral or bacterial--is not required to produce the immunogen, one does not have to be concerned that an agent will escape from the production laboratory because only a small piece of the infectious microbe is used in gene cloning. Also, one does not have to be concerned with inactivating the agent. These are important advantages especially in the case of FMD. In some countries where FMD vaccines are produced, outbreaks have been traced to the escape of the virus from the production laboratory. In other cases, outbreaks have been traced to improper inactivation of the virus. The cloned gene products also will not require refrigeration--a distinct advantage in the tropics and in less-developed countries.

The use of genetically programmed bacteria is a promising avenue to vaccine manufacturing.

REFERENCES

Bachrach, H. L. 1982. Recombinant DNA technology for the preparation of subunit vaccines. J. AVMA. 181(10): 992.

Bittle, J. L., R. A. Houghten, H. Alexander, T. M. Shinnick, J. G. Sutcliffe and R. A. Lerner. 1982. Protection against foot-and-mouth disease by immunization with a chemically synthesized peptide predicted from the viral nucleotide sequence. Nature 298(7):30.

Gilbert, W. and C. Vila-Komaroff. 1980. Useful proteins from recombinant bacteria. Scient. Am. 242(4):68.

Goldstein, G. and M. Sanders. 1983. Monoclonal antibodies in clinical medicine. Clin. Immun. Newsletter. 4(6).

Kleid, D. G., D. Yansura, B. Small, D. Denbenko, D. M. Moore, M. J. Grubman, P. D. McKercher, D. O. Morgan, B. H. Robertson and H. L. Bachrach. 1981. Cloned viral protein vaccine for foot-and-mouth disease: responses in cattle and swine. Science 214:1125.

McKercher, P.D., D. M. Moore, D. O. Morgan, B. H. Robertson, J. J. Callis, D. G. Kleid, S. Shire, D. Yansura, B. Small. 1983. Genetically Engineered Polypeptide Antigen for Foot-and-Mouth Disease: A Dose Response in Cattle. FAO Proc., European Commission for Control of FMD. Rome, Italy.

Part 6

GENETICS

21

GENETIC IMPROVEMENT IN HORSES: PRINCIPLES AND METHODS, PART 1

Joe B. Armstrong

Mankind has long been aware that differences exist among individuals, and that selection and mating systems produce genetic changes in domesticated animals. Animal breeding is actually one of the oldest sciences of which we have written record. According to the Bible, Jacob was effectively selecting for color in sheep and goats as early as about 1746 B.C. Genesis 30:32-43 implies that Jacob practiced selection among the males of Laban's flocks so as to increase the number of animals of specific colors that were to become his by virtue of an agreement with his father-in-law, Laban.

The writings of Varro (116-27 B.C.) indicate that farmers of that era were practicing selection among the animals of their flocks and herds. On the subject of selection in swine, Varro is quoted as follows:

> "A man, then, who wishes to keep his herd in good condition should select first, animals of the proper age, secondly of good conformation (that is, with heavy members, except in the case of feet and head), of uniform colour rather than spotted. You should see that the boars have not only these same qualities, but especially that the shoulders are well-developed."

Perhaps more familiar than the exploits of Laban and Varro are those of Robert Bakewell, who lived from 1725 to 1795. Bakewell of Dishley, Leicestershire, England, is perhaps the most noteworthy of early animal breeders and is often referred to as the founder, or father, of modern animal breeding. Bakewell's breeding work was with Longhorn cattle, Leicester sheep, and Shire horses. He achieved considerable success in improving his animals by combining selection with inbreeding. Bakewell's basic principles, which are still widely quoted today, were: "Like produces like or the likeness of some ancestor; inbreeding produces prepotency and refinement; breed the best to the best."

Genetic improvement in horses involves the same basic principles as genetic improvement in beef cattle, sheep, hogs, and dogs. Unfortunately, less factual genetic information is known about the horse.

In addition to a lack of concrete genetic knowledge, the horse industry faces other major deterrents in its quest for genetic improvement. Some of these deterrents are listed below.

Failure to Establish and Maintain Clear, Obtainable Goals

True "breeders" develop realistic goals for their horse breeding operations and earnestly pursue them. Unfortunately, there are few true breeders of horses. Most are "multipliers" who have no planned breeding program. If they do have a plan, it is abandoned in favor of something new every time popular opinion shifts--and popular opinion blows with the winds. Aim at nothing and you will surely hit it! It takes planned perserverance to be a breeder.

Reproductive Efficiency

The horse is generally recognized as having the poorest reproductive rate of most domesticated animals. A national average for broodmares foaling live foals generally is considered to range from 50% to 80%. The percentage of foals weaned would be smaller.

The difference between mares and other domesticated females in reproductive efficiency can be attributed partially to the unusual reproductive patterns of the mare. These patterns include 1) the long estrous period within which ovulation can occur at any time, 2) a seasonal ovulatory pattern with a prolonged estrus prior to the first ovulation, and 3) sporadic estrous behavior. The reason(s) for these unorthodox patterns is not readily apparent--genetics? environment? management?

Lack of Selection for Reproductive Efficiency

The major selection practiced in horses has been based on their ability to walk or run. Economics ensures that other farm animals (cattle, sheep, hogs, etc.) are selected heavily on the basis of reproductive efficiency.

In addition to requiring on the average two mares for each foal produced annually, lack of attention to reproductive efficiency results in an increased generation interval. Genetic improvement is made only when generations are turned. Breed the same mares to the same stallion for 15 yr and the genetic base will remain the same. The only changes will be in management (environment), except that all the mares and the stallion will be 15 yr older. To maximize genetic improvement, as soon as they reach breedable age, the superior female offspring from the above matings should be bred to another stallion.

Economics

As compared with research on other farm animals, little research has been conducted on horses. The horse, unlike cattle, sheep, and hogs, has had no firm market value in past years because it is not a meat animal. It also costs more to maintain a herd of horses. These factors, when multiplied by poor reproductive efficiency and long generation intervals, have made the horse a poor experimental research vehicle.

However, the increase in interest in horses and in the number of horses over the past 40 yr has made the horse an entity of great economic importance. Increasing research has been generated over the last 20 yr. Nutrition trials and studies can be conducted over relatively short periods, but meaningful genetic studies require generations! Therefore, time and money are major factors related to our lack of knowledge in horse genetics.

In spite of the deficiencies in our bank of genetic knowledge, most horsemen are not adequately tapping the store of knowledge that exists today. The historian Hagel is quoted as saying "The only thing we learn from history is that we do not learn from history."

BASIC GENETICS

Genetic improvement in horses is attained through 1) the use of mating systems and 2) the selection of superior individuals in a given generation to be the progenitors of the succeeding generation. **Selection** is simply determining which individuals will be the parents of the next generation. If selection is to be effective in changing a population of horses, variation--differences between individuals--must exist in the population. **Variation** is the net result of differences in heredity, environment, and the joint effects of heredity and environment; it is often referred to as "the raw material with which the breeder works." Such raw material now exists in horses and has existed through the ages!

The ingenuity and perseverance of man are required to shape this material into new and improved horses.

The gene is the basic unit of inheritance. Countless numbers of genes occur in pairs on threadlike structures called chromosomes in the nuclei of cells. The horse has 32 pairs (64 total) of chromosomes in each cell nucleus. One chromosome from each pair is inherited from each parent at the time of fertilization. While every cell nucleus has 64 chromosomes, each ovum produced by the mare and every spermatozoon produced by the stallion contains 32 chromosomes. The chromosome number is reduced from 64 to 32 during the formation of ovum and spermatozoon by the process of meiosis.

The location of a gene on a chromosome is known as a locus. Each member of a gene series is an allele and there is only one allele at a given locus on the chromosome. There may, in fact, be several alleles for each gene. Our knowledge of coat-color inheritance demonstrates this fact very visibly. During meiosis (the reduction of chromosome number from 64 to 32 in the germ cells) a chromosome pair becomes entwined and exchanges chromosome (genetic) material from one chromosome member of the pair to the other chromosome of the pair. The alleles are mixed randomly in the pair of chromosomes, and as the chromosome pair divides, each germ cell receives one member of the pair of chromosomes. This random mixing of alleles reduces the probability of exact replication of parents and among siblings. This mixing enables a foal to receive genes from all of its grandparents.

While the horse has 32 pair or 64 total chromosomes, the ass has only 31 pair or 62 total chromosomes. The resultant cross between the horse and the ass produces superior, rugged work animals (mules and hinnies), but they are infertile. Mules and hinnies have 63 chromosomes, having received 32 from their horse parent and 31 from their ass parent. The influence of chromosome pairs in the sex cells results in sterility in hybrids of these species.

Sex of the individual is determined by one pair of chromosomes. In the male, the sex chromosomes are designated as XY and are of unequal size or are unmatched. In the female, the chromosomes are of equal (matched) size and are designated XX. During meiosis, the mare can only contribute the X chromosome to her ova while the stallion contributes approximately 50% X and 50% Y chromosomes to his spermatozoa. It is therefore the male that determines the sex of the offspring. Because the Y chromosome is smaller than the X chromosome, fillies receive more genetic traits from their sires than do colts.

Qualitative and Quantitative Inheritance

As the term indicates, qualitative refers to that kind of inheritance that is controlled by one or a few pairs of genes. That which we can see, the phenotype, may be a direct indicator of the genotype of the horse. Sex of the horse, coat color, curly hair, and hemophilia are examples of qualitative traits. Breeding systems to change and(or) improve qualitative traits are reasonably uncomplicated and predictive and will not be considered in this discussion.

It is considerably more difficult to design breeding systems that will accurately and quickly change those traits that are quantitatively controlled. Quantitative traits are difficult to recognize because the phenotype is controlled by many pairs of genes. Many quantitative traits are difficult to measure or define. Examples of traits that are obviously controlled by many genes and are difficult to measure are "cow sense," soundness, and jumping ability.

Measurement of these traits is highly subjective and varies according to the person making the judgement. Quantitative traits that are more easily measured are racing speed, wither height, and weight at a given age. The stop watch and weights and measures provide definite, objective measurements.

Heritability

Most, if not all, traits are genetically controlled at least to some degree. The degree to which traits are genetically controlled is the heritability estimate for that trait. The heritability value indicates the amount of the average differences in traits between the individuals selected to be the parents of the next generation and the population from which they were selected. We expect to see these differences passed on to their offspring. With all selection, there tends to be an incomplete selection or a regression back toward the breed average rather than an averaging of the superiority of the selected parents. The degree of this incompleteness of selection is directly related to the estimated heritability value of the trait. The heritability estimate is the measure of the differences controlled by genetic influences that are measured or observed between horses.

Observed differences are made up of both genetic and environmental factors: Phenotype = genotype + environment. Because the phenotype is what we see or observe, it is important to know a horse's environmental influence if we are to be able to make accurate genetic evaluations. The more a trait is influenced by environment, the less genetic change the breeder can make in a given period of years.

Change resulting from selection for quantitative traits may be formulated as follows (change can be either in a positive or negative direction): change = heritability x selection differential. Selection differential is defined as the difference between the average merit of the stallions and mares selected to be the parents of the next generation and the average merit of the entire horse population from which they were derived. The stallion and the mare each contribute equally to the genetic makeup of their offspring. For an individual foal, each parent is equal in genetic importance. Extra emphasis is generally placed on the genetic merit of the stallion because he will normally sire 10 to 200 offspring annually compared to the mare's production, at best, of one foal annually.

$$\text{Genetic change} = \text{Heritability} \times \frac{(\text{Sire superiority} + \text{Mare superiority})}{2}$$

A simplified example of selection for a quantitatively controlled trait would be selection for wither height.

Assume the average wither height for a breed of horses is 14 hands and you wish to increase wither height as rapidly as possible. Wither height is generally regarded as being about 40% heritable. From a population that averages 14 hands (56 in.), you select a 16 hand (64 in.) stallion and mate him to mares that average 15 hands (60 in.). The stallion is +2 hands (8 in.) and the mares are +1 hand (4 in.) superior to the average of the horse population from which they were selected. Plugging their values into the formula, we would expect the offspring from these matings to have a wither height of:

$$\text{Change} = .40 \times \frac{(8 \text{ in.} + 4 \text{ in.})}{2} = 2.4 \text{ in. above the breed average}$$

Wither height of the selected offspring should average 2.4 in. higher than the average height of 56 in. of the original population of horses. The offspring resulting from this selection would measure 56 in. + 2.4 in. = 58.4 in. (14 hands 2 1/2 in.) at the withers.

The genetic change demonstrated here is permanent and will be carried from one generation to another; it may be added to or subtracted from depending upon whether or not future selections are positive or negative for wither height. It should be readily apparent that the higher the heritability value of a trait, the more rapidly a breeder can make progress in that particular trait.

Maximum progress is made when heritability is high and selection is for a single trait. Therefore, horse breeders should be discreet in their breeding programs and attempt to select for as few traits as possible in a given generation. Thoroughbred breeders have been most successful in selecting for speed over a given distance. Speed and(or) money earned are the primary selection criteria. Money earned or average-earnings-index values take into account both speed and soundness, and enable breeders to make considerable genetic progress. Attention to smaller "fancy" traits, such as length and shape of ear, etc., serve only to slow down the genetic improvement in the thoroughbred breeders' quest for faster, sounder horses. Table 1 lists heritability estimates for some traits.

It should be noted that the values in table 1 are referred to as "estimates." Because many traits that horse breeders desire to change and improve are difficult to measure or define, it should be apparent that the accuracy of the heritability estimate fluctuates in direct proportion to the breeder's ability to measure and define the trait. This is the reason that few estimates of cow sense, soundness, and reproductive traits are found in the literature today.

TABLE 1. HERITABILITY ESTIMATES

Trait	Heritability
Wither height	.25 to .60
Body wt	.25 to .30
Body length	.25
Cannon bone circumference	.19 to .30
Trotting speed	.40 to .50
Racing speed	.25 to .67
Reproductive traits	Low?
Cow sense	Medium to high?

Heritability estimates are broadly classified as low, medium, and high. Low heritability estimates range from .10 to .25; medium from .30 to .50; and high from .50 to .80. Highly heritable traits are generally easy to measure, and positive change is made by utilizing the animal's own record of performance. Conversely, traits with low heritability are difficult to select for because environmental factors (training, feeding, etc.) are responsible for most of the observed differences between individuals for the trait.

The net effect of selection is to change the average merit of the population for a particular trait. Most biological measurements in livestock populations are considered to be normally distributed. A theoretical normal distribution curve in which the values are clustered at the mid-point, thinning out symmetrically toward both extremes, is illustrated in figure 1.

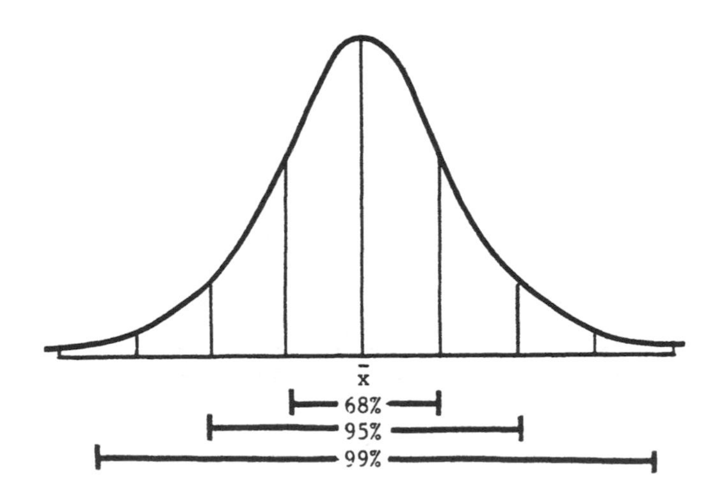

Figure 1. Theoretical normal distribution curve.

The bracketed areas indicate the percentage of the total population under a theoretical normal curve, as distributed around the average or mean ($\bar{x}$) of the population.

If selection for a trait is effective, the mean value or merit of the population will be shifted to the right. For selection to be the most effective, the breeder should choose his stallions from at least the upper 10% of all stallions and the mares from the upper 60% of all mares. The normal curve illustrates this type of selection (figure 2).

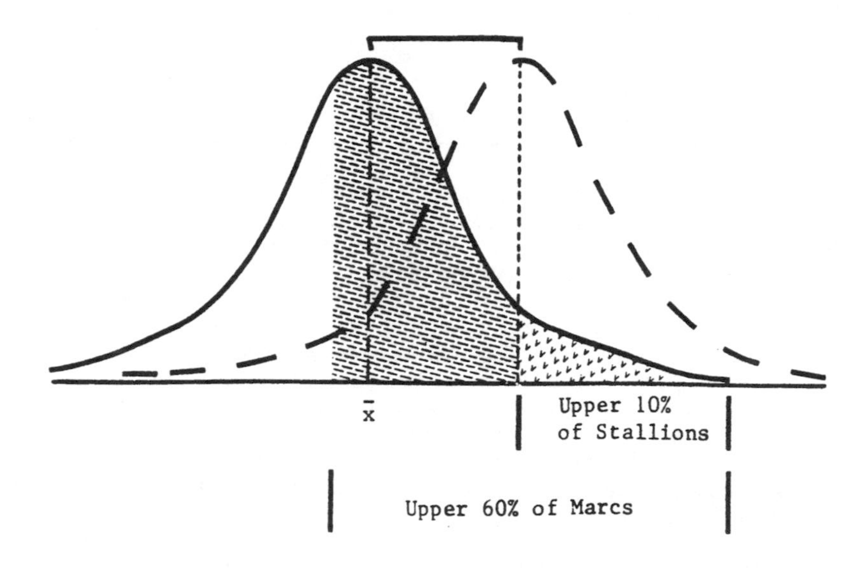

Figure 2. Genetic change.

The population resulting from this selection intensity should result in a normal curve whose mean is increased, as illustrated in figure 2 by the curve formed by the dotted line.

The genetic change that results is permanent and is passed from generation to generation. It is equivalent to building a wall with bricks and strong mortar.

Number of Traits Considered

If the breeder is concerned with improvement in only one trait, all of his selection efforts can be made for the one trait. Each time he considers another trait in his selection program, the selection pressure is lowered for both traits.

SUMMARY

Genetic improvement in horse breeding is slow, at best, and the responsibility for breed improvement rests upon the shoulders of the few tried-and-true breeders who have bred for specific goals over a period of generations. Dedication to the pursuit of breeding outstanding horses is a must. The dedicated breeder has a well-defined, obtainable goal, and does everything within his means to reach that goal in the shortest period of years. He must select and use stallions from the upper 10% of his breed, based on accurate and intelligent measures of merit. He must consider the stallion's own record; those of his parents; and, if possible, the records of his offspring (most breed associations have these records readily available to their breeders). The dedicated breeder must be extremely discriminating in the selection of his broodmares. Poor producers should be culled from the herd and replaced with young fillies that are the result of mating superior stallions to superior mares. It is best if the replacement mares have proven records of performance and production. The more knowledge one has about his breeding stock, the more efficient and rapid will be his genetic improvement.

Breeding superior horses may well be the greatest challenge in the livestock industry. There are no shortcuts. Variation, "the raw material with which the breeder has to work," is abundant.

22

GENETIC IMPROVEMENT IN HORSES: INBREEDING AND ITS CONSEQUENCES, PART 2

Joe B. Armstrong

The definition of inbreeding is slightly different for all animal breeders; a definition that most find acceptable is "the mating of individuals that are related to each other more closely than are randomly chosen individuals." When mating partners are less closely related, we refer to this as outbreeding.

Inbreeding means the mating of relatives. Related mates have common ancestors and therefore may have received some genes that are alike from these common ancestors. The offspring of these mated relatives may receive two genes at a given locus that are identical; that is, the identical genes are both copies of the same gene that was carried by the common ancestor(s).

The genetic consequences of intense inbreeding without selection have long been known to be disadvantageous. Inbreeding leads to a depression in performance of most traits; causes a deterioration in constitutional vigor and health; decreases fertility; and increases lethals and physical abnormalities.

Inbreeding depression is caused by the increased proportion of homozygous loci in the inbred individual. In other words, the inbred individual inherited a double dose of a gene (or genes) that was present in a single dose in the common ancestor of its sire and dam.

Inbreeding is generally expressed as a percentage value and can be considered to be an estimate of the percentage of the total genes that have been put in the homozygous state. Inbreeding coefficients are generally expressed as F values and the inbreeding coefficient of an individual is exactly one-half the relationship between its parents, unless those parents are themselves inbred. When the parents are inbred, there is a formula for correcting the inbreeding coefficient.

The formula for computing inbreeding coefficients is:

$$F_X = \Sigma(1/2)^{n+1}(1 + F_A)$$

where:

F_X = inbreeding coefficient of individual X

Σ = summation of all the independent paths of inheritance which connect the sire and dam of X

n = number of segregations in a specific path between the sire and the dam of X

F_A = inbreeding coefficient of the common ancestor for each path

Half-Sib Matings

Half-sibs are 25% related and therefore the progeny of half-sib matings are 12.5% inbred as the following bracket and arrow-style pedigree demonstrates (figure 1).

$$F_X = \Sigma(1/2)^{n+1}(1 + F_D)$$

$$= \Sigma(1/2)^{2+1}(1 + 0)$$

$$= \Sigma(1/2)^{3}(1)$$

$$= .125$$

Figure 1

Full-Sib Matings

The mating of full sibs represents a much more intense system of inbreeding than does the mating of half-sibs. With full-sib mating there are two common ancestors and two independent paths of inheritance to evaluate (figure 2).

Common Ancestor	n	Contributions
C	2	$(1/2)^{2+1} = (1/2)^3 = .125$
D	2	$(1/2)^{2+1} = (1/2)^3 = .125$
		$F_X = .25$

Figure 2

The above calculations assume that the common ancestor C and D were themselves noninbred. Had either C or D, or both, been inbred, the F_X value would have been slightly larger.

Parent-Offspring Matings

Parent-offspring matings are equal to full-sib matings in intensity of inbreeding. As full-sib matings, parent-offspring matings are 25% inbred as shown below (figure 3).

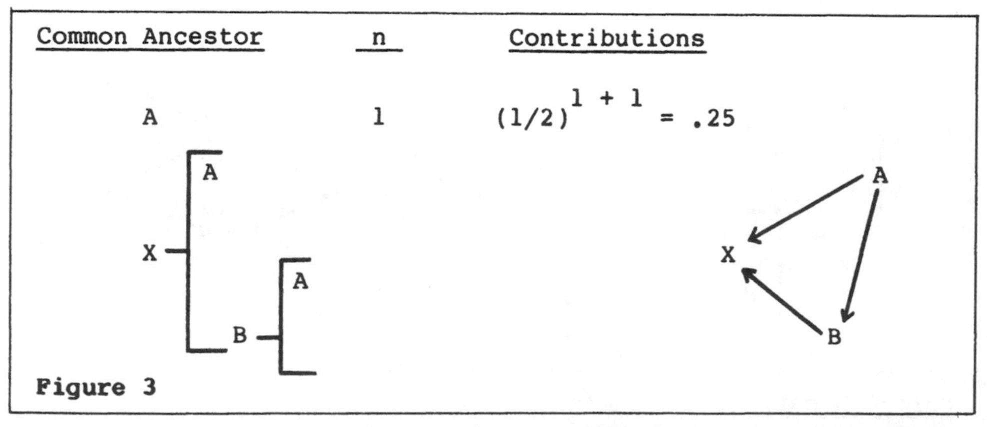

Figure 3

Computation of inbreeding coefficients becomes more complex when the common ancestors have themselves developed as the result of matings of relatives for several generations.

While inbreeding is generally thought of as disadvantageous, it should also be noted that inbreeding makes desirable and undesirable genes homozygous, impartially.

The key in any breeding program is not selection alone but effective selection balanced with culling. Each time you cull an inferior individual, you have raised the genetic merit of the breeding herd.

Inbreeding depression cannot be adequately explained without considering its opposite, hybrid vigor. The development of hybrid corn through the crossing of inbred lines is one of the major achievements in agriculture in the twentieth century. Much of our animal breeding theory revolves around these experiments. There are good experiments detailing inbreeding depression and hybrid vigor in chickens and hogs, where generations can be turned rather rapidly. Some data are available on cattle populations, but controlled experiments are nonexistent with horses.

Most investigations on the relationship between depression of performance and degree of inbreeding suggest this relation to be linear. Also, those traits that are related to fitness are sensitive to inbreeding depression, generally have a low heritability, and show large heterotic effects in crosses. Highly heritable traits generally show little hybrid vigor.

In practice, horse breeders have noticed that some families of horses can apparently withstand continued close breeding--mild inbreeding or linebreeding--with no noticeable detrimental effects, while other families cannot exist under a similar breeding program.

Causes of Inbreeding Depression and Hybrid Vigor

Homozygosity, or the fixation of genes in double doses (AA, bb, CC, dd), is the effect of inbreeding, and heterozygosity (Aa, Bb, Cc, Dd) is the distinguishing feature of hybrid vigor.

The superiority of the hybrid is generally considered to be the result of many pairs of genes in a heterozygous condition.

Hybrid vigor is generally defined as the percentage by which the hybrid progeny surpass the average of the two parents in performance. Some people would prefer that the definition be the percentage by which the hybrid progeny surpass the performance of the best parent (overdominance).

The concepts of gene action that are generally used to explain hybrid vigor are shown in an oversimplification in figure 4.

Dominance and overdominance are the two gene concepts most often used to explain hybrid vigor. The three genotypes under the bars on the graph represent only one pair of genes. The number of possible gene pairs and combinations becomes mind-boggling.

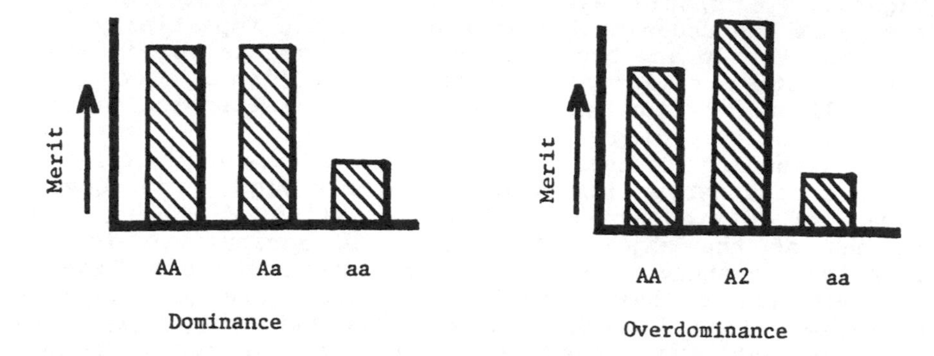

Figure 4

The overdominance concept assumes three different individuals. The first has the dominant gene A in double dose (AA), the second has A and a each in a single dose (Aa), and the third has the recessive gene a in double dose (aa). If overdominance is the correct explanation, the interaction (or heterozygosity) of a dominant gene with its recessive counterpart produces a greater effect than does either A or a when each is the homozygous condition.

The dominance explanation offers the same three genotypes but only two phenotypes because AA and Aa are identical in performance merit and greater than aa.

It is easy to see how the dominance theory of hybrid vigor is compatible with the theory of developing inbred lines where genes are fixed in a double dose and inbreeding depression occurs, and thus, how vigor is restored when the inbred lines are crossed.

The fixation of genes in double dosage occurs independently of whether they are good, deleterious, dominant, or recessive.

By changing heterozygotes to homozygotes, inbreeding brings to light many of the recessive genes that would otherwise remain hidden. Since recessive genes normally have fewer desirable effects than do their alleles, inbreeding tends to lower the average outward merit of inbred animals. By bringing undesirable genes and traits to light, inbreeding allows breeds to be purged and therefore allows for more rapid improvement of the breed.

The problem for the breeder is: At what rate to inbreed? If the rate is too rapid, too many individuals will be produced that are homozygous for some undesired genes as well as for some desired ones. When inbreeding is too mild, too many generations will be needed to accomplish much with it.

The level of merit of the breed or family in which inbreeding is to be practiced also determines the rate at which inbreeding may be carried on and still be successful.

Studies of most breeds indicate that inbreeding is at a low level, even when the breeds have started from small numbers--as have most of our horse breeds. In these breeds, and families within the breeds, genetic relationships have been maintained at fairly high levels to popular ancestors within the breed, while avoiding an increase in inbreeding. This has been accomplished through linebreeding.

In spite of the disadvantages of a slight loss in fitness, the increase in prepotency of those individuals that are not defective renders close inbreeding a tool for breed improvement when it is used by breeders who understand genetic principles and are not afraid to cull.

Examinations of the breeding programs of essentially all of the breeders of merit within all breeds and species reveal that many of these patriarchs have practiced inbreeding or, at least, strong linebreeding programs.

Even in Robert Bakewell's time, it was known and stressed that inbred animals were more apt to be prepotent and effective when used in outcrosses than were animals of equal individuality but not inbred.

23

GENETIC IMPROVEMENT IN HORSES: LINEBREEDING AND OUTCROSSING, PART 3

Joe B. Armstrong

Linebreeding is a common term among breeders of purebred livestock. Breeders often joke that if a close mating works it is linebreeding, and if it does not work it is inbreeding! In linebreeding the aim is to get or keep as much blood of a particular stallion or mare in the animals of the herd as is possible, while at the same time keeping inbreeding as low as possible. Linebreeding is accomplished by using parent animals that are closely related to the admired ancestors but are little (if at all) related to each other through any other ancestors.

Linebreeding differs from inbreeding primarily in that it is directed toward maintaining a high relationship to a chosen ancestor--and secondarily in that it is usually less intense than most inbreeding programs.

Most breeders think of the King Ranch when linebreeding is mentioned. The King Ranch used linebreeding very effectively in concentrating the blood of Old Sorrel.

Linebreeding is practiced because sires and dams generally do not live long enough for a breeder to get all the sons and daughters he wants from the admired ancestor. Each time a son or daughter is mated to an unrelated individual, their offspring receive only about one-fourth of their inheritance from the outstanding grandparents. In three to four generations of outbreeding, the original outstanding ancestor's influence is so diluted or scattered that the present offspring bear little or no resemblance to the outstanding ancestor. For example, how many times have you asked the breeding of a horse; and are then told it is a Three Bars horse, but when you inspect the pedigree you find Three Bars one time in the fifth generation?

Linebreeding takes advantage of the laws of probability as they affect Mendelian inheritance to hold the expected amount of inheritance from an admired ancestor at a nearly constant level instead of letting it be halved with each generation.

Linebreeding also builds up homozygosity and prepotency within a breed because it is a form of inbreeding. The homozygosity produced by linebreeding is generally for increased merit for desired traits because it is a slower form of inbreeding and usually is carried on in conjunction with a rather strict selection and culling program.

Linebreeding also separates breeds into distinct families within the breed. These families are closely related to an admired ancestor. Cutting- and reining-horse families provide excellent examples of linebreeding to about four or five admired ancestors, and of the effectiveness of selection within these families and between families for specific "nicks." (Nicking results when the right combination of genes for good characteristics is contributed by each parent.) In the 1981 National Cutting Horse Association (NCHA) Futurity, 11 stallions were the grandsires of 82% of the nearly 600 entries.

Linebreeding is best accomplished by breeders whose herds are superior to the average of the breed. Again, the higher the merit of the herd, the greater the gain to be received from linebreeding. After the admired ancestor dies, the relationship to it that can be attained in future animals is limited to that of its closest relatives that are still living. Look at the demand today among cutters and reining enthusiasts for daughters of King, Leo, Three Bars, and Sugar Bars.

Thoroughbred breeders avoid inbreeding because of the resulting loss in constitutional vigor. However, many successful breeders practice mild inbreeding in an attempt to keep the influence (or blood) of the great horses in their present breeding and racing stock.

Cutters and reiners practice linebreeding more vigorously than do breeders of racehorses primarily because of the heritability of the traits. Cutting and reining traits are thought to be lowly heritable. Lowly heritable traits are known to respond well to family selection and to have great hybrid vigor when families are crossed. As a trait, speed is generally considered to be medium to high in inheritance. Traits that are highly heritable respond well to individual selection (choosing sires and dams on the basis of their own performance records) and show low hybrid vigor.

Outcrossing is the crossing of family lines in an attempt to produce nicks that are superior to either of the family lines. Outcrossing is a minor part, but eventually a necessary part, of most linebreeding programs. Any inbreeding program that is carried far enough will probably fix some undesirable traits, and mild outcrossing will be needed to remedy the undesirable traits.

A popular saying among purebred cattle breeders is that "you linebreed to get famous and outcross to get rich."

When breeders begin outcrossing to produce nicks, the relationship to the desired ancestor is automatically lessened. Breeding for nicks has been responsible for the diminish of many once-prominent sire lines and female families.

Linebreeding has a definite place for the serious horse breeder, regardless of the breed or traits being selected for. To be successful, the breeder must have a clear set of realistic goals, a herd of superior merit, and a lifetime to invest in his breeding program. Few breeders have possessed these simple attributes, but those few have made great strides in herd improvement and are known and respected by their peers.

24

GENETIC IMPROVEMENT IN HORSES: BREEDING WINNERS IN CUTTING AND REINING, PART 4

Joe B. Armstrong

Cutting- and reining-horse popularity is at an all-time high. Futurities, derbies, and maturities are household words to horsemen involved in these sports. The Reined Cowhorse classes combine the best of these two exciting classes. Many trainers today are financially secure and successful specializing in training only futurity horses, etc.

The genetic picture of horses bred for cutting and reining is indeed interesting and is quite different from racing horses. Why?

The speed of the horse is one of the few performance events that is objectively measured. Racehorses run a given distance against the clock. The inheritance, or heritability, of speed is known to be medium to high. Animal-breeding principles have shown that improvement in traits that are highly heritable is best accomplished by breeding the "best-to-the-best" on the basis of individual performance. Animal-breeding principles also demonstrate that inbreeding reduces fitness and that highly heritable traits do not respond vigorously to crossbreeding or the crossing of different family lines within a breed.

For the above reasons, racehorse breeders have been highly successful in breeding the fastest males for given racing distances to the fastest females.

On the other hand, published heritability values for cutting and reining traits indicate that these traits are lowly heritable. Animal-breeding principles dictate that maximum progress in lowly heritable traits is achieved when heavy emphasis is given to family or ancestral records of performance. This is linebreeding!

Study the pedigrees of the successful cutting and reining horses and they fit a very definite pattern. That pattern is linebreeding to, for all practical purposes, four broad families of horses: King (Zantanon), Leo, Three Bars, and King Ranch.

Trace the history of the National Reining Horse Futurity. Fourteen of the last sixteen winners trace to King, and the other two winners trace to Ed Echols, which was King's half-brother. Both were sired by "The Mexican

Man of War," Zantanon. As you study the pedigrees of these horses, you find the four broad families mentioned earlier showing up again and again.

National Cutting Horse Association (NCHA) cutting-futurity finalists and winners display a similar pattern. In the 1981 Futurity, eleven stallions were the grandsires of 82% of the nearly 600 entries in the futurity.

Doc Bar was the paternal grandsire of 46% of the futurity entries and the paternal grandsire of 67% (12 of the 20) of the finalists. Additionally, he was the maternal grandsire of 4% of the total entries. Leo San, son of Leo, and Jewel's Leo Bars, son of Three Bars out of a daughter of Leo, contributed 9% and 6%, respectively, as paternal grandsires of futurity entries, and 5% and 28% of the 20 futurity finalists.

True, there are other families that have played important roles in cutting and reining contests, but the success of breeders in linebreeding for these traits has been phenomenal. This is a graphic demonstration of the animal-breeding principle that traits that are lowly heritable show their greatest response to linebreeding.

Lowly heritable traits are strongly influenced by environmental conditions. In cutting and reining, environment includes trainers and training!

In truth, cow sense and working ability probably do not have heritability as low as is indicated by the few studies in the literature. Ask any breeder who raises and trains cutting or reining horses. They will let you known in no uncertain terms that this ability is transmitted from one generation to the next.

The nature of our definition of heritability and the population from which we estimate these values are at least partially responsible for the low estimates--and they are simply that--estimates. Heritability is defined as that fraction (genetic) of the observed differences (genetic + environmental) between individuals that is passed to the next generation. If one was to study an entire breed of horses, the variation (observed differences) among horses in their ability to cut cattle or rein would be great. Some would have little or no ability, while others would have great ability, and a whole host of horses would be intermediate in ability. Our horses are identified and selected for their ability to work cattle or rein so that breeders do not introduce families that are recognized as having little or no cattle ability. Therefore, when cutting- or reining-futurity horses are measured and studied, we are dealing with a very narrow segment of the total horse population or breed. Because of the selectiveness of the group being studied, most of the variation observed is because of environment, training, riders, etc., and the resulting estimate of heritability is low. In studying pedigrees, one should be cautious about going too far back in an attempt to introduce the blood of a certain horse into his breeding program. Each horse receives a sample half of his genetic makeup from each of his parents. The diagram below shows the percentage of genes received from each parent, grand-parent, etc.

An ancestor in the fourth generation contributes only 6 1/4% genetic influence on the offspring in question. This is the reason most breeders look at the first two generations and only occasionally at the third generation. Past the third generation, there is little genetic influence.

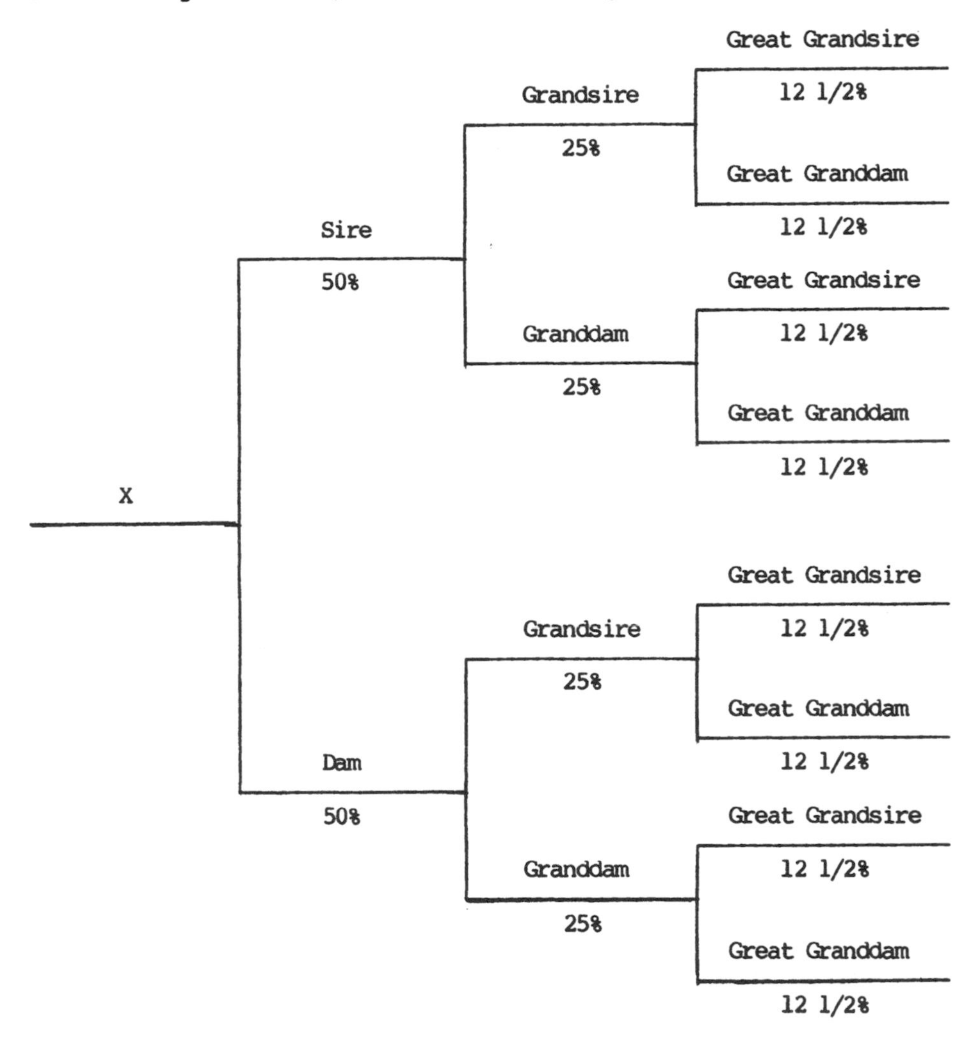

Linebreeding or mild inbreeding is an exception to the above statement. Linebreeding requires mating of animals that are more closely related than are the average of the breed. Inbreeding is the mating of closely related animals.

Both linebreeding and inbreeding, when practiced with careful culling, have the effect of increasing the desirable genes for which the breeder is selecting. While the desirable genes are being concentrated, so are the undesirable genes--if they exist in the family. This is the reason that a rigid culling program is necessary when linebreeding or inbreeding.

25

HORSE BREEDING PROGRAMS

R. L. Willham

Our common pastoral nomad heritage has been preserved for us in Greek mythology. MINERVA, in all her wisdom, caught and tamed the great winged horse, PEGASUS. She gave BELLEROPHON a golden bridle as he set out to slay the dragon, CHEMERA. He became, so the story goes, the first man to ride and to conduct mounted warfare, and he probably started the romantic legend of dragon slaying. Thrace was the legendary home of the mythological CENTAUR. The account of the battle between the CENTAURS and the LAPITHAE of Thessaly reflects the fear of settled people for the horsemen. The man and horse were one. All of the fearsome traits of people were given to the CENTAUR by the Greeks. Horsemen of the steppes of EURASIA had a profound influence on our cultural evolutions.

Today the breeding of horses is steeped in the mystique of centuries. Since the horse first participated with man in raids and then wars, the breeding of horses has been the business of nations. National defense or conquest depended, until very recently, on being well horsed. From the desert patriarch to those born to money who participate in the sport of kings, all classes of people have been successful horse breeders. Lowly grooms have risen to positions of authority at the stables by cultivating the art of horse breeding.

Since the turn of the century and the rediscovery of Mendel, the science of population genetics as applied to stock improvement has become increasingly important, especially in the domestic species having high reproductive rates and economic importance in agricultural production. Horse breeding has profound economic importance, but it is more than a business; it usually is a labor of love. Even though the design and conduct of creative breeding programs is constant across species, differing in details only, few breeding principles have been applied in horse breeding. The purpose of this paper is to relate the basic breeding principles and weave them into the design of breeding programs for horses.

BREEDING PROGRAMS

> **"It is often believed today that successful breeders have some mysterious methods of which others are ignorant. Instead, the principles of the successful breeder have been exceedingly simple.... The difficulty lies not so much in knowing the principles as in applying them."**
>
> **S. Wright (1920)**

This statement, made by the population geneticist of the age way back in 1920, is as true today as it was during the golden age of stock breeding. There is no mysterious method, even if successful breeders by their sage silence suggest that there really is. Luck, pure and simple, still makes many successful breeders, especially among those with small numbers of stock under their control.

The basic breeding **problem,** in any species, is to find parents that produce progeny that are superior, in the definition of the breeder, to past progeny. The primary question is that of providing a clear definition of "superior." What can be done to solve the problem is circumscribed by the germ plasm that can be introduced by crossing. That is, as antiquity did, one can produce mules, but not use them for breeding. In the near future, it may be possible to expand our opportunities through DNA transfer. However, for today, let us limit our considerations to germ plasm that can be incorporated by crossing.

A breeding program is a complete management system that is designed to make genetic change in a desired direction in the stock under the control of a breeder. The classic definition of a breeder is one who consistently, over time, puts genetic gain on genetic gain in a given direction. Because **selection** or the choice of parents is the only directional force available to the breeder to make genetic change, the form, types, and alternatives open to the breeder in his design of breeding programs are limited.

The **form** of all breeding programs is the same, in general; they differ in time and specific species requirements. The form is the production of a genetic sample, the evaluation of this sample, and the selection of parent stock from the sample. Then the selections are mated to produce the next genetic sample. This form is cyclic, being repeated over time in a consistent direction.

Breeding programs are essentially of two **types** in many species. The types are divided on the basis of the use to be made of the output from the breeding program. If the stock produced are to be future parents, then the kind of stock sold, and the available forces to use to make genetic change, differ from those programs in which the product is simply sold for subsequent use. Breeding stock is sold on the basis of **breeding value** or the value of the stock as a parent. Stock sold for **use only** are sold on the basis of their value for a particular use. No genetic segregation or recombination is intended for use-only stock; unlike the use-for-breeding stock which are evaluated in terms of their progeny in the stable of the buyer. Further, the types differ in the genetic forces available for use. After a breed is chosen, only animals of that breed can be selected for use. Besides using selection within breeds, herds selling animals for use only can consider using all available sources of genetic variation in the production of stock. Cross combinations between breeds are possible. The opportunities are greater for the producer of stock for immediate use than for breeders producing breeder stock.

BREEDING PRINCIPLES

Two genetic phenomena form the basis of breeding programs. The first and most important is the resemblance among relatives. The second is inbreeding depression with its converse--hybrid vigor or heterosis.

The basis of **selection** is the resemblance between parent and offspring. The sire and dam of each individual contribute a sample half (one gene or the other at random from each locus) of their hereditary material to that individual or progeny. The degree of resemblance for any observable or measurable trait indicates the influence of gene effects, not gene pair effects in producing variation among animals. That is, if a breeder can easily pick out the offspring of a particular parent for a given trait, the trait is heritable. If, on the other hand, the resemblance is low, the variation in the trait is less heritable. Using the degree of resemblance among sets of related individuals, it is possible to deduce the **heritability** of a given trait. The heritability is simply the fraction of the variation that is caused by gene effects among animals treated alike. And genes (one or the other at each locus) are what are transmitted from parent to offspring each generation. These ideas lead to the concept of **breeding value** or the value of the animal as a parent. A working definition of breeding value is--twice the difference between the progeny average of an individual and the average of the group or herd. It assumes that the other parents are a random sample from the herd. The difference is doubled because a parent transmits only a sample half of their genes to an offspring. But progeny are not the only relatives that can be used to predict breeding value. The group's own performance, sibling average performance, and progeny performance can all be used as the information about various relative groups becomes available. For example, the breeding value of an individual can be predicted by multiplying the heritability of the trait by the difference of the performance of the individual and the contemporary group average. Obviously, this prediction would not be as accurate as that from a large number of progeny.

The primary use of the heritability idea is in the design of selection schemes in breeding programs because heritability can be used to predict the response to selection. In its simple form, heritability times the selection differential (difference between parents chosen and all possible parents) predicts the average change in the group, herd, or stable. It is the same idea as that of predicting the breeding values, which can be used as selection criteria. Heritability profoundly influences the selection scheme. When the trait to be improved is highly heritable (60% to 80%), the performances of the individuals are accurate enough to be used as the only selection criterion. When the heritability of a trait is moderate (30% to 50%), then the average performance of relative groups will add to the accuracy of evaluation. And when the heritability of a trait is low (10% to 20%), average performance of related groups is essential to increase the accuracy of the selections.

Averages work this way. The common element (the fact that relative groups have genes in common from an ancestor) is constant and the plus or minus deviations from this commonality cancel each other out in the averaging. Thus, a group of 20 paternal half-sibs share (on the average) 1/4 of their genes or have them in common, so the heritability of this average is higher than that for individuals. The design of a breeding program is to develop procedures that will maximize genetic change per unit of time for the traits of importance in the breeder's chosen direction.

Performance in horse breeding involves at least three basic classes of traits: reproduction, conformation, and disposition. In all species, the reproductive complex is slightly heritable and shows important heterosis when distinct genetic groups are crossed. Conformation is highly heritable and requires little more than individual evaluation; heterosis is moderate. Disposition, or the ability to be taught and interact with man, is moderately heritable. Here the average performance of relatives can be used to some advantage. Heterosis is moderate for measures of disposition but is probably higher than that for conformation. A negative relationship exists between heritability and heterosis. In a breeding stock program, real genetic change can be made over time in conformation and disposition, but little real change can be made in the reproductive complex.

Now consider the other genetic phenomenon--inbreeding depression and its converse hybrid vigor. Inbreeding, or the mating of individuals more closely related than average in the group, is practiced by many breeders of horses for a myriad of reasons (the worst of which is that no other stud is available). Inbreeding involves the effect of chance at Mendelian segregation. It is measured by the inbreeding coefficient first deduced by Wright in 1921. The breeder has absolutely no control over the chance process. He takes the result of the draw and by selection either saves the good result or culls the bum result. He has no other power!

Inbreeding results when the mated animals have common ancestors in their pedigree. The process produces lines and sublines that become different from each other by random drift. Inbreeding influences each locus the same way; inbreeding increases the probability that the pair of genes at a locus are identical by descent or are copies of the same gene from a common ancestor. This is fine if, by chance, the "good" genes at a locus become alike in an individual, but it is a disaster if the "bad" genes become alike. On the average, over the lines, there is inbreeding depression--performance goes down, especially in the reproductive complex. But occasionally there will be one line out of many, just by chance, that will out perform the whole group. Many lines must be formed, evaluated, and selected before that one line is isolated. A large number of animals is required to involve inbreeding creatively in a real breeding program.

Breeders who luck out with one line have done so by chance, not by anything they did; but it does happen, even with 100 to 1 odds. You and I hear about the lucky breeder and try to emulate his results. We never hear about the other 99 who did the same thing and went out of the breeding game.

Linebreeding is inbreeding but for another purpose. It is selection based on the pedigree. If a breeder finds a truly outstanding parent, he can linebreed to this animal to raise the relationship of his stock to this individual. This has been a useful procedure, but the actual inbreeding effect can be minimized while the relationship of animals to a selected ancestor is increased dramatically. (Linebreeding to the Belgian stallion, Farceur, is an excellent example.)

When lines produced through inbreeding are crossed, the resulting individuals will show heterosis or hybrid vigor and will, on the average, recover that lost by the inbreeding depression. (The highly reproductive species, corn, poultry, and swine use this procedure to increase production.) The half Thoroughbred/half Quarter Horse does rather well on the track. Part of this is hybrid vigor, if the cross is sufficiently divergent genetically.

CHOICES

Because selection is the only directional force available to the breeder to make genetic change in a given direction, the breeder can choose from only three factors, direction, differences, and decisions, in the design of a breeding program. These factors are three-dimensional, and direction is paramount.

A breeder must decide first on direction. The word direction is used, rather than goal, because a breeder moves a group of animals through time in a direction; he never really achieves a goal, as such. This decision influences the others. Next, he must develop a means to evaluate differences among his stock in the traits of major importance in the direction chosen. Finally, he must develop a decision procedure to make use of the differences among his stock to move his group in his chosen direction. Selection is the decision process. How he mates the selected parents is influenced by the direction chosen and by the units among which he makes selection. The individual is the smallest unit of selection, but the units could be among inbred groups or lines or families. Most breeding programs fail because of vacillating direction or goals. So-called breeders, who select for the hot item this year and something else next year, just are not breeders.

Again, to quote S. Wright (1920):

> **"....Instead, the principles of the successful breeder have been exceedingly simple. He isolates and fixes a good type by careful selection and close breeding. If ambitious to take a greater step in advance, he crosses types with characteristics which seem to offer possibilities for a desirable combination and fixes the new ideal by continued selection and close breeding. He brings inferior stock up to higher levels by consistent use of prepotent sires of the same improved type."**

No more succinct definition of breeding principles have ever been written. The Thoroughbred breed was the result of a divergent cross of the native horse with some oriental blood, and then the population was closed.

SUMMARY

The design and conduct of breeding programs in all species of stock have common elements. Because selection or the choice of parents is the only directional force available to breeders, the form, types, and choices of the breeder are really few. Breeders must first choose their direction, then develop evaluations that produce differences among their stock, and lastly design decisions that are based on the differences evaluated to move their germ plasm in the desired direction.

REFERENCES

Wright, S. 1920. Principles of livestock breeding. USDA Bulletin No. 905.

26

THE QUARTER HORSE: THEN AND NOW

Robert M. Denhardt

The Quarter Horse has always been more specifically native American than has the Thoroughbred. Instead of consciously emulating the racing customs of England, the short-horse man chartered a new course. He felt no obligation to English-racing procedures or to English-racing bloodlines. Certain individuals, certain bloodlines produced sprinters, and it was to those stallions that he took his mares. The American short-horse man cared little what Weatherby, Skinner, Edgar, or Bruce might say about the sire of his choice. He just wanted to breed to blazing speed--and he did.

The above should not be construed to indicate that I am against registration and all that it can accomplish. What is important to remember is that it was "early speed" that was the objective of the breeder of the quarter-running horse. That is precisely why some of the great foundation sires, such as Printer (1800), Cherokee (1847), and Traveler (1900), with no known pedigree, can still be listed among the greatest. That the sires and dams of these three stallions were unknown bothered the short-horse breeder not at all.

It would be equally absurd to ignore a Quarter Horse sire because he was a Thoroughbred. If one arbitrarily eliminated registered horses that were great producers of early speed, such as Sir Archy (1805), Bonnie Scotland (1853), or Uncle Jimmy Gray (1906), part of the history of the quarter running horse would be missing. These three Thoroughbred stallions are no less important than their unpedigreed contemporaries Printer, Cherokee, and Traveler. To get a good overall view of the history of a unique equine type, the only American contribution to the running horses of the world, a person should not spend time trying to classify sprinters as Thoroughbreds or Quarter Horses.

The information available on horse breeding and racing in the 1600s is admittedly cursory. Although short races were common before the middle of the eighteenth century, I do not consider that the quarter-running horse became a distinct type until the importation of Janus, whose influence was, and to some extent still is, paramount.

One does not have to look far to find the reason why short racing was popular in the colonies. Think how much easier it was to find a short, level, straight stretch on which to match a race and how difficult it would have been to build a smooth, oval surface large enough for horse racing. The early straightaway races seldom even reached a quarter of a mile. After all, everyone wanted to see the races from start to finish. So the early races in the colonies were short, and generally only two horses raced at a time. This style of racing followed the frontier as it moved across the continent to the Pacific Ocean.

Much of the popularity of horse racing lies in the thrilling stretch run for the finish. Even among the long-horse devotees, who think a race should be a mile or more, it is not until the horses near the wire that they jump to their feet and roar for their favorites. The early jockeying for position is interesting to the knowledgeable horseman, but the finish stretch is the thriller. Quarter racing has distilled the essence of racing and bottled it all to explode in about 20 seconds of ecstatic emotion.

Sprinting has always been a "traveling" sport, and the best quarter-running horses were always on the move, looking for new competition and new money. Owners, trainers, jockeys, and horses traveled the country over looking for just the right race. In contrast, most of the better Thoroughbred stallions ran for a few years at the established race tracks and then retired to stud. The good mares were taken to them, so most of their offspring could be found in a fairly concentrated area. The quarter-running stallions, on the other hand, traveled far and wide, campaigning and breeding for 10 yr or 12 yr until their speed diminished; that explains why the blood of the better sprinters was widely dispersed during those early years.

QUARTER RACING IN THE OLD DAYS

Love of short racing does not seem to be an inherited trait: you have to acquire it. Once you have the appetite for it, however, the hunger lingers on and on. Any number of men since colonial days have lived out their lives in the sole pursuit of match racing, and who can say that they were not as happy as, or happier than, their soil-tilling brethren?

Perhaps one factor that makes short racing such an intriguing sport is the special challenge created by all the variables inherent in it. The idea is to do a better job of assembling the critical elements than your neighbor or friend does and so win the race, or the wager, and the right to crow. Today, horse racing at the recognized tracks is under the supervision of state governments, who see it as their duty to present to the public all the facts about the horses, the jockeys, and the conditions of the race and track. These variables are not immediately apparent at a match race. To make a valid match for your horse (or a sensible wager on such a race), all above factors, and more, have to be determined. Since facts may be scarce, a good deal of gossip and rumor must be evaluated before a judgment can be reached. Often intuition is as valuable as knowledge, and the successful match-race devotee cannot always say where one ends and the other begins.

One of the first lessons to learn is not to be gullible. It is necessary to accept all statements about the match with a grain of salt. You can believe only what you personally know to be true. After a few races, most of those who lay a dollar on their favorite have become adept as well as dedicated. One also can tell the fan who has paid his dues by the way he accepts the outcome of the race. The knowledgeable zealot always accepts it as something ordained, as though a superior being has determined the outcome. The verdict is no more to be questioned than the wind or the rain.

When one tears himself away from the thrill and tension of a well-run match between two good Quarter Horses, it is nothing less than captivating to watch the true devotees gather in small groups, where each explains how he knew which one was going to be the winner (even when he had bet on the other horse). The discussion also covers in minutest detail just where the trainer or jockey made the big mistake. These men are as serious in demeanor as any ambassador to the United Nations and, it could be added, probably have equal comprehension of the topic under discussion.

There is another group of short-race enthusiasts: those who bet little or not at all on the outcome of a match. For them the simple pleasure of watching two horses thundering down the raceway and flashing over the finish line at almost the same moment is reward enough. Many Quarter Horse breeders belong to this group. Their reward comes in raising a faster horse than their neighbor's. Two common human traits are the desire to compete and the desire to excel. Raising and racing Quarter Horses meet both desires admirably.

There is still another factor, one that was more common in the days before trucks, trains, and airplanes began shuffling horses from one racing establishment to another. That was the bond of affection that arose between the owner and his racehorse. On today's circular tracks the closest approximation to that relationship is the fondness that may develop between the trainer or stableman and his charge. During the first three-quarters of the 19th century the owner traveled with his horse from town to town, state to state, matching races as he could. They got to know each other well. Possibly the feeling they shared has been expressed best by Sam Hildreth. Writing about his early career with Quarter Horses, he tells how he and his father (the whole family, in fact) took Red Morocco from Missouri all the way to Texas to match a race with Sheriff Jim Brown. When they returned home to Missouri, the Hildreths were distraught because they had lost the race and had had to leave their beloved Red Morocco with the Sheriff.

Some of the lure of short racing has been mentioned earlier, but the only true way to understand it is to expose yourself to a match race--if you can find one. Pari-mutuels and state racing commissions are rapidly making the old two-horse match an endangered species. The reason is obvious (but nonetheless regrettable): there cannot be successful pari-mutuels when half or more of the people are betting on one horse. The only places where match racing has a home today are the few states that have local tracks and no pari-mutuels. Attend one. You'll like it.

RECENT CHANGES IN QUARTER HORSES

The first, and greatest, sire of Quarter Horses was an imported English stallion named Janus. The description of Janus indicates that he stood a little over 14 hands, had great bone with tremendous muscles, and was compactly built. For 200 yr, his conformation was considered to be the ideal. When the American Quarter Horse Association (AQHA) was formed in 1940, he was still considered the model Quarter Horse. His conformation was adopted as typical for the breed. By 1940, three exceptional 20th-century Quarter Horse sires had been identified that were built along the general lines of Janus: Joe Bailey, Joe Moore, and Red Dog.

Dan D. Casement, writing in the introduction to the first edition of the American Quarter Horse Association Stud Book (1940), described the ideal Quarter Horse as follows:

> **There is the small, sensitive, alert ear, his wise bright eye, the amazing bulk and bulge of his jaw which seems to betoken his bulldog tenacity and resolution. There is his short back, deep middle, and long belly, his low-slung center of gravity, and the astonishing expanse of his britches, seen from the rear, surpassing the width at the croup.**

In the same article Casement quoted William Anson, the famous turn-of-the-century Texas Quarter Horse breeder, as saying that the immense breast and chest; the enormous forearm, loin, and thigh; and the heavy layers of muscle were not to be found in like proportions in any other breed in the world. Casement ended his article with the warning that only negative and harmful purposes would be served by any attempt to refashion the shape of the Quarter Horse in imitation of any other breed. He was, of course, referring to the Thoroughbred.

It is doubtful that Johnny Ferguson with his Thoroughbred Top Deck ever worried about early descriptions of the Quarter Horse. Top Deck, Three Bars, Depth Charge, and other great Thoroughbred sires of the modern Quarter Horses produced too many horses with blazing speed. Their get are the ones who win the futurities and stake races and whose sons are syndicated for large sums of money. Their owners are not concerned that they have seven-eighths or more Thoroughbred blood and that they are at least a full hand taller than the ideal Quarter Horse.

Dan D. Casement was not the only early official of the American Quarter Horse Association who wanted a bulldog, not a greyhound. Of the first officers of the fledgling organization--Bill Warren, president; Jack Hutchins and Lee Underwood, vice-presidents; Jim Hall, treasurer; and Bob Denhardt, secretary--one, Jack Hutchins, was a race-horse man. He kept his running horses--most of them sired by a son of the Thoroughbred, Chicaro--separated from his Quarter Horses and did not try to register any of them. His principle Quarter Horse stallions were Lobo and Billy. Both of them were bulldogs. Bill Warren's principal stallion was Pancho, a heavyset 14-2-hand grandson of Little Joe. Lee Underwood had two principal stallions, Chief and Dexter, and neither one was a greyhound. My Del Rio Joe was also of the bulldog type.

To anyone with prescience, perhaps the route the Quarter Horse has taken since 1940 would have been obvious. The clues were there. To begin with, there were the "ABC" classifications. At the start, only pure, dyed-in-the-wool bulldog Quarter Horses were to be registered; however, the officers of the new breed association soon came face-to-face with the cold fact that only a few Quarter Horses of the type they wanted were available for registry. The idea of a Quarter Horse Association had been heartily supported by breeders and buyers, and many people were asking for Quarter Horses. In the end it was agreed to register a horse of the "correct" type with an "A" after its number, a horse that showed at least half-Quarter Horse characteristics was to have a "B" after its registration number, and half Thoroughbreds were to be registered with a "C." Very few breeders could boast of an "A" band of horses. Then came the annual meeting, where the majority ("C" breeders) abandoned the "ABC" classification, and all numbered horses were simply registered as Quarter Horses.

One of the promoters of the "C" type Quarter Horse was, in my opinion, the best horseman ever employed by the AQHA. His name was Jim Minnick. He frankly told the first executive committee that he personally favored a "C," or half-breed-type horse. He preferred the Thoroughbred-Quarter Horse cross; his polo-playing experience had demonstrated to him that they were the best polo ponies. Jim was hired as the first inspector, and during the first years his selections provided the base stock for the breed. His choices set the pattern. He also judged most of the early official shows. His activities should have provided some clear indication of how the association was going to go.

Other factors encouraged the move away from the bulldog. Since few breeders had horses of that type, it was popular to criticize them, and one generally heard the word "bulldog" coupled with "mutton-withered" or some other term of a disparaging nature. No doubt the criticism was often made sincerely. There are horses in all breeds that can be faulted, but that does not make every individual of the breed or type faulty, nor does it make all the "A"-type horses poor. Many, such as Red Dog and Poco Bueno, were superb individuals.

Many breeders besides Jim Minnick preferred the half-breed horse. Some were directors of the new American Quarter Horse Association. Take Bob Kleberg, of the King Ranch, for example. In the early days, the King Ranch provided a large share of the income of the new breed association, registering more horses than any individual or corporation. But the Klebergs were not bulldog breeders, nor did they want to be. The King Ranch foundation stallion, Old Sorrel, was half Thoroughbred, and the foundation dams, the Lazarus group, were all Thoroughbreds. Peppy, Wimpy, Macanudo, Hired Hand, and other basic stallions in the breeding program were not bulldogs of the Steel Dust stamp. They were ideal "C"-type Quarter Horses and great individuals; however, there were no Lobos, Tonys, Red Dogs, Joe Baileys, or Zantanons among them. Bob Kleberg explained his position in the first edition of the AQHA Stud Book: "...the Quarter Horse makes an ideal foundation on which to cross the Thoroughbred." He did it as well as anyone who ever lived.

THE MAJORITY ARE STILL QUARTER HORSES

Although most of what I have just said concerns short racing, racing actually represents but a small part of the Quarter Horse's talent. History is largely compiled from written records, and the ordinary use of the horse is seldom recorded. His brilliant racing feats always were.

Unfortunately, this is still true. Racing at Ruidosa or at Los Alamitos can be seen on TV and found in the sporting section of newspapers and magazines. Trail riding, barrel racing, roping, cutting, and general ranch work are seldom mentioned--even though it is in such fields that the true value of the Quarter Horse is found.

It is personally gratifying to me to see that these utility horses are still Quarter Horses. Few are Thoroughbreds in looks or blood, and only the exception is over 15 hands. These are the animals that justify the work put in by the founders of the Quarter Horse Association.

Part 7

MANAGEMENT AND REPRODUCTION

27
PASTURE BREEDING

Michael Osborne

Management of thoroughbred-horse farms requires understanding of many phases of biology, animal behavior, and human behavior. The author's experiences in the operation of thoroughbred farms in Kentucky and Ireland provide insights to many aspects of horse husbandry, marketing, and related technology. An overview of the author's observations is offered to assist horsemen and horse-farm managers to improve specialized phases of their horse operations.

The stallion should be placed in the pasture and the mares released to him on a gradual basis. Putting all the mares together will result in an amount of fighting to establish a pecking order of dominance among the mares. The established pecking order will be disturbed by putting the stallion into the mare herd--it appears that some mares will fight again to reestablish the pecking order in the presence of the stallion. The stallion should breed a mare immediately before being let loose into the mare herd. Mares who constantly court the stallion's favor, even when not in oestrus, should be removed because the stallion may pay exclusive attention to them. A groom should be careful in any attempt to remove the stallion from the herd because the stallion may attack the groom and may also breed the mare unnecessarily--having "one for the road." Stallions breed at dawn and dusk and at any other time when the stallion or mare herd are disturbed by man, another stallion, or pony teaser. The latter serves as a good detective because his presence in the vicinity of the mare herd will make the stallion breed a mare--evidence of which mare is in oestrus. The problem of knowing which mares have been bred is difficult for an observer. Interference by a groom apprehending the stallion and the presence of a pony stallion are the best solutions. Stallions do not cover every mare that shows oestrus. They can be most selective, particularly with infected mares, and mares treated with prostaglandin are less attractive to the stallions. The results from pasture breeding are excellent, but the system involves a high risk and constant observation. The physiological limit for an active stallion is 20 mares with good reproductive records and 15 mares with chronic problems of barrenness.

28

"HOW OLD IS HE?"

William C. McMullan

"How old is he?" This familiar question can be very important to you, especially if you are buying a horse. A 6-yr-old is worth more than a 12-yr-old, and a good horseman will know the difference. You can get an idea by judging his overall appearance, muscle tone (sagging and flabby stomach muscles), and hair coat (white hairs at temple in an older horse). Feel the jawbone; the branches of a young horse are thick and rounded, becoming thinner and sharper on their border as the horse ages.

The Arabs used to estimate age by pinching the skin (the skin of an older horse is less pliable and drier). Other methods include feeling for bony projections at the base of the tail or determining the space between the ribs and correlating these findings with average ages.

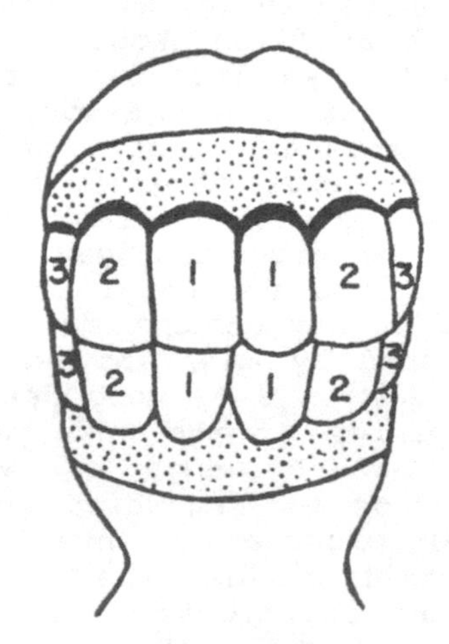

Figure 1.

All these things may be of some value at times, but a more accurate method is by examination of his teeth, especially the lower incisor teeth (figure 1). The incisors are the 12 (6 upper and 6 lower) smaller teeth easily seen at the front of the mouth if the upper and lower lips are parted. These are the teeth with which he bites off blades of grass. The lower six are really three pairs; the two teeth in the middle are known variously as I^1 (1st incisors), centrals, middle incisors, pinchers, or nippers. In the rest of this paper, they will be referred to as I^1.

The next pair (the first tooth to the left and the first tooth to the right of pair I^1) is known as I^2 (2nd incisors) or intermediate incisors. The third pair (the second tooth to the left and the second tooth to the right of pair I^1) is I^3 (3rd incisors) or corner incisors. You must know these, because all of the most accurate determinations are made from these three pairs.

Before someone finds a horse that does not fit the rules that follow, let me warn you in advance that my discussion is based on the average, not the exception. It stands to reason that a horse grazing a pasture of rocks will wear off more tooth than one grazing in a meadow.

The most accurate of all indicators is the time at which the incisors erupt (break through the gums and become plainly visible).

In the colt, eruption of the incisors is most easily remembered by the rule of the three 8s: pair I^1 is through the gum by 8 days (actually birth to 8 days); pair I^2 is in by 8 wk (actually 4 wk to 6 wk); pair I^3 is in by 8 mo (actually by 6 mo to 9 mo). The corners are through the gums by 8 mo, but the upper corner does not touch the lower corner for 8 to 10 more months.

Remember 8 days, 8 wk, and 8 mo. These are baby teeth, also known as milk teeth, temporary teeth, or deciduous teeth. On many charts they are shown as Di 1, Di 2, Di 3 (Deciduous incisor 1, 2, 3).

This brings up an important point: How can we tell baby teeth from adult or permanent teeth?

Look at the teeth of colts and horses at every chance. You will notice that the permanent teeth: 1) are larger and longer; 2) have a broad neck between the root and crown, whereas baby teeth are narrowed at the neck; 3) have perpendicular parallel grooves and ridges on the face of the incisors, 4) are darker in color, and 5) are more flat in curvature.

We have to know these characteristics or we might mistake a big 2-yr-old's mouth for that of a much older horse.

At 2 yr old, the pairs I^1 should show considerable neck. A very important change is about to take place. At 2 1/2 yr of age, the baby teeth (upper and lower pairs, Di 1) will be pushed out by the permanent teeth (upper and lower pairs, I^1). So, right in the middle, we have four bigger I^1 and smaller pairs Di 2 and Di 3 on either side.

One year later, at 3 1/2 yr, pairs Di 2 are replaced by the larger I^2; thus at 3 1/2 yr, we have eight larger teeth, pairs I^1 and I^2 on the upper and lower table. The corner incisors are much smaller.

One year later, at 4 1/2 yr, we will see the eruption of the permanent corners, pairs I^3. It will be about 6 mo (5 yr old) before the upper and lower I^3 touch, however.

Our horse now has 12 permanent incisors that will stay the rest of his life.

In most male horses 4 yr to 5 yr old, and in 2% to 3% of mares, an upper and lower tooth comes in on either side, between the corner incisor and the first jaw tooth--fairly close to where a bit would rest. These are called canine teeth or sometimes tusks, tushes, or fang teeth. Actually, only 60% to 70% of male horses have canine teeth in both upper and lower jaws; about 6% to 7% have them only in the upper jaw, and 26% to 30% have them in the lower jaw.

All the teeth are in by the age of 5 yr, so a 5 yr old is said to have a "full mouth." Between 4 yr and 5 yr of age, a horse is really busy; he cuts 4 new incisors, 4 canines, and 8 molars; a total of 16 new teeth. This explains why a 3 yr old sometimes outworks a 4 yr old.

The jaw teeth (molars, grinders) will not be discussed here since they are seldom used in determining age. Their presence (six upper and lower on each side) does indicate, however, that the horse is at least 5 yr old.

Remember key ages:

2 1/2 yr	3 1/2 yr	4 1/2 yr
I^1 at 2 1/2 yr	I^2 at 3 1/2 yr	I^3 at 4 1/2 yr

From the age of 5 yr and afterward, the main indicators are still the incisor teeth, but we must judge from the changes in the shape of the incisors and changes in the table surface (biting surface) due to wear from grazing and chewing. The lower incisors are more accurate indicators than the uppers. There are two identifying marks found on the table surface on the incisors that you must be able to recognize if you expect to be able to age a horse past 5 yr old (figure 2). These are:

- The cup. Each incisor tooth, if you look down (in the case of the lower incisors) on the table surface, will have a dark brown or black cavity in the center of the tooth, which goes across toward the edges of the tooth. This cavity is known as the cup. At age 5, the cups in the incisor teeth are 1/4 in. or more in depth. As day-to-day wear takes its toll, the black cavity or cup is worn down and becomes shallower until at 6 yr old, the cups are almost gone from lower I^1s. At 7 yr old, the cups are gone from lower I^2s and at 8 yr old, the cups are gone from lower I^3s.

  Remember: CG (cup gone) at 6, 7, and 8 yr old; I^1s at 6 yr old; I^2s at 7 yr old; and I^3s at 8 yr old.

  The cups in the upper incisors may go at 9 yr old (upper I^1), 10 yr old (upper I^2), and 11 yr old (upper I^3), or they may stay longer, especially the corner cups. Thus, if wear has been normal, a horse is "smooth mouth" by 11 yr to 12 yr old.

When the cups are first worn away, you see in their place a long oval containing the cementum. The cementum wears down to a small round spot of enamel called the enamel spot. It is so hard that in an older horse (10 yr to 11 yr old) it is elevated slightly above the rest of the table surface. It disappears from most horses by age 15.

The dental star (DS). The darker dentin of the DS fills the pulp cavity down inside the root of the incisors (figure 3). As the tooth wears, the dental star appears first as a dark yellow transverse line on the table surface on the lip side of I^1. It usually comes into view in the lower I^1 at about 8 yr. The dental star becomes oval and moves toward the middle of the table surface by age 13. By age 15 the dental star is round. Do not confuse the dental star with the enamel spot that is harder and elevated.

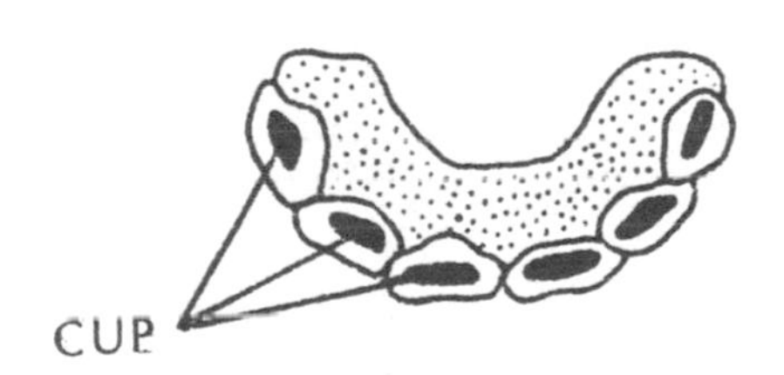

Figure 2.

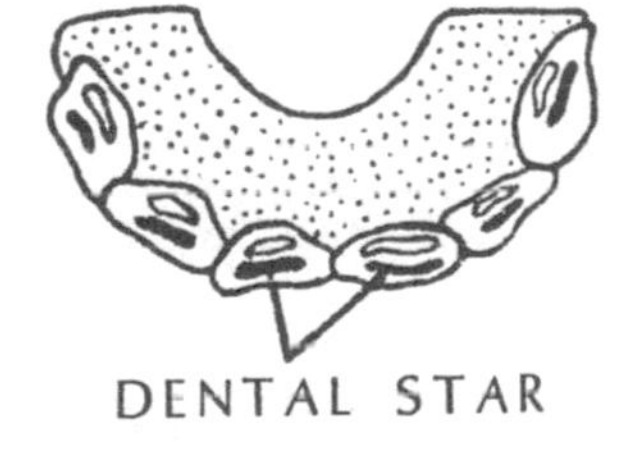

Figure 3.

Another mark that is sometimes helpful is the "7-yr notch." On each of the upper corner incisors, a "dove tail" or "hook" is formed when wear from the lower corner incisor becomes uneven (figure 4). The back quarter of I^3 is 1/8 in. to 1/4 in. longer than the rest of the tooth. It is said to appear at age 7, go away at age 8, come back at age 11, and stay until about age 18.

In the same category with the 7-yr notch (in terms of reliability) is a mark known as Galvayne's groove (figure 5). This is a shallow groove on the lip surface running up and down the length of the upper corner incisor. Cementum remains in the groove as a dark line. It is said to appear at the gum line at 10 yr old, is half-way down the tooth by age 15, and all the way down the tooth by age 20. At 25, it will be found only on the lower half of upper I^3, and by age 30 the darkened groove will be gone.

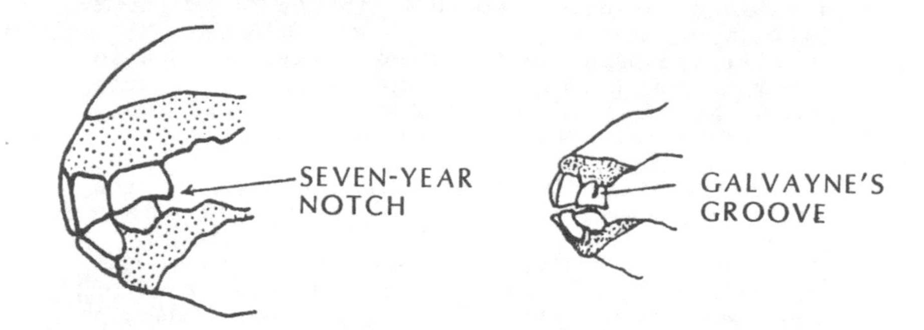

Figure 4. **Figure 5.**

These two indicators are far from being exact. Actually, all we know for sure when we see the dove tail or hook on upper I^3 is that the horse is at least 7 yr old and probably less than 18 yr. If we see Galvayne's groove, we can actually only be sure the horse is at least age 10.

Of more reliability is the relative shape of the table surface of the incisors (figure 6). At years 1 through 7, the incisor teeth are definitely oblong, with the transverse diameter (side to side) being much longer than from front to back edge. But by age 9 the diameters are about equal and I^1 is said to be rounded at age 9. I^2 is rounded at 10 yr, and I^3 is rounded at 11 yr.

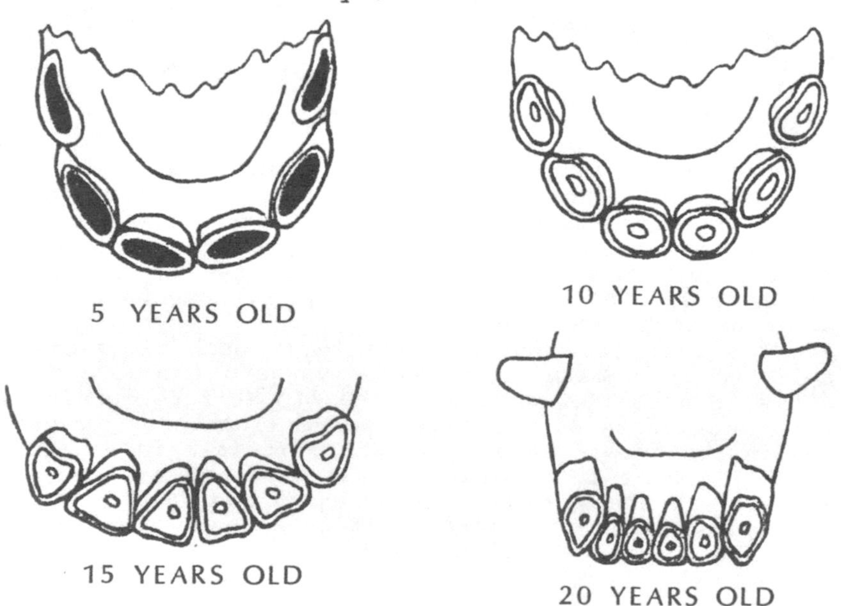

Figure 6.

As the tooth is worn further, I^1 becomes somewhat triangular at age 14 with the apex (point) of the triangle toward the tongue. I^2 is triangular by age 15 and I^3 is triangular by age 17. I^1 becomes biangular (the diameter from front to back is greater than the transverse diameter) by 18 yr, I^2 by 19 yr, and I^3 by 21 yr.

There are three other general indicators of age. The angle formed by the upper and lower incisors, when the teeth are viewed in profile, becomes more acute (sharper) with age (figure 7).

When the teeth are viewed from the front, they are seen to diverge from the median plane in a young horse, to converge in an old one, and to be parallel at middle age (figure 8).

The arcade of the incisors looking down on the table surface is a half circle in the young horse and a straight line in the old (figure 9).

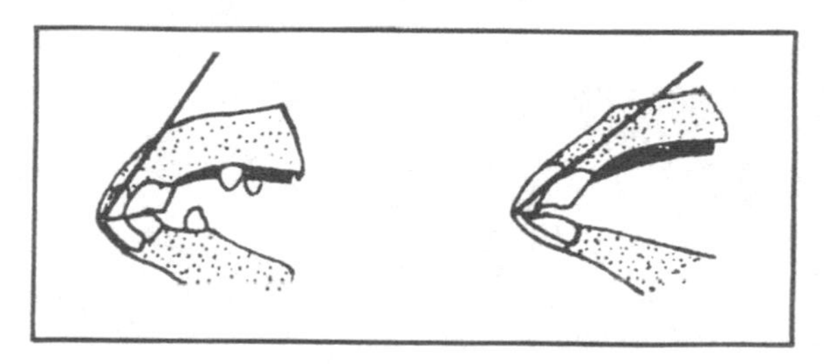

Figure 7.

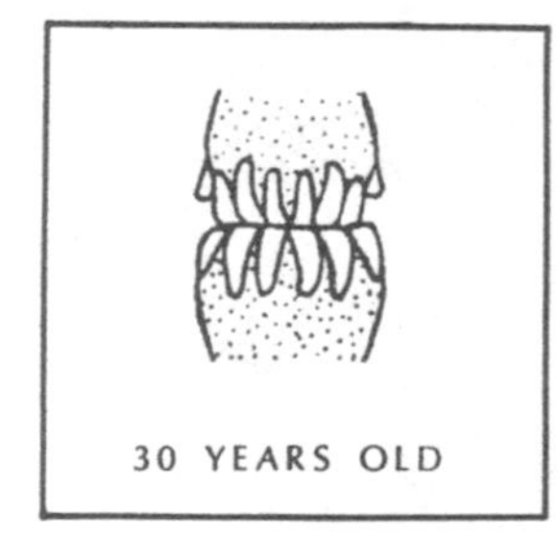

Figure 8.

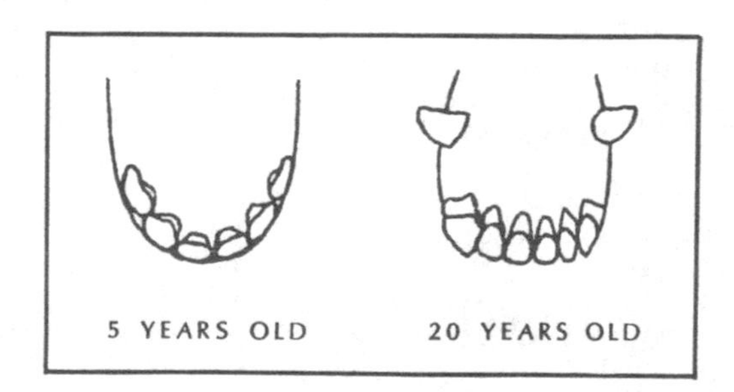

Figure 9.

Occasionally attempts are made to make an old horse look younger. One trick, known as "gypping," is to paint or dye the greying hair its original color. "Puffing the glen" is an attempt at filling the depression above the eye, which tends to sink deeper in old age. A more common method of making a horse younger is "bishoping" (figure 10). This involves drilling the incisors of an aged horse and staining the cavity so as to make an artificial cup, indicating to the novice horseman that the horse is still under 10 yr of age. The artificial appearance of the cup (it will not be lined with enamel as a natural cup will) and the shape of the teeth (round at 9 yr to 11 yr, triangular at 14 yr to 17 yr, and biangular at 18 yr to 21 yr) should be the give-away.

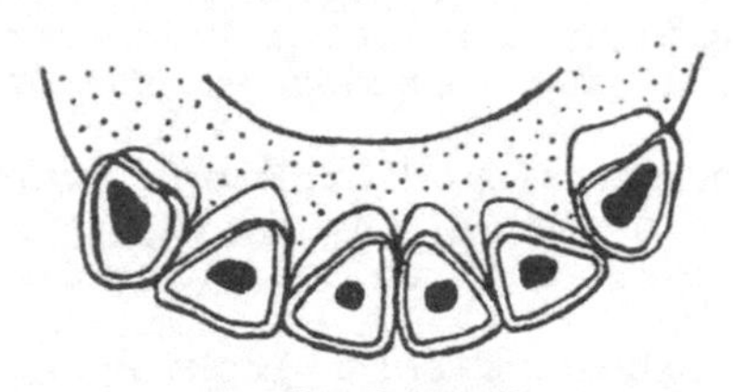

Figure 10.

These things are difficult to learn all at once. Until you have them memorized, table 1 will help. Memorize these now.

Baby teeth erupt	8 days	8 wk	8 mo
Permanents erupt	2.5 yr	3.5 yr	4.5 yr
Cup gone	6 yr	7 yr	8 yr
Round table	9 yr	10 yr	11 yr

TABLE 1. AGE INDICATORS

	I^1	I^2	I^3
Baby erupt	8 days	8 wk	8 mo
Permanents erupt, yr	2.5	3.5	4.5
CG (cup gone), yr	6	7	8
Dental star, yr	8	9	10
Table round, yr	9	10	11
Star centers, yr	13	13	13
Table triangular, yr	14	15	17
Star round, yr	15	15	15
Table biangular, yr	18	19	21
Galvayne's groove, yr			10 to 30
7-yr notch			7 and 11 to 17

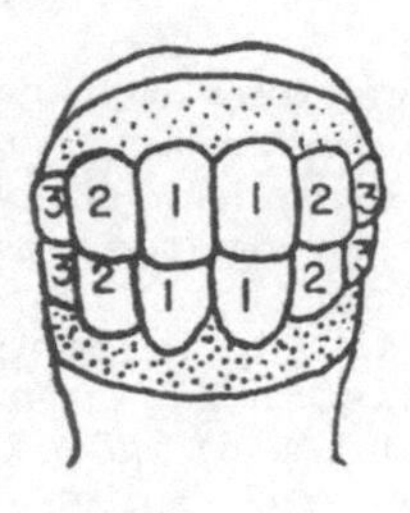

29
CONSIDERATIONS IN MANAGING THE BREEDING STALLION

Doyle G. Meadows

Many questions have to be considered and answered when standing a stallion. Obviously, one of the first considerations to be addressed is the selection of a stallion. Basically three questions must be answered to determine the worth of a superior breeding animal: 1) does the stallion have the desired pedigree; 2) does he have the desired and preferred conformational traits; 3) does the breeding stallion have the desired performance traits or has he sired superior performing individuals?

Furthermore, the stallion being considered must be able to make a contribution to an existing herd, whether it belongs to the stallion owner or a public-owned herd. In addition, the stallion must be fertile and breeding sound. A provision for a breeding soundness evaluation by a veterinarian should be incorporated into a "contract for purchase." In some breeds, color should be considered when selecting a breeding stallion.

Most stallion stations or stallion owners must depend on outside (public) mares for additional income. However, it is very important to understand some basic concepts when standing a stallion to the public. The first concept is this: only good stallions make a profit. Therefore, if one is to be successful in the breeding business it is imperative to stand a quality stallion. Next, the stallion owner must set a breeding fee and should not deviate from the established service fee. Many stallion owners get into trouble by "making deals" with friends, neighbors, and other horsemen. In the long term, these deals generally cost the stallion owner both friends and money--certainly not the original intention. This brings up another important consideration: the stallion owner must use business-like procedures in the establishment and implementation of a stallion service contract. This stallion service contract takes all the gentlemen's agreements and guesswork out of the horse breeding business. After the stallion service contract has been signed, the stallion and mare owner know their mutual expectations, which leads to a happier situation for all parties.

It is absolutely mandatory that the stallion owner not cut corners in handling the mares. Providing low-quality mare care to outside mares is a blueprint for failure in the horse breeding business. Mare owners will pay for excellent mare care but will not return to a breeding farm that has given them poor mare management, regardless of its stallion battery.

Communication is the key to a successful breeding establishment. Stallion stations must keep the mare owners informed of the current status of their mare. A weekly report to the mare owner about his mare is an ideal way to build a lasting relationship between the breeding farm and mare owner. Communication is a must to be a successful horse breeder.

Many horsemen ask questions concerning stallion usage. Here are some management guidelines that may assist in horse breeding operations. A 1-yr-old stallion should not be used to breed mares; however, most yearling stallions are producing some live viable sperm and are capable of settling a mare. Yearling stallions should not be allowed to run with a group of mares. Many 2-yr-old stallions are being used today; however, their book should be limited to 12 mares to 15 mares. It is important to note that these values are based on live cover--not artificial insemination numbers. It is very easy to overwork a 2-yr-old stallion and to cause him to become sexually depressed. Special care and attention should be given to the 2-yr-old stallion. Most stallions go to the stud barn when they are 3-yr-old. These stallions should be able to breed 25 mares (not services) and still maintain breeding aggressiveness and health. Typically, a 3-yr-old stallion should not be mated twice in one day.

The 4-yr-old stallion is an essentially mature stallion and can be bred to 35 mares to 40 mares. This stallion can be bred twice in 1 day, but should be doubled sparingly. The 5-yr-old is a mature stallion and can be bred to 40 mares to 50 mares; he can be doubled (bred twice in one day) 2 times to 3 times/wk. An old stallion 15 yr to 20 yr old definitely requires special attention; he should be treated and handled as if he were 2 yr old. It is very important to note these are average figures and should be used as a guideline or management tool for stallion managers. Many stallion managers may breed twice the number of mares noted here. In particular, one Thoroughbred stallion manager in Lexington, Kentucky, bred 75 mares to one stallion. This number exceeds the traditional number of mares typically bred at a Thoroughbred breeding farm by 50% to 60%. Ultimately, the stallion manager must determine the number of matings a stallion can handle and still maintain breeding effectiveness.

Many of the problems that are observed in the breeding stallion are a direct result of mismanagement. The following management practices should help the stallion manager overcome and eliminate many sexual behavior problems associated with stallions. Handlers should avoid unnecessary roughness when handling the stallions. This only leads to a frightened and abused stallion. At all costs, avoid over-using a young stallion. Stallion managers should avoid excessive use of a stallion as a teaser; however, some teasing may be beneficial to certain stallions. Stallions should not be housed in complete isolation. This only leads to abnormal behavioral traits in certain stallions and may decrease effective life of the stallions through reduced exercise programs.

It is very important to learn behavioral characteristics of individual stallions. If a stallion manager is to be effective with a particular stallion, it is imperative that the handler understand behavior traits of that stallion. Stallions have to be placed on routine health programs and not mismanaged just because they are stallions. In addition, the stallion must be placed on an excellent plane of nutrition. If a stallion is to perform at an optimal level, he must receive both appropriate quality and quantity of feedstuffs. Prior to the breeding season, a semen fertility test should be conducted to determine the fertility level of the stallion. If a problem exists, the stallion manager has ample time to correct the situation prior to breeding mares.

The ability to handle a stallion (or any horse for that matter), is basically an art combined with the scientific aspects of documented research. If a horseman is to gain and maintain full control of a stallion, he must understand the concepts of horse psychology and behavior, nutrition, health, and a score of other things. The same is true with the stallion breeding farm; the farm must be operated just like any other business that implements sound business-management practices. Hopefully, the previously mentioned management practices will help the horseman be a more effective horse breeder.

ACKNOWLEDGMENT

Acknowledgment is extended to the Horse Section (Dr. Gary Potter, Dr. Doug Householder, Mr. B. F. Yeates, and Dr. Jack L. Kreider) Texas A&M University for providing factual material for this paper.

30

BROODMARE MANAGEMENT: PRIOR TO, DURING, AND AFTER FOALING, PART 1

Doyle G. Meadows

Broodmare management is the subject of volumes of material with some articles written about the open mare, others about the pregnant mare, and still others about the lactating mare. This paper addresses only the basic principles related to management of the broodmare, in general.

Regardless of the stage of productivity (open, pregnant, or lactating) mares must be on a proper plane of nutrition. Horsemen often tend to overfeed open or pregnant mares, but may leave the lactating mare nutritionally deprived. The following recommended nutrient levels (NRC, 1978) are a basis for providing adequate nutrition for mares. The values are expressed on a 90% dry-matter basis for greater practical application (table 1). Many horsemen feel that a mare's nutritional requirements skyrocket from the time she becomes pregnant. This simply is not true. It is only the last 90 days of gestation that the mare's nutrient requirements are altered, and then not drastically. Here are a couple of relevant points: 1) the protein and calcium/phosphorus levels of the mare increase, and 2) the major difference will be in the quality of the feedstuffs being fed. The mare (last 90 days of gestation) should receive higher quality feedstuffs--not more total pounds of feed.

As soon as the mare foals, her nutrient requirements increase dramatically. There is a sizable increase in the percentage of protein in the diet and a large increase in the total pounds of feed intake to meet additional energy demands on her body (table 1). A large amount of high quality feedstuffs is required to adequately feed the lactating mare if optimum foal growth is to be obtained. However, as the mare progresses through lactation, her nutrient requirements are reduced (at the end of 90 days). The figures listed in table 1 should be used as a feeding guideline. Some heavier-milking mares may demand additional feed to maintain acceptable body condition. Remember, it is relatively easy to feed an open mare, but a good manager is required to oversee a good nutritional program for lactating mares.

TABLE 1. NUTRIENT CONCENTRATION IN DIETS FOR HORSES AND PONIES

	Digestible energy, Mcal/lb	Crude protein, %	Ca, %	P, %	Daily feed[a] lb
Mature horses at maintenance	.0	7.7	.27	.18	16.4
Mares, last 90 days of gestation	1.0	10.0	.45	.30	16.2
Lactating mare, first 3 mo	1.2	12.5	.45	.30	22.2
Lactating mare, 3 mo to weaning	1.1	11.0	.40	.25	20.6

Source: NRC (1978).
[a]Feed required for typical 1100 lb mature mare.

A good nutrition program cannot be overlooked when managing broodmares. Many mares are thought to be in good condition simply because they have a large, distended abdomen; however these mares may appear very thin upon foaling. Many breeding farms are criticized unduly for poor feeding management because of mares in such condition. The horsemen should look for evidence of condition (fat) along the topline of the mare and not necessarily along her underline. (Feeding the foal will be covered in another paper in this book, "Managing the Newborn Foal.")

It is especially important to evaluate the nutritional level of the mare prior to and during the breeding season. If mares are placed on an ascending plane of nutrition prior to the breeding season, they tend to move easily, cycle, and conceive. This tendency has been documented by van Niekerk and van Heerden (1972), who reported that mares that increased in body weight were earlier to ovulate in the breeding season than were those that did not gain weight. Also, Ginther (1974) reported the onset of estrus after January 1 was shorter for mares that gained weight than for those that lost weight. The greater the weight loss, the longer the breeding season was delayed, thus it is important that mares be in a gaining nutritional state prior to and throughout the breeding season.

Another essential management practice for broodmares is the establishment and implementation of a herd health program, which should be a simple, effective health program established between the farm manager and attending veterinarian.

An Oklahoma veterinarian (John Beall) uses the following herd health programs on many farms. In the spring, Beall vaccinates all mares for sleeping sickness (prior to May 10 in Oklahoma) and for tetanus. For the pregnant mare, Beall gives rhinopneumonitis vaccine at the 5th, 7th, and 9th month of gestation. Mares are dewormed on the farm every 60 days to 90 days, depending on the level of management and number of horses. Deworming medications at the farms are customarily alternated by most veterinarians so that the parasites will not develop immunity to a single medication.

Deworming on a routine basis (in addition to prevention and control measures) is an important part of any herd health program. Drugs used to control worms and to break life cycles can be administered directly into the stomach through a stomach tube, fed in combination with rations, or placed inside the mouth in the form of a paste. Deworming intervals depend on weather conditions such as rain, temperature, and humidity, as well as the season of the year. Thus, deworming programs are different in the various areas of the country. For numerous horses on smaller tracts of land, more frequent deworming practices should be implemented in all classes of horses. Exact deworming schedules for a specific area can be obtained from local veterinarians.

Many commercial deworming preparations are available, and medications should be changed periodically to prevent immunity to certain drugs. Each medication controls different kinds of worms; for example, some deworming medications will control ascarids, strongyles, and pinworms but will not control bots. An additional medication should be used that will control bots.

Internal parasites in horses can be very destructive; but with a proper herd health plan, the parasites can be controlled so that successful feeding, breeding, and performance are maintained in all classes of horses.

A mare is referred to as seasonally polyestrus; that is, mares will cycle many times during a particular season. Mares typically begin to cycle in early spring and continue to cycle through the early summer months. It is well documented that cycling starts as a response to increasing photoperiod (light). Therefore, the length of the mare's breeding season can be lengthened by altering the photoperiod. One lighting regime that works well in Oklahoma is as follows:

December 1, first week:	Light program begins with lights on from 5:00 p.m. to 6:00 p.m.
2nd week	5:00 p.m. to 7:00 p.m.
3rd week:	5:00 p.m. to 8:00 p.m.
4th week:	5:00 p.m. to 9:00 p.m.
5th week:	5:00 p.m. to 10:00 p.m.
6th week:	5:00 p.m. to 11:00 p.m.
7th week:	6:00 p.m. to 12:00 p.m.

Continue from 6:00 p.m. to 12:00 p.m. until April 15, then discontinue the light program. The lights should be on an automatic timer; do not depend on someone to turn the lights on and(or) off at the correct time. The light program is used on open mares, as well as on pregnant mares and mares that have foaled.

REFERENCES

Ginther, O. J. 1974. Occurrence of anestrus, estrus, diestrus, and ovulation over a 12 mo period in mares. Amer. J. Vet. Res. 33:1173.

NRC. 1978. Nutrient Requirements of Horses. National Academy of Sciences--National Research Council. Washington, D.C.

van Niekerk, C. H. and J. S. van Heerden. 1972. Nutrition and ovarian activity of mares early in the breeding season. J. South African Vet. Assoc. 43(4):355.

ACKNOWLEDGEMENT

Acknowledgement is extended to the Horse Section (Dr. Gary Potter, Dr. Doug Householder, Mr. B. F. Yeates, and Dr. Jack L. Kreider) Texas A&M University for providing factual material for this paper.

31

BROODMARE MANAGEMENT: PRIOR TO, DURING, AND AFTER FOALING, PART 2

Doyle G. Meadows

A filly will reach puberty at 12 mo to 15 mo of age and will begin coming into estrus (heat) on a rhythmic cycle (figure 1).

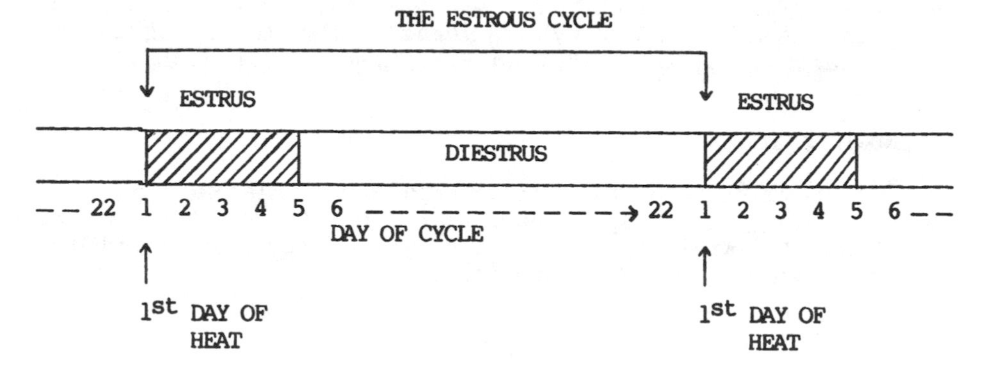

Figure 1.

Most reproductive physiologists define the complete cycle as the amount of time (days) from the first day of heat until the first day of the next heat period. The estrus (heat) portion of the total cycle refers to the period of receptivity to the stallion. During this time, typically about 5 days, a follicle is rapidly developing in the ovary and will be released (ovulation) during the latter stages of estrus (mid- to late estrus). Therefore, the time to breed in the estrous cycle is mid- to late estrus.

Diestrus is that portion of the estrous cycle in which the mare is not in heat. The anestrous period is the time of sexual rest and cessation of most ovarian activity. This is during the late fall and winter months, although location (Maine or California) may affect the length of the anestrous period.

After the horseman has a working knowledge of the estrous cycle, he can determine a mare's condition in relationship to the cycle. This can be done by rectal palpation and the use of a teaser stallion. Most breeding farms use a combination of these methods. Good records are a must in horse breeding; they are necessary in managing the mares so that as many as possible are in foal. This objective is directly related to the ultimate success or failure of the business. The following heat intensity scoring system is used by several breeding operations.

Score of Mare on Heat Intensity (1 to 5)

1. Rejection of stallion
2. Not receptive, but does not fight
3. Will accept stallion, slow to tease, no profuse excretions from vulva
4. Accepts stallion willingly, increased vulva activity (winking), urinates, profuse secretions
5. Accepts stallion, even backs up to stallion; all items listed in 4, plus breaking down (squatting)

This system is useful because the horse breeder has more precise information rather than simply knowing that the mare is in or out of heat.

Depending on results of the teasing, a mare may or may not be palpated. Mares can be teased by leading the stallion in front of their stalls and paddocks or by placing the stallion in a small pen in the center of a paddock full of mares. In the latter situation, the mares will go over to the stallion and the stallion manager can observe closely and make notes. Some mares, although they are in heat, will never challenge other mares for a place around the stallion's pen, due to their position in the dominance hierarchy of that group of mares.

Horsemen have begun to pay more attention to the horse breeding business. Many started with a pasture breeding program, advanced to hand mating, and now in many instances are using artificial insemination. Horsemen are realizing that they do not have to accept poor conception and low foaling rates as a fact of life. They are becoming more and more educated in the total field of equine science. The universities have established outstanding horse curriculums to meet the demands of horse-oriented students.

Certainly, the horse breeders have established controlled breeding programs. Using more sophisticated record-keeping systems and breeding methods, the breeders now know where they are and where they are going. For breeders, the governing factors for management decisions are knowing the cycle, knowing when ovulation occurs, and then to breed or artificially inseminate accordingly.

Mares do not always follow the cyclic pattern previously indicated. In fact, Dr. Warren Evans states that mares are "reliably erratic." Several factors may be attributed to irregular cyclic behavior. One of these factors has been mentioned, i.e., the season of the year. Mares usually cycle from early spring until late summer. A second factor is that of a persistent corpus luteum. The corpus luteum is responsible for secreting progesterone, a hormone that prevents sexual receptivity. Thus, the corpus luteum must stop functioning to decrease progesterone levels in the blood so that the next estrus can occur.

Cystic ovary is a third factor contributing to irregular cyclic behavior. This situation accounts for a prolonged heat with an enlarged ovary and without ovulation, which obviously interrupts normal cycling. Other factors causing irregular cyclic behavior in mares may include hot weather in June and July and the effects of foaling.

Variations in the heat period also have been reported, including conditions such as silent heat and heat without ovulation. Split estrus and estrus with multiple ovulations also occurs.

Some not-so-scientific suggestions that may help the breeding business include prevention of practices such as hauling the mare immediately after foaling. Many horsemen who have tried to breed on foal heat have lost a foal because they hauled the mare and newborn foal in inclement weather or under poor trailering conditions (such as with an open horse or stock trailer).

Be sure to breed the mares to stallions that will maximize the ultimate return. Generally, it does not pay to breed a low-quality, average-pedigreed mare to a world-class horse. Many times, additional expenses for board bills at breeding farms can be eliminated with a routine health and reproductive examination by a veterinarian prior to transporting the mare to the farm. If a problem is found, it can be cleared up prior to breeding.

One last thing, horsemen should be active in both breed and marketing associations to help promote their product.

ACKNOWLEDGMENT

Acknowledgment is extended to the Horse Section (Dr. Gary Potter, Dr. Doug Householder, Mr. B. F. Yeates, and Dr. Jack L. Kreider) Texas A&M University for providing factual material for this paper.

Part 8

FOAL HEALTH AND MANAGEMENT

32
FOALING

Michael Osborne

Management of thoroughbred-horse farms requires understanding of many phases of biology, animal behavior, and human behavior. The author's experiences in the operation of thoroughbred farms in Kentucky and Ireland provide insights to many aspects of horse husbandry, marketing, and related technology. An overview of the author's observations is offered to assist horsemen and horse-farm managers to improve specialized phases of their horse operations.

Foaling in the thoroughbred mare takes place under the cover of darkness in most instances. Up to 80% of foalings in the housed mare take place between 20:00 hr and 02:00 hr. Foaling rarely takes place between 10:00 hr and 14:00 hr (midday). A noticeable forward shift in time of foaling takes place on holidays or Sundays. The peak time when 80% foaling takes place is 18:00 hr to 01:00 hr. This proves that quietness among the working staff or their absence is an influence. A fall in the barometric pressure also hastens the onset of foaling. The speed of the foaling process and the ability of the newborn to get to its feet and run with its mother has ensured the survival of the species. The problems arising during foaling, such as malposition of fetus, oversize, broken ribs, convulsive syndrome, and problems of adaption, can be controlled by management. The afterbirth (placenta) should always be examined for evidence of infection and to ensure that none remains within the mare. The nonantibiotic treatment of uterine infections and contamination is a worthwhile practice.

33

MANAGING THE NEWBORN FOAL

Doyle G. Meadows

Tremendous demands have been placed on the horsemen to produce large young horses. In the recent past, it has been economically important to raise an extremely large colt or filly so that they would be more competitive in the showring and(or) racetrack. This has been accomplished partly by breeding earlier in the breeding season, feeding large amounts of feed, and using more intense management of the health and overall well-being of the young horses.

Several problems have arisen due to these maximum (generally non-optimum) growth rates, which are attributed primarily to large amounts of feed. Many horsemen have begun to improve their overall management efficiency to ensure optimum, healthy growth rates in lieu of over-feeding. However, a sound feeding program is mandatory in any successful horse operation. It is important that horse-men take advantage of every management practice to produce a healthy, marketable foal. This paper suggests several management practices that can assist the horsemen in reaching these objectives.

After the foal is born, begins to breathe, and stands up, both the foal's natural instinct and your management are important. Assuming that there are no problems with the mare and foal, it is essential that the foal nurse the mare. There will be a strong urge for you to help the foal stand and nurse, etc., but you should let the foal accom-plish this on his own. It is important that the foal nurse this early milk called colostrum because it contains anti-bodies that protect the foal against disease. The foal is able to absorb the colostrum antibodies for approximately 36 hr after birth. Should the mare die during parturition, colostrum should be obtained from a local breeding farm or veterinarian for the newborn foal.

MEDICATION AND VACCINATION

In addition, immediately after foaling, the navel stump of the foal should be dipped in a tincture of 10% iodine.

This single treatment, although it seems relatively uneventful, is very important. Without the iodine treatment of the navel stump, the foal could get an infection that would lead to navel ill. Navel ill can cause permanent damage to the lower leg structure of the horse.

During the time the navel stump is being treated, a tetanus antitoxin vaccination should be administered to give immediate protection against tetanus for 1 day to 21 days. This should be followed by a tetanus toxoid. However, it takes 7 days to 21 days to get immunity with a tetanus toxoid vaccination. If the mare had received a tetanus toxoid one month prior to foaling, the tetanus antitoxin could be omitted because the mare is able to pass the resistance on to the newborn foal.

Another important management consideration is to make sure that the foal passes the meconium that is waste material (fetal excrement) in the digestive system. The foal should pass the meconium during the first 8 hr to 10 hr after birth. Excessive ringing of the tail and straining may indicate that the foal is constipated and should be treated to enable the excretion of the meconium. This can be accomplished with a warm, soapy water enema or a commercial enema solution. The foal may need to receive more than one enema to completely pass the meconium. Many breeding farms administer an enema to the foal as part of the early management program, whether or not the foal has passed the meconium. Constipation is a problem in the newborn foal and an observant manager is required to detect the situation.

As noted, the foal should receive a tetanus antitoxin at birth, followed later by a tetanus toxoid. In addition, at 60 days of age, the foal should receive vaccinations for protection against sleeping sickness, influenza, and rhinopneumonitis. On farms with a history of disease problems, vaccinations should be given against flu, rhino, and strangles. Strangles is an acute contagious bacterial disease caused by *streptoccus equi*. The disease primarily effects young horses (foals about 3 yr of age), although older horses may contact the disease. The bacterial organism causes inflammation of the upper respiratory tract and of the adjacent lymph nodes which may take 3 days to 5 days. If the lymph nodes are not effectively treated, they will rupture in about 10 days. After rupture, nodes will discharge a thick, cream-colored pus. In severe cases, other nodes may abscess and abscesses may form on the body surface (bastard strangles). The disease is usually spread by the nasal discharge that contaminates water troughs, feed bunks, and pastures. The infected horses may spread the disease for approximately 4 wk--and the organism remains present in the environment for at least a month after infected animals are removed. However, a vaccination program is not recommended by all veterinarians. Generally, vaccination is recommended only at farms with a previous history of strangles.

Deworming on a routine basis (in addition to prevention and control measures) is an important part of any herd health program. Drugs used to control worms and break life cycles can be administered directly into the stomach through a stomach tube, fed in combination with rations, or placed inside the mouth in the form of a paste. Deworming intervals depend on the season of the year, thus deworming programs are different in the various areas of the country. As horses become more and more intensified on smaller tracts of land, more frequent deworming practices should be implemented in all classes of horses. Exact deworming schedules for a specific area can be obtained from local veterinarians.

Many commercial deworming preparations are available and medications should be changed periodically to prevent immunity to certain drugs. Each medication controls specific kinds of worms. For example, some deworming medications will control ascarids, strongyles, and pinworms but not bots; thus, an additional medication should be used that will control bots.

Generally, in Oklahoma, horsemen deworm mares every 90 days at the breeding farm; some farms deworm every 60 days depending on level of management and concentration of horses on the land. The foals are dewormed in the routine farm deworming program after they are 30 days old. However, horsemen who fit and show young horses begin deworming the foals at 1 mo of age, and then deworm every month thereafter. Many veterinarians in Oklahoma recommend worming foals on a monthly basis. When born, foals are not infected with parasites, they are infected by direct or indirect contact with other horses. The parasites discussed here are spread in the feces (manure), which contaminates the environment, feed, and water supply. Improper management leads to high parasite loads and subsequently unhealthy conditions in horses.

FEEDING

Feeding of the foal has been written about in virtually every publication available to the horseman. Composition of diets for the young horses may be obtained from many different sources. From a management point of view, it is important to allow the foal access to a creep ration within the first 2 wk after birth. The foal will not consume large amounts at first but will begin to eat more as it grows older. The creep feeder should be placed near the feeding and water area of the mares. In addition, the creep feed must be kept fresh, (feeding only what the foals will clean up each day). At 3 mo of age, the foal is receiving only half of his energy needs from the mare's milk. Therefore, creep feeding is necessary if the foal is to gain at optimal levels.

Another major factor involved in creep feeding is its effect at weaning time. Foals will wean much more easily if they are already eating a creep diet. Generally, they will not be depressed and lose weight if they have been eating a good creep diet. Other considerations at weaning time include group weaning whenever possible; foals wean more easily when grouped. It is also very important that a sick foal not be weaned; try to wean only healthy foals that are already eating.

Hopefully these management tips will enable you to produce a healthy, marketable foal.

ACKNOWLEDGMENTS

Acknowledgment is extended to the Horse Section at Texas A&M University (Dr. Gary Potter, Dr. Doug Householder, Mr. B. F. Yeates, and Dr. Jack L. Kreider) for providing factual material for this paper.

34

INFECTIOUS DISEASES OF NEWBORN AND YOUNG FOALS

Thomas Monin

Several diseases of newborn and young foals are bacterial or viral in origin. Management of the mare before and during pregnancy, and of the foal's early environment, helps to reduce the incidence of these diseases.

RHINOPNEUMONITIS (EQUINE HERPES I)

Mares that contract this viral disease while pregnant usually abort the foal 3 mo to 4 mo later. Abortion usually occurs during the last half of pregnancy; however, if it occurs near the expected foaling date, a sick foal may be delivered.

These foals are weak and have clinical signs of respiratory disease. They may not stand. Usually, they have labored breathing with fluid sounds. The body temperature is elevated. These foals usually die within 24 hr to 72 hr despite treatment.

The incidence in a herd can be lowered by maintaining a preventive vaccination schedule in the mare. Mares should be given the vaccine during the 5th, 7th, and 10th mo of pregnancy.

Use of this vaccination program will significantly lower the incidence of abortion and weak foals at birth. It also will develop a high colostral milk level of antibodies to be passed on to the foal. This will give the foal protection from the disease for up to 90 days of age.

TETANUS (LOCK JAW)

Tetanus is usually fatal and can be contracted at any time after birth. The disease usually is associated with a sanitation problem during the foaling. Route of infection in the newborn foal is usually through the umbilical (navel) cord soon after birth; however, infection may develop at any time with the organism entering through a puncture wound.

Clinical signs of tetanus are: stiffness of the neck, legs, and body muscles; increased nervousness (especially to loud, sudden noise); flipping of the third eyelid over the eye when startled (most diagnostic); inability to nurse; stiff gait; and the tail may be stiff. The disease may appear 3 days to 4 wk after the bacteria enter the body.

Tetanus can be prevented by vaccination of the mare. The mare should receive a booster during the 10th mo of pregnancy to ensure protection of the mare during foaling and to give the foal protection through the antibodies in the colostral milk.

Foaling vaccinated mares in clean surroundings and proper care of the newborn's umbilical cord soon after birth will eliminate this disease in the newborn.

Foals from vaccinated mares should be started on their permanent immunization program by 6 mo of age; a booster should follow in 9 mo to ensure good protection.

FOAL DIARRHEA

Foal diarrhea is a perplexing condition that often is overtreated; it has several causes including: diet, virus, bacteria, internal parasites, antibodies, and eating of foreign material.

Viral and bacterial infections cause damage to the mucosa (lining) of the small intestine. Foals given good supportive care may overcome this damage; the intestinal lining will regenerate itself if it isn't severely damaged by the infection or overzealous treatment.

A change in diet or quantity of diet may cause diarrhea. Mares that produce large amounts of milk may cause their foals to develop diarrhea from overeating. Foals which have been separated from their dams for a period of time may overload their stomachs with milk. Orphan foals and foals receiving supplemental milk may develop the same problems.

The cause of foal heat diarrhea, which occurs at 6 to 14 days of age, is not clearly understood. Many different agents have been blamed, such as hormonal change in the mare, vaginal secretions running down on the nipple and being eaten by the foal while nursing, internal parasites in the dam's milk, and nutritional change in the milk. In one experiment, foals taken from their dams at birth developed a diarrhea similar to "foal heat diarrhea" at about 10 days of age.

Foals eating fibrous materials may develop diarrhea. This is due to the large colon not being mature enough to digest the material.

Some foals persist in eating dirt or sand in large quantities. Sand is especially abrasive and causes inflammation of the intestine. This, in turn, will cause diarrhea.

Internal parasites may cause diarrhea. *Strongyloides westeri* infection occurs when the foal drinks milk from its parasite-infected dam. Roundworms (ascrids) and bloodworms (large strongyles) can cause diarrhea, especially during their migration through the wall of the intestine.

Oral (and some injectable) antibiotics may cause diarrhea in foals. The diarrhea may be due to unfavorable reaction to the antibiotic or its effect on the normal bacteria in the intestine.

Foals with diarrhea should be carefully evaluated to determine the cause. Foals with "foal heat diarrhea" should be watched. If they remain alert and are nursing well, nothing more than vaseline around the anus to prevent scalding is needed. However, if the diarrhea continues for 4 days, or complicating problems arise, a veterinarian should be consulted.

When foals stop nursing and start showing signs of weakness or dehydration, a veterinarian should be consulted immediately. Dehydration shows as a loss of skin elasticity, sudden weight loss, and sinking in of the eyes.

If one feels compelled to give home treatment without veterinary consultation, confine the treatment to Pepto Bismol® and Kaolin-Pectin by mouth, or mineral oil, if a stomach tube can be passed.

Good management of the foal's health and environment will eliminate many of the causes of diarrhea. Foals should be dewormed on a regular 6-wk to 8-wk basis starting at 30 days of age. This practice should be continued until they are 30 mo old, and preferably for the entire life of the horse. Avoid separating foals from their dams for longer than 2 hr; if this isn't possible, try to milk out one side of the mare's udder before returning her to the foal. Another method is to limit the foal to 10 min of nursing; then stop him for 20 min to 30 min. Basically, any method may be used that prevents the foal from overfilling his stomach in a short period of time.

Foals that develop diarrhea, other than "foal heat diarrhea," should be moved out of contact with normal foals to prevent spread of the disease in the herd.

PNEUMONIA

Pneumonia caused by a bacterial infection is the most common form. It may be the main cause, or it may be a secondary problem related to septicemia (infection of the blood), internal parasite migration, viral infection, or stress.

The most common bacterial cause of foal pneumonia is the *Streptococcus spp.* However, other bacteria may, at times, cause foal pneumonia.

Clinical signs of pneumonia include labored breathing, a deep cough, and some nasal discharge. Many foals are depressed and unthrifty in appearance. Other signs such as loss of appetite, weakness, dehydration, and diarrhea may occur with pneumonia.

Foal pneumonia due to *Corynebacterium equi* is the worst form of pneumonia. It appears at 1 to 3 mo of age. The onset of this pneumonia usually goes unnoticed until the foal is suddenly ill with the disease. Two forms of the disease are recognized in foals, subacute and chronic. Foals with the subacute (fast onset) form may appear to be normal until they develop severe signs of pneumonia. They usually die within 4 or 5 days. The chronic (prolonged) form is the most common. It occurs as an unresponsive disease that may involve intestinal and(or) joint infection. It causes abscesses in the lung that wall off and are difficult to treat. Most foals will eventually die, and those that survive have decreased respiratory capacity and are not satisfactory for work.

Good management of foals will lower or eliminate the incidence of pneumonia. Mares should be vaccinated for all the viral respiratory diseases, with boosters given before foaling. This will ensure high antibody levels in the colostral milk, which will protect the foal. Foals should be dewormed on a regular schedule. Eliminate or minimize contact with older horses other than the brood herd. Isolate new mares and foals for 30 days before introducing them into the herd. Mares and foals involved in exhibition should be kept separated from the main herd. Avoid stressful situations such as overcrowding, excessive handling, and dusty situations. Foals kept inside should have good, draft-free ventilation.

Foals with pneumonia should be removed from the herd immediately. They should be confined to small, dry, well-ventilated, clean areas with their dams until well.

Consult a veterinarian for a complete diagnosis and treatment routine; there is more to treating pneumonia than killing the causal bacteria.

OMPHALITIS

Omphalitis is the infection of the umbilical (navel) cord and surrounding tissue. In newborn foals, this is a constant threat. The infection may localize in the cord and surrounding skin. It may spread into the bladder or develop into a septicemia (generalized blood infection).

Management and sanitation in the foaling area will eliminate this disease. The umbilical cord and surrounding skin should have repeated application (once or twice daily) of a mild, nonirritating disinfectant for 2 to 3 days. Povidine iodine solution, 2% tincture of iodine, or equal parts by volume of 2% iodine and glycerine may be used. The use of strong disinfectants such as 7% iodine, lysol, and others may cause chemical burning of the skin and obstruction of the umbilical cord.

Obstruction of the open end of the umbilical cord by using strong disinfectant or tying off with string should be avoided. This may cause infection to become trapped in the cord, forming an abscess. As the abscess grows, it may rupture into the abdominal cavity, bladder, or spread into the blood.

In emergency situations, it may be necessary to tie a string around the umbilical cord to stop bleeding. However, it should be removed within 2 hr. These foals should be observed more closely for signs of umbilical cord infection.

The clinical signs of umbilical cord infection are pus and(or) swelling of the remaining umbilical stump or body wall where it attaches. This is due to abscess formation. The area around the umbilical cord may sore to touch. The skin may be puffy and soft feeling.

Umbilical stumps should dry up within 48 to 72 hr. If this does not occur, infection may be present and a veterinarian should be consulted.

Disinfection of the umbilical cord should be accompanied by good sanitation in the foaling stall. A stall, well-bedded with clean straw or grass hay, is much preferred over dusty bedding such as wood shavings, sawdust, ground paper, or rice hulls. These bedding materials tend to stick to the umbilical cord and may cause infection.

JOINT ILL

Joint ill infections are divided into two types: 1) infection of the joint capsule or its joint, or both, and 2) infection of the growth plate (physis), usually at the lower end of the long bone adjacent to the joint. Both types may be found in the same foal.

Historically, the route of infection has been thought to be through the umbilical cord. The terms joint ill and navel ill have been used interchangeably; however, joint ill has several routes of infection.

Foals may develop joint ill from infection through the mouth (digestive tract), respiratory tract, or umbilical cord. In some instances, foals may develop the infection while still in the uterus as a result of septicemia or uterine infection in the mare.

The bacteria invade the blood (circulatory system) and are circulated through the body. The joint capsule and growth plate have a unique blood supply at this stage of life. Because of this, the infective bacteria begin to collect at one or both of these locations and infection develops.

Although clinical signs will develop suddenly, the foal may have become infected 24 hr to 72 hr previously. The larger joints, such as the fetlocks, knees, hocks, and stifles, are most commonly affected; but any joint in the body can be involved. The foals usually are first noticed to have a swelling in or around the joint. This is usually accompanied by lameness. At first this may be attributed to injury such as being stepped on or kicked. Inspection of the joint may reveal heat, pain, and swelling around or in the joint. When this condition is found, a veterinarian should be consulted.

Joint ill may be accompanied by septicemia (blood infection), which can cause serious complications that may lead to death. Sometimes, humane destruction is necessary because of severe bone infection or joint damage. Foals surviving the disease may have bone or joint damage that renders them unfit to work.

Prevention of joint ill is the same as for omphalitis --good sanitation and management of the foaling barn.

Prevention of the diseases discussed is preferable to treatment. Sanitation of the foaling barn and good management, combined with good veterinary care, will help eliminate these diseases.

The following management practices are essential and economic:

- Vaccinate all brood stock and other horses on the farm against tetanus, sleeping sickness, and viral respiratory diseases.
- Give booster vaccinations to the mare 30 days to 40 days before foaling.
- Move mares to foaling locations preferably 90 days before foaling and no later than 30 days. This will allow the mare to build protection to infections found on that farm. The foal will be protected by the colostral milk.
- Foal mares in clean, draft-free, well-ventilated stalls or clean, grassy paddocks.
- Use clean straw or grass hay to bed foaling stalls instead of dusty, small-size bedding.
- Disinfect the umbilical cord with a mild disinfectant immediately after birth. This should be repeated twice daily for 2 days to 3 days.
- Do not mix mares and foals with older horses on the farm.
- Isolate newly acquired mares and foals from the brood herd for 30 days. This should prevent introduction of a new disease into the herd.
- Mares and foals taken to shows and brought home should be kept isolated from the home mares and foals.
- Remove sick foals with their dams from the herd to prevent possible spread of disease.
- Confine sick foals with their dams in clean, well-ventilated, draft-free stalls. This will lower stress on the foals and encourage rest.
- Maintain a good deworming program, starting when foals are 30 days old and repeating at 6-wk to 8-wk intervals. Good parasite control programs must be used in conjunction with the deworming.
- Avoid the stress of overcrowding.
- Avoid other stressful situations such as dusty conditions and excessive handling.

- Stressed animals are more susceptible to herpes infection. To avoid stress during weaning, separate mares and foals in weaning groups as early as possible. This allows foals to establish social order. Weaning is accomplished by removing one mare from the group every few days.
- Put as much space between pregnant mares and newly weaned foals as the farm will allow.
- Start an active vaccination program for foals at 3 mo to 6 mo of age, with a booster following at 9 mo of age.
- Vaccinate foals for rhinopneumonitis with inactivated vaccine, two injections 3 wk to 4 wk apart. The second injection should be given about 3 wk before weaning. A third injection should follow 6 mo after the second injection, or about 3 wk before the animal is sent to sale.

REFERENCES

Bryans, J. T. 1980. Application of management procedures and prophylactic immunization to the control of equine rhinopneumonitis. In: Proc. 26th Annu. Meeting, Am. Assoc. Equine Pract., pp 259-271.

Catcott, E. J. and J. F. Smithcors. 1972. Equine Medicine and Surgery (2nd Ed). Amer. Vet. Pub., Santa Barbara, CA.

Mansmann, R. A. and E. S. McAllister. 1982. Equine Medicine and Surgery (3rd Ed, Vol. 1). Amer. Vet. Pub., Santa Barbara, CA.

35

NONINFECTIOUS CONDITIONS IN THE NEWBORN AND YOUNG FOALS

Thomas Monin

There are several potentially troublesome noninfectious conditions that can occur in the newborn or young foal. A good physical examination should be done at the time of birth by a knowledgeable horsemen or veterinarian. Many of these conditions can be treated medically or surgically if noticed in time.

RETAINED MECONIUM

While the foal is growing in the uterus, it swallows fluid. The fluid becomes bowel movement call meconium. Most foals pass this after getting up and having their first colostral milk.

The meconium is usually yellow to black, and soft to formed in consistency. A foal may have difficulty in passing meconium because it has become very sticky or hard. Some foals may pass normal-appearing meconium and then develop clinical signs. This is usually due to hardened meconium further up in the colon. Retained or impacted meconium is seen more often in colts due to their smaller pelvic opening.

Clinical signs of retained meconium may show up within 3 hr after birth. At first, an affected foal will walk around elevating and switching its tail. It will strain to have a bowel movement. If it isn't successful or relieved, the foal will start lying around. The foal may appear to be sleeping, only to awaken. It may show periodic abdominal discomfort. As the condition progresses, the foal will start to roll and express signs of colic. These symptoms have been reported to start as late as 48 hr after birth due to impaction higher up in the rectum.

Some farms routinely give newborn foals an enema. A prepackaged prepared enema in a plastic applicator bottle may be used. Another method is 1 pt of warm soapy water made from an Ivory soap bar. This solution is run into the rectum using a soft rubber tube. It is important to only use a soft rubber tube or the prepackaged applicator because hard tubes may rupture the rectum. These preventive practices are good; however, on occasion, they will not reach high blockage.

Other farms will watch for passage of the meconium. At first signs of problems, they will give the enema. If this doesn't relieve the foal, the veterinarian is consulted for further examination and treatment.

PATENT URACHUS

The urachus is a tubular structure in the umbilical (navel) cord. During the growth of the fetus, it carries urine from the fetal bladder. As the fetus matures, the urachus should start to close. It should be closed before or immediately after birth. There is the possibility that infection of the umbilical cord stump may cause the urachus to reopen.

Clinical signs are straightforward. The foal will urinate from the umbilical cord. This may not be noticed at first in colts. However, the presence of wet hair around the navel should alert one.

The immediate danger of this problem is the possibility of umbilical cord or bladder infection with subsequent septicemia (blood infection).

A veterinarian should be consulted because the foal should be catheterized to be sure the urethra is open. The umbilicus shouldn't be tied off with string because this will cause serious problems. Treatment is usually medical; however, if this is unsuccessful, then surgical closure is performed.

RUPTURE OF THE URINARY BLADDER

Rupture of the urinary bladder occurs in less than 1% of the foals born. The condition may occur in either sex but is more common in the male. Rupture happens during birth.

The fetus has a distended bladder at the time of birth. As the foal passes through the birth canal, the urethra becomes compressed. Rupture occurs as the mare presses to deliver the foal. Incomplete closure of the bladder during fetal growth has been suggested as another cause.

Foals seldom show signs of the condition before 48 hr. They will make repeated unsuccessful attempts to urinate, and as urine accumulates in the abdominal cavity, some urine may pass out the urethra. After 72 hr without the foal urinating, its abdomen (stomach) becomes distended (full), breathing increases, and the body temperature starts to rise.

The foal develops pale gums and stops nursing. It becomes sluggish as the condition progresses, and may show signs of being colicky. After 48 hr, sloshing fluid sounds may be heard if the foal's stomach is shaken.

This condition must be surgically repaired. With early detection and veterinary care, a high percentage of foals will recover.

CLEFT PALATE

Cleft palate is the incomplete closure of the hard and(or) soft palate. It is seen in a small number of foals soon after birth. The cleft may involve the hard palate and the soft palate, but in some cases, only the soft palate may be involved.

The first signs of cleft palate involve milk running out of the nostrils. Some foals with this problem have trouble nursing and usually develop foreign body pneumonia from inhaling milk. Diagnosing cleft of the hard palate can be made by looking in the mouth. Cleft of the soft palate is hard to diagnose because the foal must be restrained. On occasion, it requires endoscopic examination to see the soft palate cleft.

Foals with a soft palate cleft may go unnoticed, especially if the cleft isn't large. However, these foals may be unthrifty. Prognosis for cleft of the hard palate is poor. These foals usually have foreign body pneumonia to complicate the surgical risk. Surgical repair isn't very successful and may require several attempts. Even after several attempts, reconstruction of the hard palate may not be successful. These foals are subject to anesthetic death due to lung damage and the long time required on the surgical table.

There are several causes listed for cleft palates. Among them is genetic inheritance. Serious consideration should be given to the genetic possibility before retaining these horses in a breeding herd.

PARROT MOUTH

Parrot mouth is congenital and may be inheritable. In this condition, the upper jaw is longer than the lower jaw. The degree will vary from slight to severe.

To judge severity, the table surfaces of the upper and lower incisor (front teeth) should be viewed from the side. In a normal mouth, the table surfaces should rest evenly on each other. Affected horses may show 10% to 90% of the table surface of the upper teeth. In severely affected animals, the lower jaw may be so short that the incisors don't touch. Severely affected animals usually can't maintain themselves on pasture.

The incidence of inheritance is high, and these animals should be considered unsuitable for breeding purposes.

TYMPANY OF THE GUTTERAL POUCH

Tympanites (air) of the gutteral pouch occur when air is trapped in the pouch. The gutteral pouch is a part of the eustachian tube. An opening into the throat normally allows air to flow back and forth during breathing. The condition is more commonly seen in the young foal but also may be seen in older horses.

The condition may be noticed within 24 hr of birth when a swelling starts developing under the ear behind the jaw. One or both sides may be involved. When one pushes on the swelling, it feels like a partially filled balloon. As the pouch continues to fill, it may cause partial obstruction of the throat. The foal will make a snoring sound as it breathes.

Tympany of the gutteral pouch is due to a birth defect of the eustachian tube or inflammation. Inflammation causes tissue to swell and obstruct the tube. Surgical correction is required and in some cases may not be successful. In cases which aren't successful, the foal should be watched closely to prevent possible suffocation. These foals may eventually recover because growth allows the opening into the throat to enlarge.

UMBILICAL HERNIA

"Hernia" is the term used to describe the protrusion of an abdominal organ--such as the intestine, bladder, and pregnant uterus--through an abnormal opening in the body wall. An umbilical hernia occurs at the site of the umbilical cord. Umbilical hernias present at birth are called congenital. Those appearing 3 wk to 4 wk later are called acquired.

Congenital umbilical hernias are considered to be hereditary. Acquired umbilical hernias may be due to rough handling of the umbilical cord at birth. Excessive straining during urination or bowel movement and infection of the umbilical cord may also lead to acquired umbilical hernia.

It is best to let the umbilical cord break naturally. However, if manual breaking becomes necessary, a set procedure has been established that requires two large artery forceps. One is placed on the umbilical cord approximately 1 in. (2.5 cm) from the body wall. The other forcep is placed approximately 1 in. below the first. The cord is pulled apart without pulling on its body wall attachment.

A convenient way to measure the size of an umbilical hernia is to place as many fingers as possible side by side in the wall opening. The palm of the hand should face the person as he stands beside the foal. The size of the hernia ring can then be given in number of fingers.

Small umbilical hernias, one- and two-finger size, usually close on their own 6 mo to 12 mo of age. Large hernias, four fingers plus, are better closed while the foal is small. These large hernias seldom close by themselves. Umbilical hernias should be observed closely for changes in size or consistency. If these foals start to colic, they should be seen immediately by a veterinarian. Changes in size or swelling attended by firmness of the hernia should also be checked by the veterinarian.

On occasion, a portion or entire loop of the intestine may become entrapped in the hernia. This may cause that portion of intestine to form adhesions or die. Adhesions will cause the hernia sack to become firm to the touch. Infection of the hernia will also cause it to become firm but also accompanied by swelling of the body wall around the sack. Most foals will show signs of colic if the portion of the intestine in the sack is dying. A gradual increase in the size of the hernia may mean more intestine is falling into it. A sudden increase in size accompanied by colic or lying around more than usual should be viewed with alarm. A veterinarian should be consulted at once.

Foals developing acquired umbilical hernia should be examined by a veterinarian to determine the cause. Abscesses of the umbilical cord may appear similar to hernias. They are usually firm and painful like entrapped hernias.

Examine the hernia several times weekly to keep track of its size and consistency. Observe foals on a daily basis to avoid missing a change in appearance or the presence of colic.

Hernias are repaired by several different methods. One method is to open the hernia and sew the body-wall opening shut. Another method is to use hernia clamps. The attending veterinarian determines the method used.

SCROTAL HERNIA

Congenital scrotal hernia is seen in a small number of male foals born each year. The condition is considered inheritable. Scrotal hernia usually involves only one side but has occurred on both sides.

Many of these hernias will correct themselves as the muscles become stronger. The decision to surgically repair these hernias is based mainly on their size. However, if a foal becomes colicky and the scrotum is involved, then surgical repair should be done immediately.

The use of a hernia truss or bandages to hold the intestines in the abdominal cavity seems to aggravate the condition. Small scrotal hernias may self-correct by 6 wk to 12 wk of age. If surgical repair isn't done immediately, then these foals should be watched closely for signs that the hernia is increasing in size or that the foal is colicky. Foals with congenital scrotal hernia should be castrated at the appropriate time regardless of whether they self-correct or not.

GASTRIC ULCERS

Foals under stress from severe illness or injury may develop stomach ulcers. This condition has been noted on two occasions in two normal foals stressed by overcrowding on the breeding farm. Cortisones and nonsteroid antiinflammatory drugs used over time may also be involved.

Intermittent grinding of the teeth may be the first sign of ulcers followed by slobbering while grinding. If the condition isn't recognized early and treated, these signs increase in severity. Foals will stop nursing and begin to lose weight and become unthrifty.

Prevention of the condition is the best treatment. This is done by placing foals on preventive treatment at the onset of diseases or injuries that require extensive treatment and(or) healing time. If overcrowding is the problem, this can be handled by moving mares and foals off the farm.

LIMB DEFORMITIES

Foals born with varying degrees of crooked limbs, contracted limbs, relaxed tendon, etc., are a common occurrence. Because of the complexity of the cause and how it should be treated, it is recommended that a veterinarian be consulted as soon as the problem is noticed.

FAILURE OF PASSIVE TRANSFER OF COLOSTRAL ANTIBODIES

The foal's consumption of adequate colostrum cannot be overemphasized. Failure of colostral antibody passage from the mare to the foal is the most common condition in the equine industry. This failure of transfer causes great economic loss to the industry.

Foals depend on the colostral milk for their entire source of antibodies to protect them from disease. The intestinal tract is receptive to absorbing these antibodies from the colostrum in the first 8 hr to 12 hr of life. Absorption of antibodies after 12 hr drops drastically and ceases at 24 hr of age.

Failure to absorb sufficient antibodies from the colostrum may be due to several reasons:

- The foal doesn't drink enough colostrum. This may be seen in orphans, rejected foals, weak foals, and foals removed from the mare for other reasons.
- Low antibody content in the milk.
- Failure to absorb sufficient antibodies from the colostrum.
- Premature loss of colostrum before the foal is born.

Some mares will lose (stream) colostrum from the teats several weeks before the foal is born. When this is noticed, several pints should be collected and frozen to await the foal's birth. The frozen colostrum can be thawed and fed to the foal soon after birth.

Fortunately, failure to absorb sufficient antibodies can be detected by two simple tests. After a foal is 24 hr old, it normally should have an antibody level of 30% or better of its dam's antibody level. The zinc sulfate turbidity test is a good quick field test, but the slowed radial immunodiffusion (RID) test is a more accurate test. These tests are run on the foal's plasma.

Foals should have 500 mg of antibodies per cc of plasma. Levels below this should be treated. Treatment consists of giving the foal plasma in the vein from a suitable donor--such as a disease-free stallion, gelding, or an unbred filly.

Suitable donors should be identified and blood tested well before the foaling season. A veterinarian can help send the appropriate blood samples to a laboratory. Once donors are identified, they should be kept in the peak of health.

Some farms have their veterinarian check every foal 24 hr to 36 hr after birth in hopes of identifying those foals with low antibody levels. This management tool may seem expensive until the value of lost foals is added up at the end of the season.

REFERENCES

Mansmann, R. A. and E. S. McAllister. 1982. Equine Medicine and Surgery (3rd Ed., vol. 1). American Veterinary Publications, Santa Barbara.

Oehme, F. W. and J. E. Prier. 1974. Textbook of Large Animal Surgery. The Williams and Wilkins Co., Baltimore.

Perryman, L. E. and T. B. Crawford. 1979. Diagnosis and management of immune system failures of foals. In: Proc. 25th Annual Meeting, Am. Assoc. Equine Practitioners. pp 235-243.

Walker, D. F. and J. T. Vaughn. 1980. Bovine and Equine Urogenital Surgery. Lea and Febiger, Philadelphia. pp. 163-164.

Part 9

NUTRITION AND FEEDING

36

THESE HORSES GOT SICK BECAUSE OF WHAT THEY ATE

William C. McMullan

"When I got back home from an overnight trip and went out to feed, I was startled to find the feed room door open. It was obvious the two geldings had been in there--a 50-lb sack of sweet feed and a sack of laying mash were open, scattered and about half was eaten. I breathed a sigh of relief when I found the nags and they were O.K." That was the history I got from the owner the next day when he called to report his geldings didn't seem to want to move--kind of nailed to the ground. When urged, they moved stiff legged and carefully, landing exaggeratedly on the heels of both front feet. When I got there, I found that they were also sensitive to what would have ordinarily been routine taps with the shoeing hammer. The digital arteries were pounding away harder and faster than usual (they can be most easily palpated at the inside and outside back corners of the fetlock on level with the ergot). These signs add up to founder (laminitis), one of the most important diseases of horses. We can learn from this unfortunate and often-repeated true story:

1. Be sure the feed is secure!
2. Ground feeds, especially finer ground starch feeds like chicken and hog feeds are more likely to cause founder.
3. Signs of founder usually do not appear until 24 hr to 72 hr after a starch overload, so if a horse has eaten twice as much as he usually gets in a day (25 lb or more most of the time), be sure he gets a gallon of oil and intravenous butazolidin as soon as possible. Once signs appear, each hour delay in starting treatment reduces treatment success by about 20%. In the early stages, founder is a true emergency.
4. Remember that founder can also be caused by moldy feed or by maintaining high planes of nutrition over time, not to mention nonfeed-related causes.

This story could have had a different twist. Sometimes the result of overeating, of eating spoiled or contaminated feeds or excessively coarse roughage is colic. The pawing, kicking the belly, rolling, uncomfortable pain signs may be noticed in a few hours (spoiled or contaminated feed) or in several days (rough hay impactions). Toxicity buildup from severe colic may result in founder 1 day to 2 days later. While close to 75% of colic is thought to be related to damage by blood worm larvae, the other 25% can be greatly reduced by feeding the right amount of a clean, quality horse feed at regular 8-hr to 12-hr intervals. Avoid any abrupt changes in feed or schedule.

Insufficient and/or unpalatable water is an important contributor to impaction colic. Look at the water, smell it and see if you would taste it. In the North, a water heater is cheaper than a vet call. Some people say that pellets are worse to cause colic in a horse than sweet feeds. I think it is debatable, depending mainly on the quality and amount of each and whether or not a changeover from one to the other is involved. Horses do best when they are fed the same amount of the same kind of good quality feed at the same time every day. A little boring perhaps, but very healthy.

Colic is by far the leading cause of death in our present-day horse society. More horses die from colic than from all other diseases added together. And, like founder, early, expert treatment makes all the difference in the world; without it, the problem can mean a hopeless cripple or another statistic. 'Nuff said!

An important disease of the respiratory system which may be feed-related is heaves, also called "emphysema" but better termed "chronic bronchiolitis." Dust in hay or feed, bedding, and barns irritates the lining of the lungs and in itself can cause coughing and shortness of breath especially when exercised. Allergens present in hay (down feathers from chickens roosting in the hay; molds, especially in alfalfa hay) are inhaled and cause further damage; then come opportunistic bacteria and then more dust, etc., in a vicious cycle. This is one good reason why equine nutritionists always recommend clean hay. This disease can be cured if the offending agents are removed at the start of the disease and proper treatment and special diets are instituted early. Otherwise there may be only a treatment for the duration.

The previous three problems are unusual in horses less than 2 yr old. In youngsters, feed-related problems most often affect the musculoskeletal system. Epiphysitis (growth plate inflammation), contracted tendons and osteochondrosis (O.D., bone cyst, young horse cartilage disease) have increased in number far faster than the population. In the weanling, growth is especially rapid at the fetlock. Inflammation of the growth plates (epiphysitis) above and below the fetlocks can be recognized by firm swelling, producing an hourglass appearance. If severe, pain and lameness will be present. Growth in the yearling is rapid at the growth plate of the upper knee, which will be abnormally enlarged when epiphysitis is present.

Contracted tendons should be suspected when the pastern bones are almost vertical and the fetlock buckles forward; this indicates contraction of the superficial flexor tendon. If the front of the hoof wall is nearly vertical but the pastern is reasonably normal, it is deep flexor contraction.

Osteochondrosis is a disease of young animals in which there is delayed and imperfect maturation of cartilage into bone. This defect may be recognized by joint swelling. But often you just have a lame yearling or 2-yr old and you have to find the spot with nerve blocks and X-rays. Part of the cause of these diseases is that we are selecting for rapid growth rate. But the compounding factor is that too often we give a weaning colt a trough full of calories. This gives growth rate a big stimulus, but the right building blocks in the form of protein, calcium and phosphorus are usually deficient; occasionally an excess of one or the other is present also, which can cause problems.

If you are pushing for halter class or early performance, I urge you to have your diet checked. Very few, if any, commercial concentrate feeds, when mixed with the proper amount of hay for a weanling colt, will supply enough phosphorus, even if alfalfa is the hay fed. Alfalfa, however, when added to a foal grain mix or pellet, will ensure that the calcium and protein requirements are met. If a colt is being fed oats and average quality coastal hay, I guarantee he will be low in protein, calcium and phosphorus. If at the same time, he is fed 2-lb to 3-lb concentrate for each 100-lb body weight, he will take in about twice as many calories as required and be a good candidate for one or more diseases.

Wobbler syndrome may be added to the list of nutrition-related diseases in the next year or so. A similar syndrome has been produced in dogs by diet manipulation.

Another nutrition problem is that of Overvitaminosis. This overdosage of vitamins and(or) minerals is the result of looking for the competitive edge, not following directions, or feeding more than one supplement. It is easier than you think to exceed 10 minimum daily requirements (MDR) of vitamin A, and if this is done by injection this may be enough to cause a problem. The ones most often overdosed are cobalt (used to treat anemia--in excess it can cause anemia); iodine (when pregnant mares were given 12 times MDR iodine, they had deformed foals); vitamin E, vitamin A, and vitamin D. This is a little ridiculous when you consider that 2 hr sunshine/day supplies needed vitamin D, and 6 hr/day on green grass for 4 wk will supply enough vitamin A for 3 mo to 6 mo.

One last beware: beware of feed for other species like cattle or other animals that contain feed additives. We have found out the hard way what some of them do. One rancher called in to his vet and said "I think you better come quick. About half my horses are either dripping wet with sweat, colicking or staggering around like a herd of wobblers." It turned out that by mistake some Rumensin (a cattle growth stimulant) had gotten mixed in with the horse feed. Death losses and humane destruction amounted to almost 50%. While cattle tolerate and benefit from this drug, a horse can die if he gets as little as 1 g (1/5 tsp) a day in his feed. Urea in the diet gives the opposite effect. The horse is much more tolerant of it--5 times more. A horse will not get enough off lick-wheels to hurt him, but it will not do him much good either. Horses do not digest urea quickly enough to absorb it efficiently. Experimentally, urea fed at 5 oz/100 lb body weight killed seven of eight ponies. That would amount to 3.3 lb urea for a 1,000-lb horse, but this would take 100 lb of most feeds (3% urea) or 25 lb of a protein supplement.

Starvation is occasionally a problem when ignorant or destitute or ruthless owners are involved. If a mature, idle, 1,000-lb horse is getting the thumb-rule minimum, 1/2 lb grain and 1 lb hay/100 lb its body weight, it is not a starvation case--look for some other reason such as parasites, bad teeth, etc.

A very serious problem, enterotoxemia, occurs in young horses fed in groups. The largest and most aggressive foals eat two shares. They may be okay at feeding time and found dead shortly after. Sometimes no signs are observed; in others, bloat, colic, difficult breathing and air under the skin are seen. *Clostridium perfringes* type D (close kin to black leg in cattle and the tetanus organism) is the likely cause. This "sudden death" syndrome is more common on the weekend when the regular help is off or in a hurry.

Another less significant effect of overfeeding, especially of protein, is "protein bumps"--a nodular, hive-like skin problem seen most often in sales yearlings. Reduce the protein intake and the nodules go away.

From these situations, one should realize that an equine needs a prescribed diet in correct amounts, all tailored to the age, workload and physiological situation (the pregnant mare needs no increases until the last 3 mo, and then a 15% increase in calories and a 25% increase in protein). At foaling, her caloric intake needs to be doubled because of lactation, and the total amount of protein in the ration raised to 14%.) The feed should be clean, fresh and offered on a regular schedule.

I'll wind up with a few "don'ts":

1. Don't feed seed oats--fatal poisoning has resulted from mercury seed treatment.
2. Don't let your horse nibble on ornamental plants --for example, decorations at the horse show. At one show a horse ate some Japanese yew while waiting for his class and fell over dead a little while later.
3. Don't feed the same amount of corn if you are switching from oats. Corn is much higher in energy and can cause colic if overfed.

4. Don't feed alfalfa if it has any large insects which might be blister beetles--1/4 oz of beetles will kill a horse.
5. Don't allow baling twine or wire or other hardware items to be consumed with the diet. Every small piece can puncture a gut or be the foundation for a fecal ball which may get large enough to block an intestine (enterolith) and cause severe colic.
6. Toxic weeds are not usually a problem if a horse has a choice of something better to eat, so don't make him forage in a wasteland.
7. Don't feed straight wheat. Wheat gluten is sticky and increases the likelihood of impaction.

And remember:

1. Horses do not have nutritional wisdom.
2. A full tank does not make a car run faster.
3. Excess energy is stored as fat, not high spirits.
4. You can't feed a $500 horse into a $5000 horse.

37

COMMON SENSE AND UNCOMMON SCIENCE IN HORSE FEEDING

Melvin Bradley

In your previous experience with ration checking and balancing, you may have had to deal with megacalories in the metric system, International Units in vitamin recommendations, and parts per million in trace mineral discussions. But take heart, there is a better way! Scientific jargon and feeding precision are necessary in nutritional research and data reporting but are not well accepted and understood by the average horseperson, feed dealer, and horse feeder who communicate in pounds, parts, and(or) percentages. Instead of trying to feed your medium-working adult, 500-kg horse megacalories of energy and grams of protein, why not feed him 12 to 15 lb of commercial or custom-prepared grain ration that contains correct percentages of energy, protein, minerals and vitamins, along with 20 lb of quality hay?

My presentation is designed to help you 1) evaluate rations that you may now be feeding your horses and 2) balance these and other rations, both by hand and by computer. The subject is divided into:

1. The art of horse feeding
2. Nutrient sources and requirements
3. Special feeding situations
4. Computer use in ration checking and balancing
5. The feeders' mineral quiz
6. Custom ration building
7. Chemical testing of horse feeds

You can profit by following this step-by-step procedure in studying and practicing ration "checking" and balancing while working toward a thorough understanding of these two important functions for horse feeding and management.

"PERCENTAGE" FEEDING

Instead of discussing detailed ration balancing and feeding based on the classical definition of "the amount of feed to feed an animal for a 24-hr period," let's follow the lead of swine and poultry feeders who group animals into similar weight and age groups to be fed a specific amount (lb) of bulk feed formulated in percentages for that specific group. For example, swine producers have breeder, creep, growing, and finishing feeds with appropriate percentages of energy, protein, calcium, and phosphorus for each stage of production. Horse feeders with animals of mixed ages and stages of work and production should use two or three different grain rations and both legume and nonlegume hay in their feeding programs. Access to these feeds allows the feeder to offer young stock high protein and mineral grain mixtures with legume hays, while limiting intake of obese, idle, adult horses by omitting grain and feeding mostly nonlegume hay with sufficient legumes offered to supply needed minerals, vitamins, and protein.

THE ART OF FEEDING

Although the science of ration balancing and chemical checking for accuracy cannot be overemphasized, the scientifically fed horse herd is in trouble without good judgment on the part of an experienced feeder. This is because biological systems do not honor hard and fast man-made rules.

Two horses of the same weight may have energy requirements to perform the same task that differ by 50% (table 1). Feeds, especially hays, that look alike may vary 100% in energy and protein yield to the horse, yet would be given the same credit in ration formulation (table 2). Finally, "picky" appetites, dominance hierarchy, and speed at which horses eat greatly affect group-fed animals.

Some general guidelines for consideration in feeding horses are as follows:

- Feed only quality feeds.
- Feed balanced rations.
- Feed half the weight of the ration as quality hay.
- Feed higher protein and mineral rations to growing horses and lactating mares.
- Feed legume hay to young, growing horses, lactating mares, and out-of-condition horses.
- Use nonlegume hays with adult horses.
- Regulate hay-to-grain ratio to control condition in adult horses.
- Feed salt separately, free-choice.
- Feed calcium and phosphorus free-choice.
- Keep teeth functional. Horses 5 yr and older should be checked annually by a veterinarian to see if their teeth need floating (filing).
- See that the stabled horses get exercise; they will eat better, digest food better, be less likely to have colic, and have better stable habits.
- Feed according to the individual needs of the horse. Some horses are hard keepers and need more per-unit weight.

TABLE 1. DAILY NUTRIENT NEEDS FOR A 1,100 LB MATURE-WEIGHT HORSE[a]

Class	Total feed, lb	TDN,[b] lb	CP[b]		Ca		P	
			%	lb	%	g	%	g
Mature horses at maintenance	18	9.10	8.60	1.54	.30	26	.20	16
Mares, last 90 days of gest.	18	10.20	11.10	1.85	.50	38	.40	34
Lact. mare, 1st 3 mo	25	15.80	14.00	3.30	.50	56	.40	34
Lact. mare, 4 mo to weaning	23	13.60	12.25	2.70	.45	46	.30	30
Creep feed, supplemental	-	-	18.00	-	.90	37	.60	22
Foal, 3 mo	12	8.35	18.00	1.85	.90	37	.60	22
Weanling, 6 mo	14	9.90	16.00	1.95	.70	38	.50	28
Yearling, 12 mo	16	9.50	13.30	1.85	.60	34	.40	24
Long yearling, 18 mo	17	9.70	11.10	1.75	.45	31	.33	21
2-yr-old, light training	18	9.20	10.0	1.55	.45	28	.33	19
Mature working horses:								
Light work	20	13.30	8.60	1.55	.30	26	.20	16
Moderate work	30	20.00	8.60	1.55	.30	26	.20	16
Intense work	40	24.50	8.60	1.55	.30	26	.20	16

Source: NRC (1973; 1978).
[a] As-fed basis.
[b] TDN—total digestible nutrients; CP—crude protein.

- Feed by weight, not volume. A gallon of one grain may vary 100% in nutrient yield as compared to a gallon of another grain.
- Minimize fines in a prepared ration. If a ration is ground fine, horses will be reluctant to eat it, and the chances of colic will increase.
- Offer plenty of good water twice daily, no colder than 45F. Free-choice water is best.
- Change feeds gradually over a period of a week or more when changing from a low-density (low-grain), high-fiber ration to one of increased density.

TABLE 2. AVERAGE ANALYSIS OF SOME COMMON HORSE FEEDS

	TDN, %	CP, %	Ca, %	P, %
Grains				
Barley	70	12	.05	.32
Corn, No. 2	80	9	.02	.26
Molasses, liquid	54	3	.0	.0
Oats	66	13	.09	.32
Wheat bran	60	16	.10	1.30
Hays				
Alfalfa				
Early cut	50	12	1.4	.2
Average	49	11	1.2	.2
Late cut	48	10	1.0	.2
Timothy				
Good	50	7	.3	.2
Poor	48	6	.3	.2
Mixed Hays	50	9	.6	.2
Minerals				
Defluorinated rock phosphate			30	13
Dicalcium phosphate			22	19
Limestone			33	0
Commercial				
No. 1			15	14
No. 2			12	8
No. 3			10	5
No. 4			20	5
No. 5			6	2
Protein supplements				
Cottonseed meal	65	41	.26	1.00
Linseed meal	62	35	.38	.78
Soybean meal	76	44	.30	.64

Source: University of Missouri Columbia Forage Lab.

- Start on feed slowly. Horses on pasture should be started on dry feed gradually. Start this on pasture if practical, and gradually increase the feed to the desired amount in a week to 10 days.
- Do not feed grain to tired or hot horses until they have cooled and rested, preferably after 1 hr or 2 hr. Instead, feed hay while they rest in their blankets or away from drafts.
- Hungry horses should complete their eating at least an hour before hard work.
- All confined horses should be fed at least twice daily. If horses are working hard and consuming a lot of grain, three feedings are mandatory.

- Half of the hay allowance should be given at night when the horses have more time to eat and digest it.
- Keep internal parasites under control to maximize health and optimize feed efficiency.

NUTRIENT CHARACTERISTICS OF SOME COMMON HORSE FEEDS

Grass is nature's way of feeding horses. When not dormant, most grasses supply an abundance of protein, energy, vitamins, and minerals for all but very young growing horses and those at hard work. The palatability and abundance of grass in early spring may founder fat horses and its high moisture content does not allow sufficient intake to sustain horses at hard work.

Good nonlegume hays average about 50% digestibility (TDN), 10% crude protein, .3% calcium and .2% phosphorus.

Grasses show extreme variation in quality, depending on stage of maturity, curing process, and variety. Calcium and phosphorus levels remain relatively constant in damaged nonlegumes, but carbohydrates (energy), protein, vitamins, palatability, and digestibility suffer badly.

Legumes are substantially higher in protein, calcium, palatability, and vitamins than are nonlegumes, but they may contain equal amounts of phosphorus and only slightly more TDN than nonlegumes. Because of their heavy foliage, early-season maturity, and susceptibility to molds in the curing process, legumes should pass a quality test before use with horses. If abundant mold is present, horses will become colic, heave, and develop hay allergies when forced to eat moldy legumes.

Grains are high-energy feeds, ranging up to 80% TDN. They are medium to low in protein and phosphorus and are extremely low in calcium. Oats, corn, and grain sorghums are palatable, with oats the choice of most horses, and horse feeders, but oats is far from the perfect horse grain. It varies in quality from "excellent" to "unfit," has few vitamins, and has a very bad ratio of calcium to phosphorus.

Corn is more uniform in quality, is higher in TDN, and has more vitamin A than oats but has less protein, is less palatable, and tends to increase the incidence of digestive upsets because it has less fiber.

Grain sorghum compares well in nutrients with corn but, because of its small size and hard seed covering, is less digestible, and some varieties are unpalatable. It should be processed for horses (but not ground into dust).

Barley is a good horse grain when it does not exceed 50% of the grain mixture. Some horses find it unpalatable at higher levels.

Protein sources are the oilmeals, legume hays, and wheat bran. Of the oil meals, soybean oil meal has a higher quality protein because of a better balance of essential amino acids.

Except when feeding young foals, horse feeders can usually meet their horse's protein needs economically by feeding various amounts of legume hay. In addition to proteins, legumes supply significant amounts of minerals, vitamins, and energy.

Table 2 shows the composition of these common horse feeds.

SUPPLYING ENERGY

Of all the nutrient requirements energy sufficiency or excess is easiest to estimate. If the horse is fat, it is consuming too much; if thin, too little. Stand to the side of your horse, 10 ft to 12 ft away, and look for ribs. If you see none, especially back ribs, the horse may be too fat; if more than two or three ribs are observed, it needs more feed (energy). Thin, group-fed horses, low in the "pecking" order and offered a palatable, high-energy feed only once or twice daily, will not get their share and must be separated from the group for individual care. This is especially true of mixed ages, because yearlings and 2-yr-olds cannot compete with adults. Further, they need higher percentages of protein and minerals than do adults whose muscles and bones are mature.

SPECIFIC FEEDING SITUATIONS

It has been established that horses have different nutrient requirements at different ages and stages of reproduction, growth, and work. It stands to reason that a variety of different hays and grain mixtures are needed in a stable of many horses. Ideally, hay storage facilities will accommodate a half year's supply of three types of hay: good legume, good nonlegume, and average-but-not-moldy-nonlegume. In addition, two types of grain mixtures are essential: 1) a 12% protein, .3% calcium (Ca), and .2% phosphorus (P) for adults, and 2) a 16% protein, .7% Ca and .55% P for weanlings and early yearlings (NRC, 1973 and 1978; Bradley, 1983). Creep rations with higher protein and mineral levels should be used for foals.

Weanlings may consume a high density (energy) ration at a rate of up to 3% of their body weight, but half of it should be excellent quality legume hay. Idle, adult horses with obese tendencies may stay in good condition on 1 1/2% to 2% of body weight with no grain and average hay.

Feed according to condition, regardless of amount.

THE OBESE HORSE

Fat, adult horses--like fat humans--are not very athletic. It is easier and less traumatic to condition a thin horse than reduce weight on a fat one. In many cases, the owner does not realize how fat the horse really is (Henneke et al., 1983). Short, low-set, thick-necked horses with hard, heavy chests are in danger of developing laminitis. To reduce risk, energy intake should decrease and exercise increase--conditions that may be difficult to manage. Confined horses, if exercised, may be satisfied on reducing (less energy) diets, but they are the exceptions rather than the rule because skimpy amounts of feed are quickly eaten, and hunger and idleness drive the horse to wood-chewing, cribbing, and sometimes coprophagy (eating feces).

A less traumatic management practice is to feed 4 lb or 5 lb of a good legume hay daily to ensure adequate protein, minerals and vitamins, plus free-choice feed of an average nonlegume for bulk and the remaining energy requirement.

FEEDING GROWING HORSES

Foals should be creep fed an 18% to 20% protein ration, along with all the excellent quality legume hay they will eat. If closely confined, some foals may overeat grain and develop swelling in the ankles, erect pasterns, and shifting lameness. In these rare cases, they may require isolation and a reduction of grain to 1 lb daily/100 lb body weight. Be absolutely sure that the level and ratio of Ca to P is correct in their diets.

Weanlings may be fed on creep feed the same as are foals but with lower percentages of protein, Ca, and P in their grain diet (table 1).

Both foals and weanlings are more easily fed with half their diet consisting of quality legume hay. Heavy grain feeding creates many problems (see Mineral section).

As growing horses become long yearlings and 2 yr olds, less grain and more hay and(or) grass can be used until they begin to work and more energy is required.

FEEDING WORK HORSES

Racehorses, distance horses, and young dressage horses are in a class by themselves in energy requirements. The challenge is to find a high-nutrient-density grain with sufficient palatability to ensure large volume intake without the horse "going off feed" or experiencing digestive disorders. These rations may contain a grain ratio of four oats to one corn, liquid molasses, and sufficient minerals and vitamins for good balance. When fed in three or four feedings/day, with uneaten grain removed and feed troughs cleaned after each feeding, horses have less tendency to go off feed and lose weight and strength.

Good hay is essential to the diet of these hard-working horses. At weights of 900 lb to 1,100 lb, horses in the Tevis Cup 100-mile ride were eating 12 lb to 16 lb of grain, plus equal amounts or more of hay. Half of the group ate straight alfalfa. Young racing horses have been shown to consume similar amounts.

Nonlegume hay probably will reduce digestive upsets, but skilled feeders can feed either type of quality hay with minimum risk.

Supplemental feeding (either by injection or orally) is hard to justify scientifically, but is practiced by many feeders. Vitamins are toxic at higher levels and should not be used indiscriminately.

Mares in late pregnancy need more and better-quality feed than is required earlier, and their energy and protein needs increase sharply in early lactation.

Idle, adult horses can subsist indefinitely on grass or hay alone. If not parasitized and if not old or with bad teeth, grain feeding is unnecessary.

FEEDING SHOW HORSES

Show horses should be sleek and attractive, have "blooming" hair coats, and should be at the peak of health and physical condition. Start conditioning 2 mo to 3 mo prior to the show season with attention to parasite control, exercise, and nutrition. Protect the hair coat from abrasion and sunburn by stabling and blanketing the horse.

Feed to the upper limits of energy, protein, and vitamins without contributing to excess fatness, looseness of bowels, or vitamin toxicity. Adult horses may consume 10 lb to 12 lb daily of a 14% crude protein ration while consuming 12 lb to 15 lb of free-choice legume hay.

Midwest grains have been shown to be marginal in linoleic acid, an energy nutrient necessary for hair coat bloom. Addition to the daily ration of 1 oz of regular "grocery store" corn oil has shown beneficial effects in some research trials.

Don't forget the grooming! Brisk brushing twice daily is essential for a good hair coat along with routine exercise.

Know the energy requirements for the type and kind of work your horse is doing and feed accordingly. Observe how he eats his feed and train yourself to detect small body-weight gains and losses before they become serious. There is little need to chemically analyze grain rations for energy, but an accurate appraisal of hays is essential (see the following).

PROTEIN FEEDING

Protein requirements of horses vary from birth to maturity but are little affected by amount of work. Some research indicates that excess protein adversely affects calcium availability.

The importance of different protein levels in diets of horses is emphasized in table 1. Three-month-old foals required diets with 18% crude protein compared with a 8.6% requirement for mature horses. Obviously foals that eat grain rations for adult horses from their mothers' feed boxes are not well fed as compared to those consuming a well-formulated, foal creep ration. Unless the owner has a large number of foals, commercial creep rations are recommended because of convenience, quality, and cost that would be incurred in mixing a custom ration that might be stale before it was used.

Weanlings grow well on 16% crude protein rations; unfortunately, few get them. The standard-quality oat and timothy hay diet does not supply sufficient protein or minerals for weanlings. Weanlings are best fed a formulated (home or commercial) ration of 16% crude protein, .70% Ca, and .50% P, with equal amounts of legume hay and free-choice, high-phosphorus minerals.

COMPUTER RATION "CHECKING" AND BALANCING

The optimistic nutritionist has reason for rejoicing because he sees unlimited opportunity for routine personal-computer use in ration "checking" (for adequacy) and formulation (for balance) by horse owners, feed stores, veterinarians, and county extension agents. Indeed, computer service is likely to be available to most horsemen in a decade or less. The feed shopper will be able to ask the dealer's computer about how the grain feed he is considering will meet the requirements of his animals (based on age, production, reproduction, etc.) when fed with home-grown roughages. In answer, words will race across the screen to compare nutrients furnished in the feed with those required by the horse, with excesses or deficiencies underscored. If deficiencies surface, the computer will respond (upon request) by offering suggestions on how to correct them.

Almost all state extension specialists and many county extension offices will soon have this capability for use by clientele, either by a visit to their office or by telephone. When extension workers or veterinarians suspect ration deficiencies on a farm or ranch, they can call from the horse owner's phone and get an immediate response from a computer in their office. This service is now offered free to horse owners in Missouri and on a consulting basis out of the state.

The pessimist will ask, "Why will a chain-smoking horseman be concerned about his horse's diet when he never takes advantage of free blood pressure checks and his genetically susceptible wife avoids routine physicals?" He can also point out that "a little knowledge is a dangerous thing" and that a computer magnifies this danger. He is correct on both counts. Further, the computer can only make calculations that have been known for decades yet have been little used by horsepersons and educators.

Growers and finishers of production-food animals have taken advantage of scientific feed formulation for years to remain competitive. These animals are asked for peak performance for only a small part of their potential life span, but horses are expected to remain sound and perform well for a lifetime. Attention to proper feeding is now easier to practice and will probably grow steadily more important as horsekeeping costs increase.

The major advantages of computers are the speed of calculations and convenience in ration work. Pounds, parts, or percentages of feeds fed can be easily entered into the computer for a quick "check" for adequacy. Many small computers now in the home have check capabilities with simple software programs. However, machines capable of both checking and balancing are more expensive and nutritional judgment is necessary for ration formulation.

A thorough study of the composition of various feeds, nutrient requirements of different horses, and practice in ration checking and balancing will make you a better feeder and result in sounder, better performing, and longer-lived horses.

CUSTOM RATION BUILDING

You can usually formulate a ration as good or better than what you can buy in your locality, if your feed dealer has a variety of feed ingredients and adequate machinery with which to prepare them.

You can give yourself the following quiz, mostly about minerals, to judge whether you want to learn more about checking and balancing.

THE HORSEPERSON'S MINERAL-FEEDING-PRACTICES QUIZ

Under practical farm and ranch conditions, many young horses are subjected to poor mineral-feeding practices. This is because minerals are not plentiful in standard feeds that young horses consume, requirements are high for young horses, mineral ratios are often incorrect, and minerals are often excessively and(or) improperly supplemented by the horse feeder. Further, the effects of poor mineral feeding are not readily apparent and often result in irreversible damage to bones that appears as lameness in later life.

Finally, most owners think "my horses are in good condition or even fat, so their mineral needs have been met" when in reality the fastest-growing horses may be in the most trouble because of high-carbohydrate, low-mineral diets.

In traveling from farm to farm and evaluating different feeding programs, I have encountered wide variations in mineral feeding practices. These are often supported by strong feelings, high costs, and great confidence that each practice used is essential to good performance, normal health, and well-being of the horse. To reinforce these beliefs, about 25% of horse magazine advertising is devoted to products that "give your horse that extra edge" over its competition.

Listed below are 20 good and bad practices often found on horse farms and ranches. Give yourself this quiz and compare results to answers immediately following the questions.

QUIZ (Answer "True" or "False.")

1. ____ Quality oats and excellent timothy hay make a good ration for stabled weanlings.
2. ____ A bright legume hay (alfalfa, red clover, etc.) fed with oats as the grain would more nearly meet the mineral needs of weanlings than would oats and timothy hay in question 1 above.
3. ____ Adult horse grain feeds from reputable companies, fed with good nonlegume hay (timothy, brome, orchard grass, etc.), are sufficient for weanlings when a free-choice mineral block is fed.
4. ____ Horse feeders, like swine and poultry feeders, need a different grain ration for different age groups, i.e., creep ration, growing ration, adult ration.
5. ____ Since weanling grain feeds are expensive, it is a good practice to "cut" them in half (add) by mixing with homegrown oats or corn.
6. ____ It is a good idea to feed mixed hay (half legume, half nonlegume) or straight legume hay with equal parts of commercial and homegrown grains.
7. ____ Horses of all ages usually eat whole, homegrown grains readily. A hard-working adult horse would be well fed with one-third by weight nonlegume hay and two-thirds equal parts of oats and corn.
8. ____ Mixed-grain rations require approximately 20 lb of feeder-grade limestone per ton to balance the mineral ratio.
9. ____ Most agree that "The larger the weanling, the better it shows and sells". To raise a larger weanling, heavy grain feeding is a must, i.e., a ratio of 4 parts grain to 1 part hay, essentially fed free-choice.

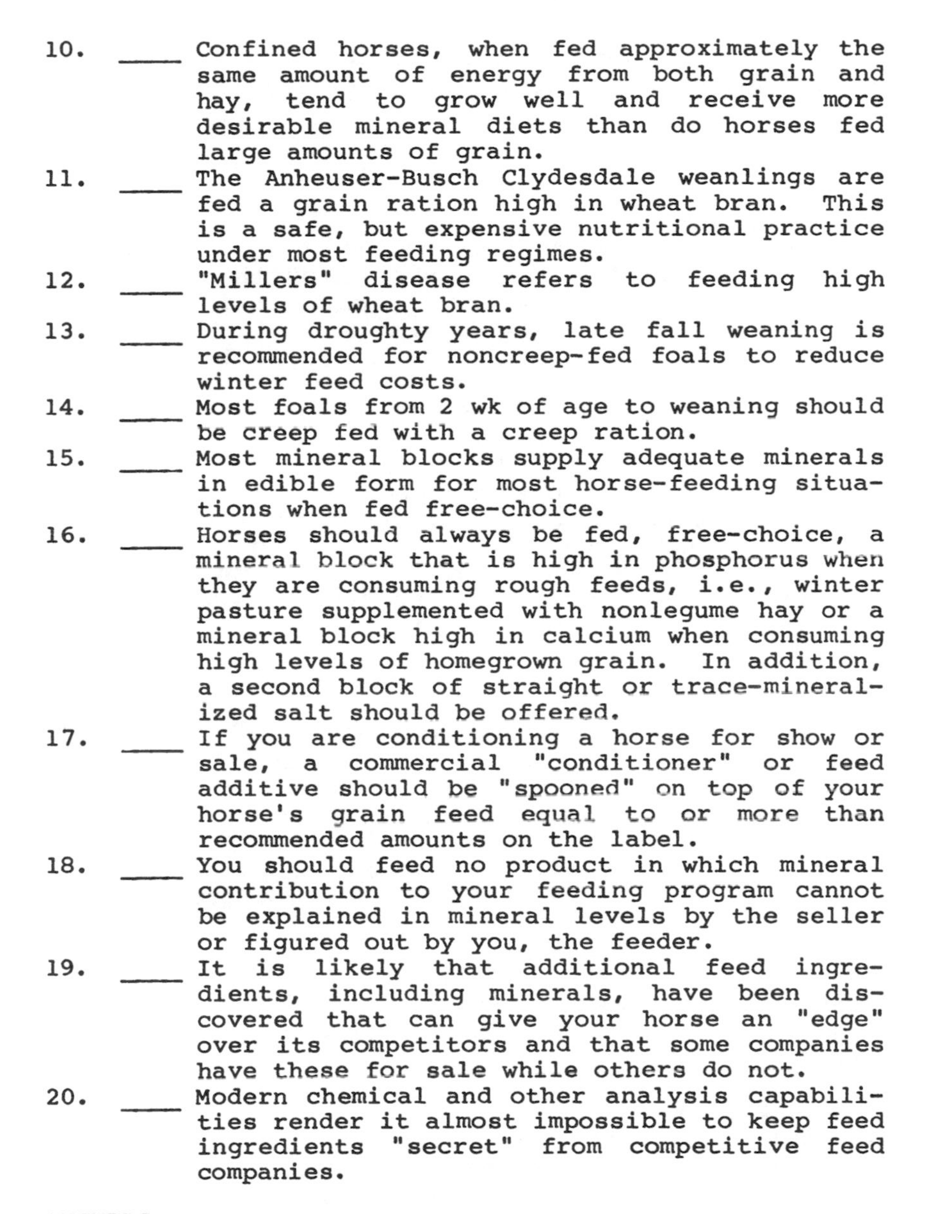

10. ____ Confined horses, when fed approximately the same amount of energy from both grain and hay, tend to grow well and receive more desirable mineral diets than do horses fed large amounts of grain.
11. ____ The Anheuser-Busch Clydesdale weanlings are fed a grain ration high in wheat bran. This is a safe, but expensive nutritional practice under most feeding regimes.
12. ____ "Millers" disease refers to feeding high levels of wheat bran.
13. ____ During droughty years, late fall weaning is recommended for noncreep-fed foals to reduce winter feed costs.
14. ____ Most foals from 2 wk of age to weaning should be creep fed with a creep ration.
15. ____ Most mineral blocks supply adequate minerals in edible form for most horse-feeding situations when fed free-choice.
16. ____ Horses should always be fed, free-choice, a mineral block that is high in phosphorus when they are consuming rough feeds, i.e., winter pasture supplemented with nonlegume hay or a mineral block high in calcium when consuming high levels of homegrown grain. In addition, a second block of straight or trace-mineralized salt should be offered.
17. ____ If you are conditioning a horse for show or sale, a commercial "conditioner" or feed additive should be "spooned" on top of your horse's grain feed equal to or more than recommended amounts on the label.
18. ____ You should feed no product in which mineral contribution to your feeding program cannot be explained in mineral levels by the seller or figured out by you, the feeder.
19. ____ It is likely that additional feed ingredients, including minerals, have been discovered that can give your horse an "edge" over its competitors and that some companies have these for sale while others do not.
20. ____ Modern chemical and other analysis capabilities render it almost impossible to keep feed ingredients "secret" from competitive feed companies.

ANSWERS

Odd numbers are false, even numbers true. If you missed no more than three questions, your mineral feeding program is probably good. If you missed five or six, you passed, but if you missed ten or more, your horse is probably in trouble and you will profit most from the following discussion.

Each of the even-numbered questions relates to the previous odd-numbered question, thus they are discussed together.

DISCUSSION OF QUESTIONS

Questions 1 and 2:

1. Oats and timothy hay furnish only 30% of the calcium and 50% of the phosphorus that weanlings require for healthy bone growth. Further, the Ca to P ratio is reversed, yielding only 81% as much Ca as P, when it should be 1 1/2 or 2 times more Ca than P.
2. Substitution of alfalfa for timothy will meet the Ca need and correct the Ca:P ratio (3:1), but will leave the ration below the optimum P level. Free-choice feeding of a high P mineral supplement would help, but some foals would not eat enough to meet their needs. Better results would be gained from feeding "built-in" P in a commercial or custom grain mixture or by feeding 20% bran (by weight) in the grain mixture.

Observations:

- Grain is a rather good P source but disastrously low in Ca with a reverse Ca:P ratio (5:6).
- Nonlegume hay, regardless of quality, is just an average source of both Ca and P--low in both for weanlings but sufficient in both for adult horses. A ratio of 1.5:1 to 2:1 is good.
- Alfalfa is a good source of Ca and an average source of P, making it an ideal supplement for grains. Its Ca:P ratio is about 6:1.
- Mineral requirements for weanlings, especially those from larger breeds that exhibit rapid bone growth, are high (some good horse nutritionists use 1% Ca and .9% P) with correct ratio crucial for normal bone growth and density.

Questions 3 and 4:

3. Many horse owners feed the same grain mixture with nonlegume hay to horses of all ages. The combination of these two feeds is too low in both Ca and P for weanlings. (See next answer.)
4. Swine and poultry producers routinely use different grain mixtures for different age groups and consider the practice essential. This is because grain feeds for young animals are formulated with higher protein, Ca, and P levels than are adult grains. When feeding horses of different age groups, horse feeders should follow the lead of swine and poultry feeders by using at least two types of grain mixtures from a reputable company

for their horses, or they can have custom rations mixed.

Questions 5 and 6:

5. "Cutting" or adding homegrown grain in equal parts or more to commercial horse grain is standard practice in many stables. "Cutting" the commercial ration with corn reduces protein and Ca levels and unbalances the Ca:P ratio.
6. The use of good legume hay, instead of nonlegume hay, with "cut" commercial grains restores the Ca and most of the protein lost as a result of this practice.

Observations:

- Almost any grain or commercial grain mixture fed with equal parts legume or nonlegume hay will meet the protein, Ca, and P requirements for adult horses; but unless grains are specially formulated for weanlings, most feeds fall far short of all three of these nutrients.
- "Cutting" of commercial or custom-mixed grains reduces protein and Ca but leaves P essentially unchanged, resulting in a bad Ca:P ratio.
- Feeding a good legume hay largely corrects the effect of "cutting."

Questions 7 and 8:

7. There is nothing wrong with feeding homegrown grains so long as they are supplemented with Ca. However, this ration would be very low in Ca and would have a negative Ca:P ratio. Mineral-poor diets for adult horses are almost always the result of high grain intake with nonlegume hay. Supplementing the grain mixture with limestone or feeding mixed hay easily corrects the deficiency.
8. Twenty pounds of limestone will not always raise the Ca levels in a ton of mixed grains to meet foals' needs, but it will always improve its Ca and P ratio.

Observations:

- Unless you feed homegrown grain mixtures with mixed or legume hays, 20 lb of limestone per ton should be added to them. If this grain mixture is to be fed to young horses with higher Ca and P requirements, 30 lb of limestone and 15 lb of dicalcium phosphate will improve it. However, do not feed excessive levels of these minerals to young or adult horses.

Questions 9 and 10:

9. Heavy grain feeding poses many unique problems; because Ca, trace minerals, and (to a degree) protein come from hay, these nutrients often are

deficient. Fiber, the stimulant for normal digestion, may be insufficient, resulting in colic. A high carbohydrate (grain) diet causes excessive weight from fat storage in some foals--often more than immature bones can carry.

10. While more condition can be gained quickly with high levels of grain feeding, the risk is seldom worth it. When horses of all ages receive half of their energy from good hay or grazing, few nutritional problems arise. Unless intake is a problem with young horses or those working hard, 2 lb of good hay are almost always nutritionally better than 1 lb of good grain.

Questions 11 and 12:

11. A few people are of the old "bran feeding" school. Wheat bran is an excellent horse feed when appropriately used. It probably can be used best with stressed horses to aid normal digestion. Under most feeding regimens, adequate calcium supplementation is neglected.
12. In the "old" days, millers of flour were often given the bran portion of the wheat grain that, when fed to their horses, resulted in "Millers" disease or "big head" (nutritional secondary hyperparathyroidism). Low blood-calcium levels trigger the parathyroid to withdraw Ca from bones. The connective tissue that replaces the calcium in facial bones causes a visible enlargement.

Questions 13 and 14:

13. Early weaning is indicated in a drought season because mares are essentially dry after midsummer. Forages are also very low in quality.
14. Almost all foals should be creep fed because grains are inexpensive and the foals experience less setback when weaned. Some mares never milk well and their foals need supplemental feed.

Questions 15 and 16:

15. Mineral blocks vary in composition and Ca:P ratios. Some have high percentages of salt and others contain high levels of protein to encourage consumption. Select them carefully to fit your horse's needs.
16. This question describes conditions under which different blocks are fed. Remember, the three important minerals are Ca, P, and salt. A sweating horse may lose 2 oz to 3 oz of salt daily at hard work. It should not be forced to eat from a salt source that also contains Ca and P.

Questions 17 and 18:

17. Spooning "something magic" on top of horse grain feeds has been estimated to cost over $100 per show and racehorse per year. Results are hard to measure, but research with horses eating good diets shows no beneficial effects of the "magic" feed.
18. Randomly spooning or force-feeding trace minerals is dangerous because excess iodine, zinc, manganese, and copper are toxic to horses. Supplemental trace mineral feeding seldom is needed when horses graze or eat quality hay.

Questions 19 and 20:

19. About 1 in 10 pages in major horse magazines advertises feed products that give your horse an "edge." The cost of using these products on a daily basis ranges from 10% to 25% of the feed cost of your horse. The real "edge" comes from a sound basic-feeding program.
20. Secrets are easy to solve with modern chemical analysis methods.

RATION BALANCING

Of course the easiest way to check and balance horse rations is by computer (as explained earlier) or by use of a trained nutritionist. However, learning the process by hand calculation gives one an understanding not to be found by any other method. If you learn and practice the mechanics of ration balancing and routinely chemically analyze the feed that your horses eat, both you and the horses will profit from this knowledge and practice.

Steps in percentage ration balancing:

- Tabulate the nutrient requirements of horses you will feed.
- Choose the feeds you will use.
- Calculate the pounds of grain and protein supplement to give the desired grain-protein ratio.
- Calculate quantities of Ca and P present in the grain-protein and add minerals to meet the requirements.
- Add vitamins.

Examples:

Let's build a custom grain ration for weanlings whose requirements (table 1) show a need for 9.90% of TDN, 16.0% crude protein (CP); .70% Ca; and .50% P.

When equal parts corn and oats are used with soybean oilmeal, how many pounds of each must you tell your feed dealer to mix together to yield a 16% CP grain mixture?

The Pearson Square for Grain-Protein Ratio

- Average the protein in corn and oats (table 2), then use the Pearson Square to get the correct grain-protein ratio.

	CP %
Corn	9
Oats	13
Total	22
Avg	11%

- Get the percentage protein in soybean meal (44%) from table 2.

Set up the square and solve as follows:

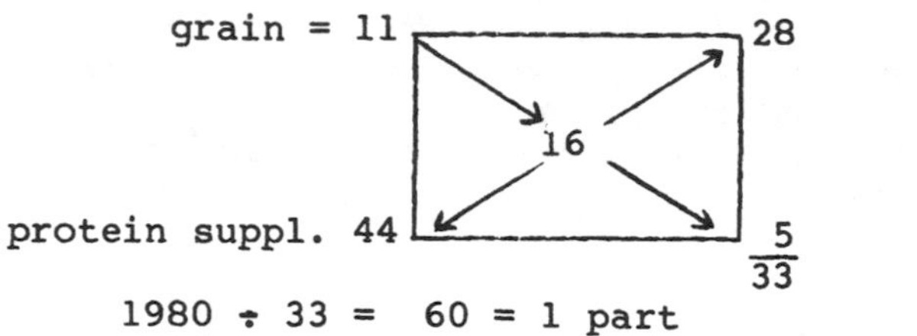

1980 ÷ 33 = 60 = 1 part
5 x 60 = 5 parts of soybean meal or 300 lb
28 x 60 = 28 parts of oats and corn or 1680 lb
1680 ÷ 2 = 840 lb each of corn and oats

Explanation: The average protein (11%) in equal parts corn and oats was placed at the upper left of the square, protein in soybean meal (44%) at the lower left, and desired protein percentage (16%) in the middle of the square. Cross subtract in the direction of the arrows to get the feed ratios (16 - 11 = 5 or five parts of soybean meal in the final mixture; and 44 - 16 = 28 parts oats and corn). The ratio of 28 parts oats and corn and 5 parts soybean meal will give a 16% crude protein grain mixture. Since horse feeds are usually mixed in ton mixers, consider using 1980 lb in your mixture. It could be any amount. When 1980 is divided by 33 parts, you get the figure of 60. Thus, of the 1980 lb, 60 x 5 parts = 300 lb must be soybean meal to yield a 16% CP feed; 60 x 28 = 1680 lb of corn and oats or 840 lb of each grain.

While this is an accurate way to determine the amount of protein supplement to add to grain, it does not account for an enormous deficiency of calcium and, to a lesser extent, of phosphorus.

If you do not supplement the Pearson Square calculations with these two minerals, it can be a dangerous method. Study the following information on calcium and phosphorus supplementation.

Determining pounds of Ca and P in the grain protein: Consult table 2 for percentages of Ca and P in the three grains. Change percentages to pounds by moving decimals two places left. Set up and multiply as below:

		% Ca	Lb	% P	Lb
Corn, 840 lb	x	.0002 =	.168	.0026 =	2.184
Oats, 840 lb	x	.0009 =	.756	.0032 =	2.688
SBM, 300 lb	x	.0030 =	.900	.0064 =	1.920
			1.824		6.792

Ca:P Ratio = 6.792 ÷ 1.826 = 1:3.720 (negative ratio)

The calculations yield actual pounds of Ca (1.82) and P (6.79) in 1980 lb of this mixture and underscore the shortage of calcium and the bad Ca:P ratio.

Correct Ca:P ratios are 1.1:1 minimum, 3:1 maximum, and 1.5:1 probably optimum for weanlings, depending upon the type of hay being fed.

In no case should intake of P exceed Ca in horse diets. Following are some symptoms often seen in horses with mineral deficiencies or imbalances.

Symptoms of Ca and P deficiencies or imbalances:

- Shifting lameness
- Erect pasterns, swollen ankles
- Crooked bones
- Slow epiphyseal closure
- Nutritional secondary hyperparathyroidism (big head)

When 20 lb of feeder-grade limestone are added to a ton of grain, the Ca:P ratio is greatly improved--but let's get it right by determining the weanling's needs and how many pounds of dicalcium-phosphate and limestone (table 2) are required to meet them.

- How many pounds of Ca and P in 1980 lb will yield .70% Ca and .60% P?

Ca = 1980 x .007 = 13.86 lb
P = 1980 x .005 = 9.90 lb
Both Ca and P are deficient in the ration.

- How much dical is required to yield a .50% ration?

$$P = \frac{\text{need - in feed}}{\text{\% P in dical}} = \frac{9.90 - 6.79}{.19} = \frac{3.11}{.19} = 16.37 \text{ lb}$$

16.37 lb of dical added to the 1980 lb of grain and soybean meal yield .50% P.
(The reason P was figured first is because its source, dicalcium phosphate, contains Ca as well as P; and the Ca source, limestone, contains no P.)

Ca in 16.37 lb (of dical) x .22 = 3.60 lb

- How much limestone is required to yield a .70% ration?

Ca in grain and protein	=	1.82 lb
Ca in dical	=	3.60
Total		5.42

$$Ca = \frac{\text{need} - \text{in feed}}{\text{\% in limestone}} = \frac{13.86 - 5.42}{.33} = \frac{8.44}{.33} = 25.58 \text{ lb}$$

- Molasses or none?

  Liquid molasses helps control "fines" and keeps mineral and vitamin additions to the ration in suspension. Molasses mainly contributes carbohydrate to the diet. If small amounts of grain are fed, horses eat it well without molasses. However, with foals molasses is well received.

  If under 5% of molasses is used, good mixing equipment is required to get distribution, and over 10% may cause loose bowels. About 7.5% is often recommended, which totals 150 lb/ton of grain.

- Adding vitamins

  Most rations should be fortified with fat soluble vitamins, especially vitamins A, D, and E; from 3,000 to 5,000 units of vitamin A per pound of mixed feed is standard; or 1 lb to 2 lb to the ton of an A, D, and E premix. B vitamins are synthesized in abundance in healthy horses and may be omitted, except for horses under stress.

- The ration is completed by .5% salt (10 lb/ton)

- According to our calculations, the following amounts and kinds of feeds will yield approximately 16% CP, .70% Ca, and .50% P.

Ingredient	Lb
Oats	840
Corn, cracked	840
Molasses, blackstrap	150
Soybean meal, 44%	300
Dicalcium phosphate	16.37
Limestone	25.58
Salt	10
Vitamins	2
(4,000,000 IU Vit. A/lb)	2,184

The scholar will observe that we have been adding to our original 1980 lb of grain and protein supplement and have actually reduced the percentages of protein, Ca, and P by "dilution" that is, by raising the weight to 2,184 lb. Without adjustment, this method is insufficient for research or comparison of feeds or for animal-growth rates. However, it is quite adequate for practical horse feeding because horses are not fed grain at maximum intake, and more or less quantity can be used depending on need of the animal.

If you prefer higher accuracy, add 10% for molasses, minerals, and vitamins and solve on that basis; i.e., if you add 10% to a ton, solve for percentages of 2,200 lb of feed instead of 2,000 lb.

CHEMICAL ANALYSIS

The final step in the sophistication of horse-feeding management is the judicious use of chemical analysis. A well-drawn sample, analyzed by a reputable laboratory, will identify errors in 1) formulation, 2) feeds used, and 3) feed formulation procedures. Or, it will reassure the horse owner that the feeds used meet the nutritive needs of his animals.

Random grouping of grains and hays seldom fully meets the requirements of young, growing horses. Inadequate diets should be identified by chemical analysis before the damage is done. Feeds vary in nutrients by up to 100%, depending on quality, location, and care in preparation and handling. To purchase or use them without knowledge of their composition may be uneconomical, unsatisfactory, and even risky.

Finally, most reputable companies have good quality-control standards, but the best of companies may experience machine or human error in feed formulation. If you have reason to doubt, check it.

Good ration analysis services are not available to all horsemen. Check with your county extension service or state extension horse specialist about such service. If there is a horse nutrition consultant available, he or she will know of such service. Finally, contact this author through Winrock International and information on chemical laboratories will be supplied at no cost.

The horse plagued by disease, parasites, and humans has enough problems. When he is well fed, many of his problems diminish.

REFERENCES

Bradley, M. 1983. Feeding growing horses for soundness. In: F. H. Baker (Ed.) Stud Managers' Handbook, Vol. 18. p 256. A Winrock International Project published by Westview Press, Boulder, CO.

Bradley, M. 1983. Quality and quantity control in horse feeding. In: F. H. Baker (Ed.) Stud Managers' Handbook, Vol. 18. pp 274-277. A Winrock International Project published by Westview Press, Boulder, CO.

Henneke, D. R., G. D. Potter, J. L. Kreider, B. F. Yeates and J. Householder. 1983. Influence of body condition on reproductive performance of mares. In: F. H. Baker (Ed.) Stud Managers' Handbook, Vol. 18. pp 133-144. A Winrock International Project published by Westview Press, Boulder, CO.

NRC. 1978. NRC Nutrient Requirements for Horses. National Research Council, Washington, D.C.

NRC. 1973. NRC Nutrient Requirements for Horses. National Research Council, Washington, D.C.

Part 10

GENERAL HEALTH AND MANAGEMENT

38
THE SUMMER ITCH

William C. McMullan

The most common skin disease in horses is the "summer itch." Also called the "Spanish itch," it is often referred to by owners and veterinarians alike in unprintable language. Part of the problem is failing to realize that it is more than a single disease. There are several causes, any one of which can be more important to a certain horse at a certain location.

A typical case goes as follows: at 4 yr to 5 yr old, (but sometimes as early as 2 yr) you notice hair coming out at the mane, ears, face, under the belly, and in some cases at the tail head--not in all these places and not all at once, but in small patches, some of which when rubbed may get raw for a few days and then scab over. The horse will gnaw at some of the places and rub the others. As the summer wears on, the problem gets worse; so bad that by July you are ashamed to take him to the show. Treatment is most often not very effective, even though several things were tried. Finally, cool weather comes and the itching is less severe and hair comes back in most of the places except under the belly. To your disgust, the problem comes back again the next summer, usually worse than before. After two or three bouts, it may even last into the winter, especially a mild winter. And by now there are places which do not grow hair anymore, and the skin on the neck is thickened and wrinkled.

Since this condition is so seasonal, there are several things which logically are possible causes. Most important are biting insects. Biting gnats (midges, punkies, no-see-ums, and sandflies are synonyms in various localities) are small 1 mm to 3 mm insects which bite almost exclusively at the mane, under the belly, ears, face, and at the tail head (certain species only). To most horses they are just a nuisance, but to a small percentage they cause a serious allergy. In Australia where it was first discovered, the allergy is called the "Queensland Itch." Usually only one or two horses in a herd will be affected, despite the fact that the other horses frequent the same places and are bitten frequently.

These pests are so small that they can go through screens unless the screens are treated with a repellent. They are so small that many people fail to realize they can be a problem for that certain horse allergic to their bite. When a horse is allergic to a certain insect, it does not take a swarm to cause trouble. One or two are enough to start generalized itching in places the fly did not bite.

Another family of biting insects that can cause the "summer itch" and in rare instances death, is represented by the black fly *(Simulium* spp.) and buffalo gnat. The southern buffalo gnat *(Cnephia pecuarum)* is an annual spring pest in the lower Mississippi Valley. Once during a "swarm" year, thousands of mules and cattle were killed in Louisiana, Arkansas, and Mississippi. An animal must be attacked by a swarm of black flies for symptoms of the heart and respiratory systems to show up. Most of the time, skin lesions are produced by the bite and poison injected. The udder, scrotum, prepuce, inner surface of the thighs and forelimbs, the chest, insides of the ears, underbelly, and natural body openings are favored biting sites. The bites cause itching, and with scratching and rubbing they become weeping sores with small raw spots mixed in the swollen skin. Dried flakes of blood may be seen in the ears.

Breeding places for the black flies are usually the edges of moving water, but some species have adapted to still water. New hatches and abundance of flies are often related to heavy rains and river overflows that occur several times a year.

Horn flies *(Haematobia irritans)* attack in clusters, always feeding with their heads pointed towards the ground, a habit which helps to identify them. They like the neck and withers, causing the hair to fall out in patches, the size of which is determined by the number of flies. Their other favored sites are so unusual that lesions there tell you horn flies are the culprit. For reasons unknown to me, they line up like columns of soldiers feeding on a strip 1 in. to 2 in. wide at the very middle of the underbelly or down the middle of the back from the loins to the withers.

Stable flies *(Stomoxys calcitrans)*, also known as dog flies, have such a painful bite that they have been a formidable deterrent to the tourist development of beach areas. They bite on the legs and occasionally under the belly. Horses stomp their feet a lot when attacked. The bite sometimes becomes infected by bacteria or fungi; but most often there is little hair loss, just a high nuisance factor.

Horse flies and deer flies, though large, are not thought to be a common cause of skin lesions. They are important in the transmission of disease, but usually just cause a temporary whelp at the site of a bite on the neck or over the ribs. Rarely the whelp may turn white.

Face flies *(Musca autumnalis)* are present in almost all states except in the very far South and Southwest. As their name implies, they feed on mucus at the nose and around the eyes, causing inflamation of the mucuous membranes lining the eye (conjunctivitis). They also feed on fresh bites of bloodsucking insects anywhere on the face. This irritation leads to rubbing and loss of hair. Along with horn flies, their breeding site is in fresh cow manure.

Many methods of treatment for the face fly are available. In addition to the usual sprays, wipe-ons, gels and fly sticks (like deodorant sticks), fly bonnets and halters with string curtains in front have been designed with the face fly in mind.

A rare cause of summer itching in horses, though common in cattle and dogs, is mange. One species attacks primarily the legs (chorioptic); another primarily the mane, tail, forelock and feathers at the fetlock (psoroptic); and two have a more generalized distribution (sarcoptic and demodectic). These are diagnosed by seeing the mites under the microscope in a sample of skin scraping. I repeat--mange is rare in U.S. horses.

The best defense against the "insect" type of "summer itch" is to keep the horse stabled day and night in stalls with 32-mesh screens. Check with your county agent as to the choice of insecticide spray for the stall and screens. Generally, application of residual type sprays at 2-wk intervals is necessary. What worked last year may have lost its effectiveness this year. An individual battery-operated stall mister (like the ones you see in restaurants) will help. A mechanical timer pushes a lever on a spray can to release a dose of short-acting pyrethrins about every hour or so. This technique (about $40) is most effective in stalls protected from the wind. Large barns may have a pipeline system instead of the individual "boxes."

Net-type stable sheets sold as fly sheets are somewhat helpful and can be improved with the addition of solid cloth at trouble sites (for example, over the mane).

Cattle insecticide ear tags can be braided into the mane and tail or attached to hair with "twistems." Some owners put them around the horse's neck, using a breakable nylon string. The tags are inexpensive and are of some benefit.

Daily removal of manure and soiled bedding is extremely beneficial in controlling those species which breed there, such as stable flies and houseflies. It is best to spread manure on a vacant field for rapid drying. Application on a pasture in use will add to the parasite load and is not advised. If a compost pile must be used, it should be covered or sprayed daily. A "feed-thru" larvicide is available commercially. Stirofos is mixed in the feed and 24 hr later kills fly larvae in the manure. Another technique used in controlling the "manure breeders" is predator flies. These miniature, stingless wasps spend most of their life searching manure piles for fly larvae, their main diet. In fact, they seldom leave the manure and are thus of little consequence to man or beast. The predators should be scattered before the fly season and reinforced a couple of times during the following months. Insecticide can't be used close by, but fly traps and electric "zappers" work well with this program or with any other.

Repellents applied to the horse would be the answer if only they lasted long enough. Most of the commercial formulas (pyrethrin, piperonyl butoxide, and butoxypoly propylene glycol) are very similar and will not protect against gnats *(Culicoides)* for more than 6 hr. Sprays and daily repellent applications at riding time are adequate for the average horse, but not the fly-sensitive horse. One word of caution: a few horses are allergic to pyrethrins; so, spot-test a 6 in. x 6 in. patch on a non-tack contact area and wait 24 hr before an all-over application. Also be wary of sprays designed for beef cattle; some of these will blister a horse. As always, use as directed!

A new insecticide is available, a synthetic pyrethroid called "permethrin," possessing a wide range of activity and a wide safety margin. On the average horse and barn, spray as directed. Duration of action is longer than most. For the allergic horse, I use additional daily spot-treatments of trouble areas (like mane and tail, for instance).

Despite your best efforts against the pests, the allergic horse may turn out just like the dog allergic to fleas--one bite will make him itch for a week. And it's next to impossible to eliminate 100% of your insect problem. Thus for the allergic horse that must be maintained in a "show coat," the most effective treatment is to block the reaction caused by the bites with an orally administered cortisone such as Prednisolone. It is given daily a.m and p.m. until the itching stops and then is reduced to the smallest every-other-morning dose that will prevent itching. This treatment schedule can be used safely all summer long. Injectable forms of cortisone are available, but their use is more likely to cause side effects unless closely monitored. Specific dosages should be tailored to your situation by a veterinarian. Cortisone drugs should not be used in the face of infection, serious wound healing or fracture, or pregnancy.

In an experimental treatment program now underway at Texas A&M, a series of injections of ground and purified *Culiciodes variipennis* is given in hopes of desensitizing the horse by the formation of blocking antibodies. This is fairly successful in people and dogs. I am not expecting a high success rate, but when it works it certainly will be more desirable than drug treatment.

Another form of "summer itch" is caused when microscopic forms of the parasite *Onchocerca cervicalis* die in the skin and set off a reaction which produces itching. The locations are much the same as those caused by insects--the face, upper neck, withers, and underbelly. This is not unusual when you consider that part of their life cycle depends on them being picked up by biting gnats.

At least half of the horses in this area have microfilarie, e.g., *Onchocerca* spp, in their skin, but most of them do not have the itch. Why? Several factors influence which horses will be affected: the degree of infection, exposure to heat and sunshine, and individual allergy. A similar situation exists in dogs; many have mange mites in their skin, but only a few have skin lesions.

The lesions caused by the death of microfilarie may resemble ringworm or fungus infection. They are worse in summer and usually clear up in winter. At present, there is no safe drug that will kill the adult worm except possibly Eqvalan, the injectable wormer. We'll know more about that next year. Diethylcarbamazine and livamisole will kill the microfilarie in 4 days to 7 days, but more migrate to the skin in a few weeks. Consequently treatment is only temporarily effective. There will still be periodic episodes of itching. These can be relieved by cortisone treatment.

Fungus infections--primarily girth itch--though much more common in the fall and winter, can be a cause of summer itch. I would not expect it to recur each year. "Fungus" is greatly overdiagnosed.

Irritant weeds like small-headed sneezeweed, spider nettle, and others cause a dermatitis of the face and muzzle. The skin is thickened and chapped and may crack open if the irritation is severe. Itching gives way to pain as the dermatitis progresses. People and dogs have skin allergies caused by diet and exposure to various pollens and other materials. These are not well-documented in horses, but undoubtedly they exist. Bermuda pollens bother some dogs, why not some horses? Weed pollens bother people and dogs, why not some horses? Dietary-related skin problems are difficult to prove, but changing to another diet may bring relief. Horses have been reported to be allergic to wheat, oats, barley, bran, clover, molasses, and other feed stuffs.

Allergy skin testing is just beginning to be done in horses. This offers hope in working on this problem.

Intense heat and humidity contribute to the "summer itch" and little can be done about this, short of air-conditioning or moving to Montana.

39

THE TREATMENT OF NAVICULAR DISEASE

William C. McMullan

If you have a horse with navicular disease, you will be interested in this paper; if you do not have one, wait until next year--it's the most common cause of lameness.

The navicular bone (figure 1) is small (2 1/2 in. long, 1/2 in. wide, 1/4 in. thick) and somewhat boat shaped--in fact, the literal translation from Latin and French means "little ship." It is located in the heel of the foot between the coffin bone and the deep flexor tendon. There is a bursa (cushion) between the navicular bone and the tendon. (People have a navicular bone, but it is located at the wrist, which corresponds to a horse's knee.)

DIAGNOSIS

An affected horse usually is only mildly lame at first; some days he is sound, only to be lame the day after a hard workout. Stumbling is a nuisance factor in navicular disease. It happens because the horse is sore in the heels, and so he tries to land the foot, toe first, instead of heel first, as is normal. In addition, in advanced cases, adhesions between the navicular bone and deep flexor tendon restrict normal extension of the toe and contribute further to the stumbling problem. Most of the time, both front feet are affected; although rarely, this disease may occur in the rear feet. If both feet are equally affected, the horse will have a shorter stride, less flexion of his joints and will put his feet down more carefully (it is called "egg walking" if his pain is bad). If, instead of trotting him in a straight line, you put him in a circle to the left (counter clockwise), he will favor his left front foot. You will see his head go down when the right front foot hits the ground and then come up when the left front foot hits. If you trot him in a right circle, he will favor his right front foot.

He will flinch when hoof testers, which look somewhat like a pair of giant channel-lock pliers, are applied with pinching using maximum force. One of the pincers is placed at the middle of the frog and the other about half-way up the wall at the heel-quarter junction. If the horse continues to work, with relief afforded by pain relievers such as phenylbutazone, the disease will progress until finally he will limp despite his "pills." A very lame navicular horse may make you think he is foundered because of his reluctance to trot and his stiff way of going.

Navicular disease is diagnosed by a combination of appropriate history, clinical signs, response to "heel blocks," and x-ray findings. In some cases, one or more of these diagnostic features is absent and a firm diagnosis cannot be made.

TREATMENT

The first treatment attempted should be corrective shoeing. Many horses with the disease (or horses just sore in the heels that act as if they had the disease) are low in the heels by nature or design. This causes the tendon to squeeze harder up against the bursa and navicular bone. Trimming excess toe and(or) adding a wedge pad or shoe with swelled heels will help such a horse. Rolling the toe of the shoe allows easier breakover and reduces pain.

Horses with underslung heels (i.e., those with the heel portion of the weight-bearing wall at least an inch forward of a vertical line dropped from the hairline at the bulbs) will probably benefit from an eggbar shoe with a rolled toe.

Horses with average heels are usually shod with another type of bar shoe and rolled toe. The bar (1 in. to 1 1/2 in. wide x 1/8 in. thick) is placed under the middle of the frog, directly below the navicular bone to protect it. The bar should be thinner than the shoe so as not to touch the frog or the ground surface when bearing weight. Do not allow dirt to pack between the bar and the frog; clean it out with a pick or old hacksaw blade. If the horse is very lame due to sole bruising, a pad usually is necessary in addition to the bar shoe.

A lot of mild and(or) early cases will respond to shoeing but, generally, only about 25% of all cases will become sound as a result of shoeing alone.

Additional months or years of service can be obtained by using pain relievers, but you should make every attempt to correct any predisposing factors like low heels; hard, dry feet; shoes that do not allow expansion (nails in the heel, clips, too small). And you should realize that the horse is nearing the end of its useful services.

Juxtabursal injections of Palosein into the fossa (hollow) between the bulbs of the heels at 7-day intervals for 3 to 4 doses resulted in sound horses in 50% of the cases in a trial at the University of Missouri and 33% of the cases in a trial here at Texas A&M. It is a safe treatment but must be repeated. This treatment, as with the pain relievers, does not stop the progression of the disease and eventually will be unable to block the pain.

Feeding of yucca plant extract has been tried as a treatment but documentation as to success rate is not available. I suspect it is very low. The extract is marketed and advertised by Farnam and Millers; it is supposed to work by a cleansing action in the intestine, decreasing absorption of toxins that may cause arthritis and other joint pain.

A navicular vaccine was advertised in the Quarter Horse Journal last year by a veterinarian in New Mexico. He claimed to have discovered a disease-producing bacteria, which he named *Hemophilus navicularis*. For several hundred dollars (depending on distance) he would fly to your address and vaccinate your horse with a "vaccine" that he made from this "bacteria." The "grapevine" had it that some horses did get sound. But when he went into a state in which he was not licensed to practice, it was discovered by laboratory analysis that his "vaccine" was in fact Strain 19 calfhood brucellosis vaccine. This brucellosis vaccine is supplied in many states at no cost to veterinarians for control of brucellosis or Bang's disease of cattle. It is known that the *Brucella* bacteria can localize and cause inflammation in the bursa at the withers (fistula of the withers) or in the bursa of the upper neck (poll evil). Suspicion that *Brucella* might also affect the navicular bursa and cause lameness was reported 40 yr to 50 yr ago in the Veterinary Record published in England. None of the navicular horses I have checked had high (greater than 1:100) blood titers to *Brucella* antigens, which indicates that the presence of this disease was unlikely. I seriously doubt that the "vaccine treatment" would benefit noninfected horses. And because of the secrecy and deplorable, money-gouging unprofessional manner in which the whole matter was conducted, we cannot rely on the success rate claimed.

The injection of corticosteroids (cortisone) into the navicular bursa has been used with some success, but the injection technique is difficult and relief is usually for only a matter of weeks.

Nerving is a surgical procedure whereby the palmar digital (heel) nerves are cut and a length of nerve removed, which numbs the entire heel area including the navicular bone. This should be a last-resort procedure on horses not intended for breeding and for which there is no strong owner-pet sentiment. This is because most of the time the end result, after a year or so (if the horse is worked), is that the horse must be destroyed. The various causes for poor results are:

- In a small percentage of cases, the blood supply collapses and the hoof comes off.
- The deep flexor tendon tears apart when pain no longer helps protect the damaged tissue.

- Puncture wounds in the heel can become advanced before you recognize it because the horse feels no pain and cannot show you.
- About 5% to 20% of the time, a painful nerve tumor (neuroma) develops where the nerve was cut so that the horse becomes lame again shortly after surgery.
- Sometimes the nerves grow back and lameness recurs.
- Surgery is not completely successful because, instead of one big nerve, some horses have many small branches--some of which cannot be found to cut.

Some veterinary surgeons claim a horse stumbles more after his heel is numb, making the horse unsafe--especially for children or use on mountain trails, etc. (While I do not think this is 100% true, it must be considered when you are trying to decide on nerving.) With all these drawbacks, why even consider it? When other treatments have failed and you have a gelding who could perform successfully for your pleasure and(or) profit for another year, then it is a consideration.

Freezing a nerve, called cryoneurectomy, is similar to nerving, except that it is not really surgery. A super-cold probe is held on the skin over a nerve until both are frozen and destroyed. In some racing states, this technique is advantageous in that a horse does not have to be held out of training for the skin to heal, and the horse also can be raced in states that do not allow routine surgical nerving. The main disadvantage of freezing a nerve is that pain is relieved for only a matter of months and the relief is not as complete as with surgery.

"Alcohol blocks" are done at the same sites where cryo or surgery are done. Ethyl alcohol is injected into and around the heel nerves, two nerves on each foot. This numbness will last only weeks to months, if successful at all, thus this is the least recommended of the three "nerve jobs."

I have treated a dozen or so navicular horses with acupuncture. In three of these, the results were almost miraculous; in three others improvement was about 75%; and in the rest of the horses, there was little or no improvement. Needles are placed at certain acupuncture points and these are connected to a battery-operated stimulator. The stimulator is set at a low frequency and rate and used for 15 min, 3 to 4 time a week, for 2 wk to 3 wk or longer, if improvement is slow. Acupuncture treatment of navicular disease is safe but not highly effective; and, as with most other treatments, must be repeated.

After hearing the report by Colles of England (personal communication in December, 1981), we treated 42 horses with warfarin, a blood thinner better known as a rat killer. It kills rats by causing a fatal hemorrhage. To use it in treatment, considerable laboratory evaluation and careful monitoring are essential. Many commonly used drugs cannot be used concurrently without risk of hemorrhage, i.e., drugs such as phenylbutazone, certain wormers and antibiotics, and most pain relievers used in colic. Our success rate with warfarin was 60%, however, the medication must be given for as long as the horse is to be sound. One of the 42 horses died of hemorrhage even though dosage and monitoring were "by the book." Race horses and pregnant mares could not be treated.

Although warfarin was the most effective drug used on navicular disease to date, and it stopped the progression of the disease, a safer drug was sought.

At the British Equine Congress in November 1982, Rose of Australia (personal communication) reported that he was having greater success with a drug called isoxsuprine, which did not thin the blood (thus risking hemorrhage); rather it caused the blood vessels to dilate (vasodilator) and allowed better circulation to the navicular bone. Rose says his success rate was 88% in the first 44 horses treated, with no side effects and no deaths, and other drugs can be given concurrently with this drug. Isoxsuprine is available for human use in tablet form in the U.S. We crush the aspirin-sized tablets, mix with syrup, and give by mouth 30 min before feeding. This is done twice a day for 6 wk, then once a day for 6 wk, and finally every other day for 6 wk. Some horses have remained sound for 2 yr after this treatment, but some became lame again within 2 mo and were successfully retreated.

The results of isoxsuprine treatment of 57 horses at Texas A&M during the first 6 mo of 1983 are unavailable at this writing but will be reported at the Stockmen's School in January 1984.

The use of isoxsuprine looks like the best treatment yet. It should be combined with proper shoeing and a reasonable exercise program that promotes good circulation.

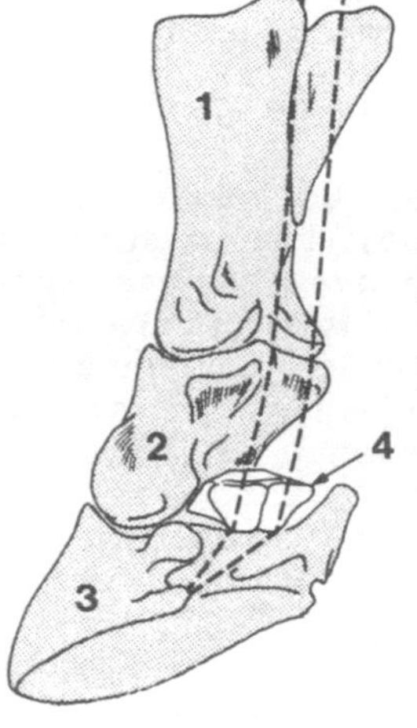

1 - long pastern bone
2 - short pastern bone
3 - coffin bone
4 - navicular bone

Dotted lines indicate position of deep flexor tendon

Figure 1.

40

APPLICATION OF ACUPUNCTURE IN TREATMENT OF ANIMAL INFERTILITY

Qin Li-Rang,
Yan Qin-Dian

Infertility among domestic animals is one of the great economic problems in animal production, and its incidence is increasing with the intensive livestock husbandry. Solving the infertility problem is a challenging task for the veterinarian. For the treatment of animal infertility, there are several methods. But the Chinese traditional acupuncture has its unique effects and is worth trying. Since the 1970s, Chinese veterinarians have made some new developments based on this ancient technique.

TRADITIONAL CHINESE CLASSIFICATIONS OF INFERTILITY

Before we discuss the Chinese traditional method, I would like to introduce the classifications of animal infertility. According to traditional Chinese veterinary medicine, animal infertility can be classified into four types.

Feeble type. Affected animals appear emaciated and dull; the skin loses its elasticity and the body is covered with a rough hair coat. The temperature is normal or below normal. The animal is in anestrus with no visible signs of heat. Though rectal palpation may reveal an atrophic ovary, significantly reduced in size, without follicle or with persistent corpus luteum. The animals are generally emaciated due to undernutrition, resulting in insufficient energy and blood. The animal's uterus is with cold syndrome due to yang deficiency. The former causes the irregular estrous cycles and ovulation, while the latter causes failure to conceive.

Fatty type. Anestrous animals appear plump and respire rapidly as they move. Rectal palpation may reveal an ovary that is normal or smaller in size with solid nodules on the surface. The follicle and corpus luteum cannot be detected. The uterus becomes much smaller and flabby.

Senile type. Animals are old and are approaching a nonreproductive state. Sexual desire has ceased and animals fail to conceive. Rectal examination indicates a reduced ovary size; no follicle or luteinized follicle can be detected. The uterus becomes flabby.

Diseased type. Affected animals have a history of inflamed genital organs or generalized chronic diseases due to treatment by unskilled obstetrical workers. Rectal palpation reveals ovary, uterus, and oviduct lesions, such as a persistent corpus luteum, ovarian cyst, chronic endometritis, and salpingitis.

The factors responsible for these types of infertility are numerous but may be of a structural, functional, or infectious nature. According to meridian doctrine, acupuncture can connect the exterior and interior of the body and rectify derangement of the defensive and constructive energy. But there are differences in acupuncture's therapeutic effects. In general, better therapeutic effects are obtained with filiform needle and electroacupuncture on infertility caused by functional disorder.

In China, acupuncture treatments for horses, cattle, and swine infertility are as follows.

POINTS AND METHOD OF ACUPUNCTURE

Horse Infertility

Yanzi

- Location. Make a perpendicular line from the tuber coxae towards the dorsal median line. The point is at the lateral one-third of the line. Bilateral.
- Method. Puncture with a filiform needle perpendicularly 12 cm to 13 cm deep.
- Response. Fibrillation of the loin and buttock muscles.

Baihui

- Location. In the depression between the processus spinosus of the last lumbar vertebra and the processus spinosus of the first sacral vertebra. One point only.
- Method. Puncture with filiform needle perpendicularly 9 cm to 11 cm.
- Response. Arching of the loin. Fibrillation of the loin and buttock muscles.

Houhai

- Location. In the depression to the tailhead and dorsal to the anus.
- Method. A filiform needle is inserted in a dorsal and cranial direction 12 cm to 18 cm.
- Response. Contraction of anus.

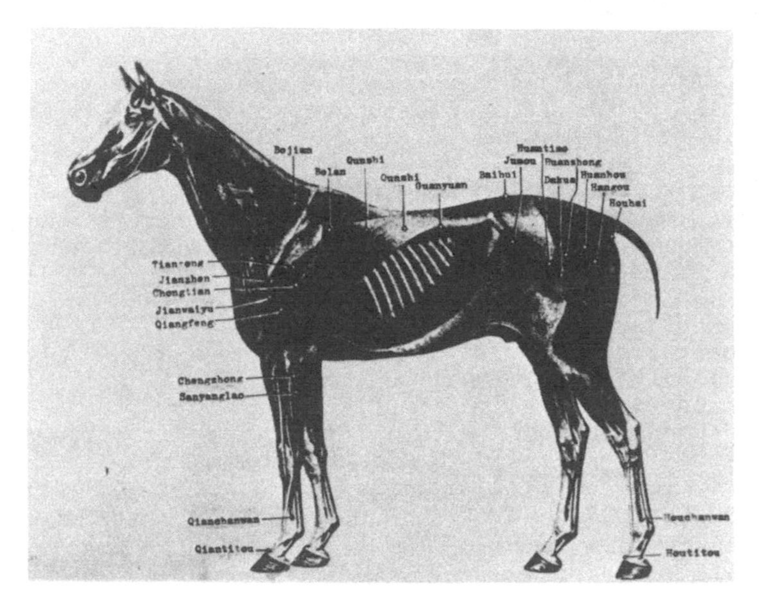

Figure 1.

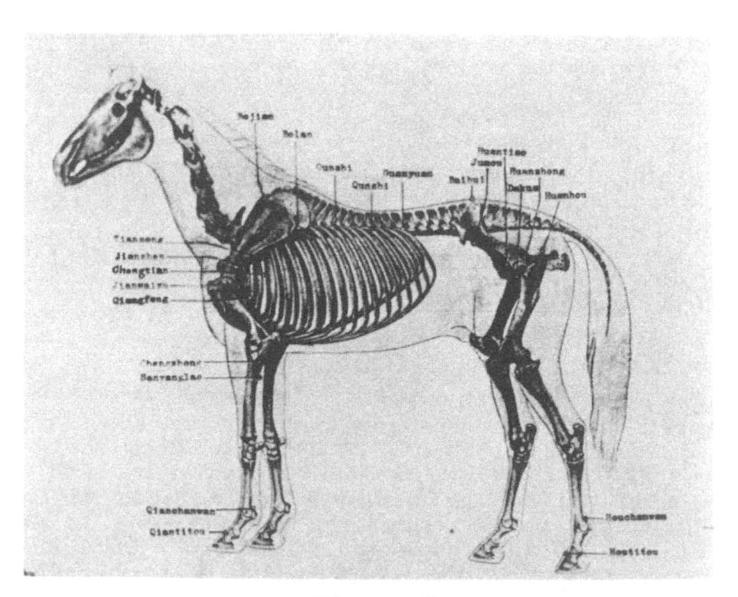

Figure 2.

Cattle Infertility

Qianjin
- Location and response are the same as the horse Baihui, but the needle puncture depth is 7 cm to 9.5 cm.

Jiaochao
- The location and response of this point is the same as the horse Houhai, but the puncture depth is 24 cm to 30 cm.

Shenshu (left Guiwei and right Guiwei)
- Location. At the middle of coxae tubae and Baihui line, which is the muscle farrow about 8 cm lateral to Baihui. Bilateral.
- Method. Puncture with a filiform needle perpendicularly downward to 7 cm to 12 cm.
- Response. Fibrillation and contraction of loin and buttock muscles.

Pig Infertility

Baihui
- Location and response are the same as the point Baihui of horse, but the needle puncture depth is 3.5 cm to 6 cm.

Jiaochao
- This point is the same as the horse's Houhai and the cattle's Jiaochao, but puncture depth is 10 cm to 15 cm.

Cuiqing
- Location. A vertical line is drawn from median of back to tuber coxae. Locate a point on this line one-third of the distance from back. Bilateral.
- Method. Puncture perpendicularly and slightly inward 10 cm to 12 cm deep.

EQUIPMENT AND METHOD OF MANIPULATION

Equipment

Needle. A filiform needle is made of stainless steel. The diameter of the needle shaft is .64 mm to 1.25 mm; the shaft is 10 cm, 15 cm, 20 cm, 25 cm, and 30 cm in length.

Electrostimulator. The electrostimulator is the major tool for acupuncture therapy. The quality of this device is closely related to the therapeutic effects. Only those with controlled output of current and pressure to meet the therapeutic intensity will be used. In our college clinic, the

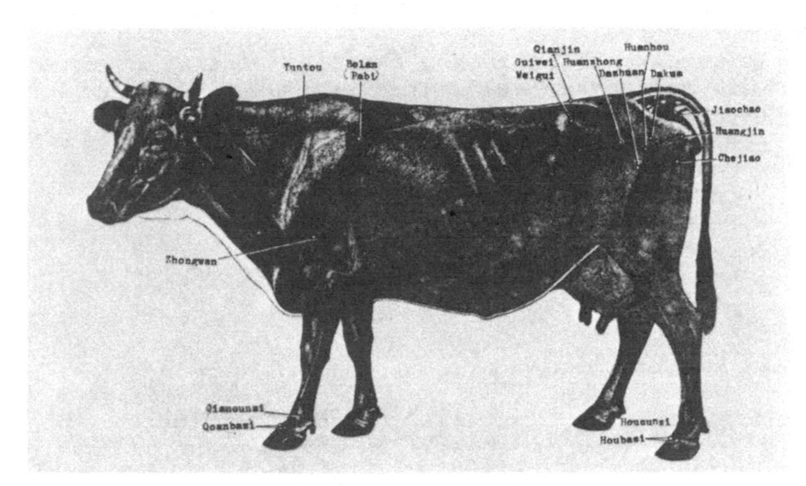

Figure 3.

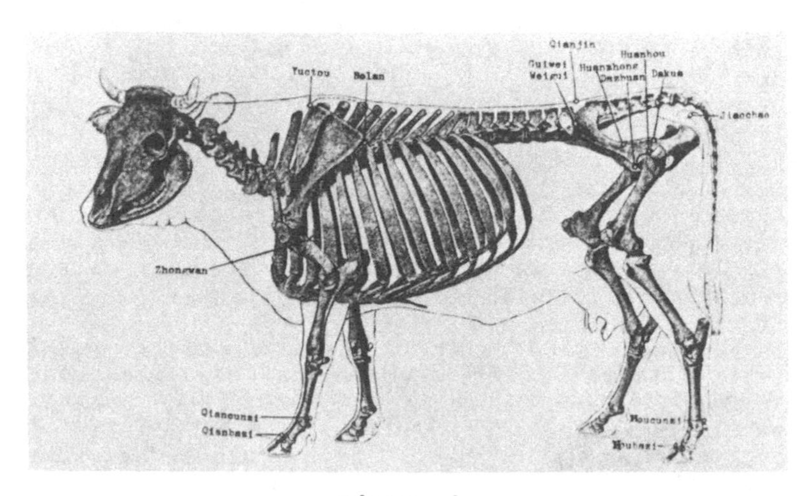

Figure 4.

71-2 type electrostimulator, which is manufactured by Tianjing Radio Factory, is used. This is a dual-purpose machine that can be used for therapy and anesthesia. The type 73-10 electrostimulator, made by Hainan Electro-Machine Factory, also is used very well. The latter type has the following technical features:

- Voltage. 18 V (direct current) battery powered with twelve 1.5 volt D batteries.
- Output wave form:
  - Biphasic spike wave
  - Continuing consistent wave
  - Interrupting wave
  - Disperse-dense wave
- Output intensity:
  - Load 200Ω, peak to peak voltage value 95 V
  - Load 200Ω, peak to peak voltage value 45V

Method of Manipulation

Choice and prescript of acupuncture points. The four points are usually divided into two groups: Baihui or Qianjin and Houhai or Jiao are in one group, while Yanzi or Shenshu are in another group. In the pig, it is the same as in the horse and in cattle--one group in Baihui and Jiaochao; another group is the bilateral Cuiqing.

These two groups of points can be used simultaneously as in endometritis. For most cases of infertility, one group point is used in each treatment.

Hand manipulation. This is the conventional method of acupuncture. At the chosen points, a filiform needle is inserted into it. When the needle approaches the desired depth, it will elicit a certain response from the animal. The needle will be left in the point for 20 min to 30 min (whichever you choose, hand or electroacupuncture).

During the time the needle is left in the acupuncture site, it should be rotated along its long axis clockwise and counterclockwise, pushing in and pulling out a little bit, or swinging with the finger to increase its stimulation for three times. It lasts for about 1 min each time. At the Baihui or Qianjin point, the needle should not be pushed or pulled. The purpose of these manipulations is to rectify the stimulation for treatment.

Electroacupuncture manipulation. The filiform needle is inserted into the point, as with the hand manipulation. A metal clamp is used to connect the implanted needle with the electrostimulator, then the switch is turned on. Great care must be taken while the voltage is slowly increased to the maximum that the animal can tolerate.

The animal shows a kind of rhythmic muscular fibrillation around the acupuncture points. By the time the frequency is about 80 Hz/sec to 129 Hz/sec, the handle of the needle will exhibit a slight movement. Interrupting and disperse-dense wave form is alternately applied and exchanged every 5 min. The operation from high frequency and low voltage to low frequency and high voltage is conducted for 5 min and then the operation is reversed for another 5 min. To prevent the animal from adapting to the stimulation, it should be kept under rather strong stimulus through the whole acupuncture period. The interval of electroacupuncture is 2 days to 3 days, and the course of treatment is the use of acupuncture 3 times. After each manipulation, the voltage should be gradually reduced to "0." Withdraw the needle as with the hand method. The needle hole should be disinfected with tincture of iodine.

CURATIVE EFFECT

In the past decade, Chinese veterinarians have treated many cases of animal infertility. It is not possible to include these cases here; however, I will list its application and some infertility effects.

TABLE 1. HAND ACUPUNCTURE EFFECTS IN CATTLE INFERTILITY

Name of disease	Point	No. of treated animals	Effective[a]		Cured[b]		Ineffective[c]	
			No.	%	No.	%	No.	%
Persistent corpus luteum	Qianjin Jiaochao	6	4	66.7	-	-	2	33.3
Cystic ovary	"	12	2	16.7	10	83.3	-	-
Chronic endometritis	"	4	-	-	1	25.0	3	75.0
Salpingitis	"	1	1	100.0	-	-	-	-
Tuberculosis	"	2	-	-	-	-	2	100.0

[a] After acupuncture, animals manifested heat within 1 mo and had regular heat later on. But conception was delayed in some animals.
[b] After acupuncture, animals showed normal heat and conceived.
[c] After treatment, animals did show heat but failed to conceive.

TABLE 2. ELECTROACUPUNCTURE EFFECTS IN CATTLE INFERTILITY

Name of disease	Point	No. of treated animals	Effective[a]		Ineffective[b]	
			No.	%	No.	%
Persistent corpus luteum	Qianjin Jiaochao	30	28	93.3	2	6.7
Cystic ovary	Shenyue	24	24	100.0	-	-
Mild endometritis	Qianjin Jiaochao	16	10	62.5	6	37.5
Chronic endometritis	Shenyue					
Chronic endometritis	Qianjin Jiaochao	8	2	25.0	6	75.0
Salpingitis	"	8	4	50.0	4	50.0
True anestrus	"	6	6	100.0	-	-
Tuberculosis	"	4	-	-	4	100.0

[a] "Effective" means that after electroacupuncture, the animals manifested heat and normal ovulation in one or two estrous cycles.
[b] "Ineffective" means that after electropuncture, the animals still had no manifestation of heat.

TABLE 3. ELECTROACUPUNCTURE EFFECT IN MARE INFERTILITY

Name of disease	Point	No. of treated animals	Effective		Ineffective	
			No.	%	No.	%
Cystic ovary	Yanzi	3	3	100.0	-	-
True anestrus	"	31	30	96.8	1	3.2
Anestrus after parturition	"	14	12	85.7	2	14.3
Persistent corpus luteum	"	10	9	90.0	1	10.0
Multifollicle	"	5	5	100.0	-	-
Large follicle	"	5	5	100.0	-	-
Alternate development of follicle	"	5	5	100.0	-	-
Delayed ovulation	Baihui, Houhai	5	4	80.0	1	20.0
Endometritis	Baihui, Houhai, Yanzi	10	8	80.0	1	20.0
Salpingitis	"	3	1	33.3	2	66.7

TABLE 4. ELECTROACUPUNCTURE EFFECT IN SOW INFERTILITY

Name of disease	Point	No. of treated animals	Effective		Ineffective	
			No.	%	No.	%
Feeble	Baihui, Jiaochao, Cuiqing	80	61	76.3	19	23.7
Fatty	"	31	20	64.5	11	35.5
Senile	"	7	5	71.4	2	28.6
Disease	"	20	10	50.0	10	50.0
Postparturition anestrus	"	57	47	86.0	8	14.0

CONCLUSION

Acupuncture is not a "cure-all" for animal infertility, but it does have satisfactory curative effects on cystic ovary, true anestrus, persistent corpus luteum, and feeble type infertility. In the treatment of endometritis, it varies with the severity of the disease. It seems that it needs to use multipoints to achieve a satisfactory effect. In 1980, Tokutaro Onkawa used acupuncture to treat 192 cases of sow infertility and 78.6% of them restored their reproductive function. L. Watanabe applied electroacupuncture to treat ovaries that had ceased to function (23), ovary aplasia (8), ovary atrophy (3), inactive estrus (49), and cystic ovary (2) with curative effects of 56.5%, 62.5%, 33.3%, 73.4%, 40%, and 55.6%, respectively.

Acupuncture brings new light to an old problem--infertility. It possesses many advantages, i.e., the method is relatively simple, no expensive equipment is needed, and the effect is satisfactory. However, intensive, careful study of this method is not available. Much more research work is needed.

REFERENCES

Veterinary Acupuncture, Agricultural Press, Beijing

Observation on acupuncture to female animal infertility. Veterinary New Therapeutic Method, Vol. 2. Shanxi Animal Science and Veterinary Medicine Research Institute.

Sun Ji. 1981. Observation on electroacupuncture to female animal infertility. Journal of Chinese Veterinary Traditional Medicine. No. 4.

Observation on electroacupuncture to dairy cattle infertility. Fujian Agricultural College, Fujian Animal Science and Veterinary Medicine Research Institute.

Zhang Yu-Sen, et al. 1981. Hand Acupuncture to treatment of animal infertility. Sichuan Animal Science and Veterinary Medicine College.

Yang Wei-Zhen, et al. Study on electroacupuncture to the treatment of persistent corpus luteum in dairy cattle. Ningxio Agricultural Academy. Ninxio Animal Science and Veterinary Medicine Research Institute.

Lo Cheng-hao. Study on electroacupuncture of dairy cattle infertility. Xinjiang Shihexi Agricultural College.

Tokutaro Onkawa. 1980. Observation on acupuncture to swine infertility. Journal of Veterinary Medicine 708, p 393.

Lyoufu Watanabe. 1982. Application of electroacupuncture to treatment of reproductive disturbance of mare. Journal of Veterinary Medicine 726, p 21.

41

ACUPUNCTURE IN TREATMENT OF IMPACTION

Qin Li-Rang,
Sun Yong-Cai

Impaction (intestinal obstruction) caused by undigested fibers is a common digestive disorder in horses and mules. According to the local animal clinic statistics of Hunan province, this disease occurs in 46% of the digestive diseases and 62% of the colic cases. If the affected animals are not properly treated they may die, causing heavy economic losses.

In the long history of China, Chinese veterinarians have accumulated extensive knowledge on diagnosis and treatment of impaction. The well-known Yuan Heng Liao Ma Ji (the Treatise on Horse) stated that impaction is an excessiveness symptom complex, retention, or stagnancy. It also stated that the excessiveness symptom complex is a common term and requires different treatment. It can be treated by acupuncture, or by drugs, or by hand. The classical veterinarians had described the acupuncture treatment of impaction by "drawn blood from Sanjian, Damo, and Titou." Since the 1960s, there have been some new developments in the acupuncture of impaction. Use of these new points and methods has proved effective, convenient, and economical. (For illustrations of the "acupuncture points" the reader should review the paper entitled "Application of Acupuncture to the Treatment of Infertility" in this book.) These new achievements have greatly enriched our traditional veterinary medicine. These can be divided into three methods which will be described below.

ELECTROACUPUNCTURE

Pre-acupuncture Preparation

An electrostimulator and a set of filiform needles are needed. There are different types of electrostimulators with various effects. From a clinical point of view, the electrostimulator should be square wave or bidirectional pulse wave type. The output frequency can be adjusted continuously from 2 cycles/sec to 300 cycles/sec. The output of electro-intensity is 0-160 V peak value.

Filiform needles are made of stainless steel (as described in my paper "Acupuncture in Treatment of Animal Infertility").

The acupuncture point is Guanyuanyu, which is located bilaterally in the depression at the intersection between the ventral margin of the longissimus dorsi muscle and the caudal edge of the last rib (figure 1).

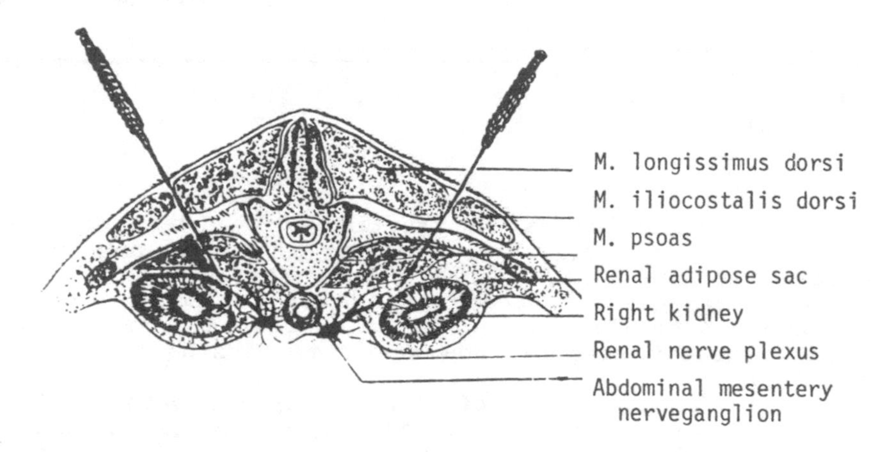

Figure 1. Cross section of Guanyuanyu point.

Method of Acupuncture

A 7.5 cm filiform needle is used. Firmly hold the anterior part of the needle shaft and expose about 1.5 cm of the needle tip with the right hand thumb and index finger. With the left hand thumb, firmly press on the skin with the nail at the site point. After a quick initial movement through the epidermis, adjacent to the nail, insert the needle obliquely 45 degrees inward and downward 5 cm to 6.5 cm deep. Then connect the needle holder with the lead to the electrostimulator. Switch on and adjust the output voltage gradually from low to high until the animal can tolerate the maximum voltage. Retain at this maximum voltage for 2 min to 3 min. Then adjust the voltage from high to low. Repeat this operation for several times each 10 min. At the same time adjust the frequency from 40 Hz to 100 Hz (hertz) and reverse it from 100 Hz to 40 Hz. This also should be repeated several times. After acupuncture, the output and frequency switch should be turned from high to low to "0" before withdrawing the needle. Electroacupuncture for each treatment takes about 20 min to 30 min.

Animal Response to Acupuncture

When the electroacupuncture is begun, the animal's bilateral abdominal muscles show rhythmatic contractions parallel with the frequency. As the frequency increases from low to high, the muscle contractions become violent, even tetanic. Care must be taken to avoid the appearance of tetany or the restlessness of the animals.

Animals being treated gradually increase bowel sound and intestinal motility with the passage of flatus. Abdominal pain is obviously relieved. When acupuncture treatment is finished, some animal patients often start defecating at once. Most animal patients recover within 10 hr after treatment. If the condition does not improve within 10 hr, treatment of the severe form necessitates another one or two series of acupuncture. In reports of 2141 cases of impaction with acupuncture treatment, effective treatment was seen in 2095, with only 46 ineffective (a curative rate of 97.8%).

AURICULAR ACUPUNCTURE

Pre-acupuncture Preparation

The animal patient is stabilized in a standing position. The site of the acupuncture point is routinely shaved and disinfected. A 5 cm to 7 cm long, sharp (round) needle or a stainless nail is used as the acupuncture tool.

Auricular Acupuncture Point

Erding point is located at the fossa upper brim in front of ear. Bilateral. Dakong is located just above Erding about 2 cm. Bilateral.

Figure 2.

Method of Acupuncture

Puncture with needle as soon as the site point is sterilized. Use thumb to press the ventral fossa in front of ear and use index finger to press rostral fold and lower root of rostral ear margin intersection of the internal ear. Hold the needle with other hand. At thumb nail, quickly insert the needle through the external and internal ear skin. Care must be taken to penetrate at the site of Erding accurately.

Puncture of Dakong

Use thumb, pressing from internal ear at the rostral fold and the rostral ear margin intersection. With other hand, hold the needle and quickly insert the needle just above the intersection at the depression through the external and internal ear skin.

Erding and Dakong have a total of four points. During acupuncture therapy, you may choose any one of these points. The needle is left in the points until the animal is cured, then the needle is withdrawn. As a clinical consideration, we often choose two points to achieve better curative effects.

Acupuncture Effect

Within 15 min after treatment, bowel sounds may reappear. Abdominal pain is gradually relieved. If the condition is not improved, it is doubtful whether the acupuncture site is correct. Readjust the site and insert the needle again. Except for advanced and severe cases, improvement will be seen after reacupuncture.

Of 2536 cases of impaction treated by auricular acupuncture, the curative rate was 94.1%. Different sites of impaction have varying curative effects. In general, the simple impaction of the small and large colon is best treated with this kind of treatment. The curative rate is high (96.8% and 94.9%, respectively). For impaction of cecum and multisegmental impaction, the curative effects to this treatment are 81.8% and 74.5%, respectively.

HAND MANIPULATION

Preacupuncture Preparation

Affected animals are stabilized in standing position. The site of the acupuncture point is shaved and sterilized as routine. A round sharp needle should be used.

Acupuncture Point

1) Qunzhi--main point. (Mass wisdom). Location: left in the last 7th intercostal space at the ventral margin of M. longissimus dorsi. Right in the last 8th intercostal space at the ventral margin of M. longissimus dorsi. 2) Qunzhi--accessory point. (Mass therapy). Location: in the last 3rd intercostal space at the ventral margin of M. longissimus dorsi. Bilateral.

Method of Acupuncture

During treatment, the acupuncture order is first the left main point then the right corresponding point. If the main point puncture is not sufficient to achieve a cure, the accessory point should be punctured as with the order of the main point. Acupuncture must be fast and accurate. There is no need to leave the needle in the point. After the needle is withdrawn, strike at the needle hole several times with a fist. The depth of insertion depends on the severity of the animal's sickness. Usually, 4.5 cm to 6.5 cm deep is adequate. Generally, the animal patients need only one treatment. If the impaction is more severe, another one or two treatments are required.

Response

The response of acupuncture is as follows: 1) increased intestinal motility. The atonic intestinal peristalsis becomes motile. On ausculation, bowel sounds can be heard within 5 min to 10 min after acupuncture, even during the course of treatment. Some of the animals defecate within 10 min to 15 min. 2) relief of abdominal pain. The abdominal pain is quickly relieved after treatment. Some of the animal patients show a sign of relaxation, with shivering of their hair coat. 3) increased secretion of digestive fluid. After acupuncture, the dry oral cavity becomes moist with saliva. In 1115 cases treated by this method, the curative rate has been 93%. Most of the affected animals require only one treatment for recovery.

SOME EXPERIENCES ON ACUPUNCTURE OF IMPACTION

The curative effect of the acupuncture in impaction is satisfactory, but there are many influencing factors such as the location and composition of the impacted mass and the degree and duration of the intestinal obstruction, which are important to its effect. However the accuracy and skill of manipulation is of utmost importance to the curative effect. The animal should be treated as early as possible. For the sake of a better curative effect, acupuncture can be combined simultaneously with other treatments such as fluid therapy, enema, etc.

Electroacupuncture Guanyuanyu, auricular acupuncture Erding and hand acupuncture of Qunzhi are effective in impaction and also have a good effect on the treatment of indigested gastric-dilatation and spasmodic colic.

Acupuncture in treatment of impaction has not shown any bad side effects. Under appropriate stimulus, acupuncture does not cause abortion, secondary enteritis or intestinal torsion. Individual cases may show rapid respiration and muscular tremor, but these side reactions subside within 30 min.

According to the Chinese traditional veterinary medical theory, impaction is an excessiveness symptom complex. This complex should be treated with the method of purgation and reduction. Therefore, when electroacupuncture, auricular acupuncture, and hand acupuncture are applied, a strong stimulus is required. This can achieve the purgation and reduction, excessiveness symptom complex is removed, and the animal recovers and is restored to normal physiological balance. At the same time, the chosen points possess bidirectional specific action. That means that after acupuncture, according to the animal's condition, it will elicit its adjustive effects on the body. The constipated animal can be purged, otherwise the acupuncture can be antidiarrhetic. Reduced digestive fluid secretion can be reversed by acupuncture. This is the reason why acupuncture has curative effects on impaction. It seems to concur with our modern medical theory. When a horse and mule are constipated, the dynamic equilibrium of the autonomic nerves is disturbed. The sympathetic nerves are in a state of hyperexcitation, whereas the parasympathetic nerves are in an inhibitory state. So the intestinal peristalsis becomes weak or disappears, and the secretion also is depressed. In the acupuncture such as that at Guanyuanyu, the needle point reaches the renal adipose sacs that are richly supplied by the sympathetic nerve. A strong stimulus causes the exciting sympathetic nerve to be appropriately inhibited. When the inhibitory parasympathetic nerve gradually recovers its excitability, the normal intestinal function of the animal is restored.

42

ACUPUNCTURE IN TREATMENT OF SPRAIN

Qin Li-Rang,
Cben Ci-Lin

Sprain is a common problem of horses and cattle. The commonest causes are improper management, including working until too fatigued and working too fast on muddy, rough, or hard road surfaces and in fields. Such mismanagement causes overextension of muscles, tendons, or joints, and aseptic inflammation or nerve paralysis. On clinical examination, the affected animals manifest marked lameness and functional disturbances. If the sick animals are not cured in time or are improperly treated, the disease may become severe or could become a chronic inflammation. Finally, the animals cannot be used as work animals, causing great economic losses.

Since 1960, Chinese veterinarians have applied acupuncture to sprains with satisfactory results. This paper provides a general description of this technique. (For illustrations of the "acupuncture points" the reader should review the paper entitled "Application of Acupuncture to the Treatment of Infertility" in this book.)

ACUPUNCTURE POINTS FOR TREATMENT OF SPRAIN

Forelimb

Main point

Qiangfeng (horse); Zhongwan (cattle)

Location: In the fossa behind the humerus, on the posterior border of the deltoideus and in between the long head and lateral head of the triceps brachii. Bilateral.

Method: Puncture with a filiform needle or a round sharp needle, perpendicularly, 3 cm to 6 cm deep.

Response: Contraction of the muscles of the same area and lifting of that limb.

Qianchanwan (horse); Qiancunsi (cattle)

Location: On the medial and lateral digital veins in the upper border of the fetlock joint. Two points on each forelimb.

In cattle, in the depression lateral and medial to the dewclaw.
Method: A small, wide needle is inserted .5 cm to 1 cm deep.

Accessory point

Bojian (Yuntou in cattle)
Location: In the fossa beside the junction scapular cartilage and the anterior angle of the scapula. Bilateral.
Method: Puncture with a filiform needle or a round sharp needle along the medial border of the scapula slightly downward 7 cm to 10 cm deep.
Response: Fibrillation of the muscle of the withers.

Bolan
Location: In the fossa at the junction of the scapular cartilage and the posterior angle of the scapula. Bilateral.
Method: Puncture with a filiform needle or a round sharp needle along the medial border of the scapula forward and downward 7 cm to 10 cm deep.
Response: Same as Bojian.

Chongtian
Location: In the sulcus muscularis 7 cm above and posterior to Qiangfeng. Bilateral.
Method: Puncture with a filiform needle or a round sharp needle, perpendicularly, 8 cm to 10 cm deep.
Response: Lifting of forelimb, contraction of shoulder muscles.

TianZong
Location: The point between the middle third and the ventral third segments of a line joining the middle of the upper margin of the scapular cartilage and Qiangfeng. It is located in the muscle cleft at the caudal margin of the scapula.
Method: Puncture with a filiform needle perpendicularly, 5 cm deep.

Jianzhen
Location: Cranial to Chongtian in the muscle cleft at the caudal margin of the scapula. Bilateral.
Method: Puncture with a filiform needle, perpendicularly, 5 cm to 7 cm deep. A small wide needle is inserted 5 cm to 6 cm deep.

Jianwaiyong
Location: Cranial and ventral to Jianzhen in the depression at the caudal margin of the shoulder joint.
Method: Puncture with a filiform needle, perpendicularly, 5 cm to 7 cm deep.

Sanyanglao

Location: In the muscle groove, 6 cm below the tubercle of the radius for the lateral ligament, lateral to the forearm. One point in each forelimb.

Method: The needle is inserted at an angle of 15° to 20° from the skin at Sanyanglao along the posterior margin of the radius slanting medially and ventrally toward Yeyan. The needle points should be felt subcutaneously at Yeyan, but should not penetrate the skin.

Hengzhong

Location: On the outer surface of the forearm, in the depression below the lateral tuberosity of radius, in the groove of the common exterior tendon. Bilateral.

Method: Puncture with a filiform needle or a small wide needle, perpendicularly, 1.5 cm to 2 cm deep.

Qiantitou

Location: About 2 cm lateral to the median point of the coronary border of the hoof. Bilateral.

Method: Puncture with medium wide needle, quickly, 1 cm deep.

Loin and Hind Limb

Main Point

Baihui (same as "Application of Acupuncture in Treatment of Animal Infertility.")

Shentang

Location: On the saphenous vein about 10 cm below the midpoint of the root of the inner surface of the thigh. Bilateral.

Method: Puncture with a small wide needle, 1 cm deep.

Huantiao

Location: In the depression at the anterior border of the hip joint. Bilateral.

Method: Puncture with a filiform needle, perpendicularly, 5 cm to 7 cm.

Huangjin (cattle)

Location: In the muscle cleft between the biceps femoris muscle and the semitendinosus, 5 cm dorsal to the tuber ischium.

Method: Puncture with a filiform needle or a round sharp needle, perpendicularly, 2.5 cm deep.

Accessory Points

Jumou

Location: At the depression of the inferior margin of the gluteal muscles and anterior to the medium trochanter of the femur.

Method: Puncture with a filiform needle or a round sharp needle, perpendicularly, 2.5 cm to 4 cm deep.

Huanzhong

Location: Cranial and dorsal to the medium trochanter of the femur and in the furrow of muscles, at the midpoint of a line joining the posterior margin of the thigh.

Method: A filiform needle is inserted 4 cm to 5 cm.

Huannan

Location: Cranial and dorsal margin of the greater trochanter of the femur, and caudal and ventral to Huanzhong.

Method: A round sharp needle or a filiform needle is inserted perpendicularly 2.5 cm to 4 cm.

Dakua

Location: In the depression cranial and ventral of the medium trochanter of the femur.

Method: A filiform needle is inserted 2.5 cm to 4 cm.

Hangou

Location: Above 7.5 cm ventral to Huangjin in the same furrow. It forms an equilateral triangle with Huangjin and the middle trochanter of the femur.

Method: Same as Huangjin.

Dazhuan

Location: In the depression cranial 6 cm to the os femoris trochanter major and in the furrow of muscularis tensor fasciae latae, muscularis gluteus medius and muscularis rectus femoris. Bilateral.

Method: Puncture with a small wide needle or round sharp needle, perpendicularly, 3 cm.

EQUIPMENT AND METHOD OF ACUPUNCTURE

Veterinary Acupuncture Needle

Filiform needle. This needle is generally used at points with well-developed musculature. It is the same as described in "Application of Acupuncture in Treatment of Animal Infertility."

Round, sharp needle. This kind of needle is similar to filiform needle, but the diameter size is bigger. In veterinary practice, a 3 cm to 13 cm long, round, sharp needle is used.

Wide needle. This needle is large, has a spade-shaped tip with a low crest along the midline. The tip widths are also three sizes: 3 mm, 5 mm, and 6 mm. This kind of needle is used primarily for drawing blood.

Method of Acupuncture

Preparation for acupuncture. Before acupuncture treatment, the animal patient should be diagnosed to determine the site of the sprain and the course of the disease. Then prescribe the points and the method of acupuncture.

Hand manipulation. Hand acupuncture of filiform needle is the same as described in "Application of Acupuncture in Treatment of Animal Infertility."

A blood needle is used with good results for acute sprains. In general, a wide needle is applied parallel with the blood vessel direction. Insertion should be fast and accurate. Care must be taken not to insert the length of the needle perpendicular to the vessel. The amount of blood to be drawn depends on animal size, characteristics of the disease, and seasonal factors. For example, in medium-sized animals with a diseased part on the upper limb, the amount of blood to be drawn is 300 ml to 500 ml. In the lower limb, the amount of blood is 200 ml to 300 ml. When the color of the blood changes from dark to brilliant red, enough blood has been drawn. After blood is drawn from acupuncture points, bleeding will usually stop; if it continues, it should be stopped.

Electroacupuncture. Electroacupuncture method and equipment are the same as described in the paper on animal infertility.

CLINICAL APPLICATION

In China, acupuncture is widely used as a routine method for treating sprains. Table 1 and table 2 summarize 797 cases of large-animal sprains.

Note: The sprain of the forelimb's fetlock uses the points of 1) Qianchanwan, Qiancunzi, and Qiantitou. When the sprain is on the hind limb's fetlock, the corresponding points will be used. 2) In large animals, the commonest sprain sites are shoulder and forearm. Fetlock and hip-joint sprains are the next most common sites. The sprain of the carpal joint and loin are also encountered. If this does occur, Baihui and Senyu as main point and Yaoqian, Yaozhong, and Yaohou as accessory points are used as the treatment for loin sprains. In general, the chosen point nearest the afflicted site is the principal for true treatment.

TABLE 1. SUMMARY OF SPRAIN TREATMENTS

Symptom	Animal species	Method of acupuncture	Number of treated animals
The shoulder is carried loosely with its affected joint. In standing, the affected limb extends forward or backward. The animal is disinclined to move. If treatment is delayed, the shoulder muscle atrophies.	Horse	Conventional method	115
	Mule	Electroacupuncture	65
	Cattle	Conventional	112
		Electroacupuncture	227
The acutely affected joint shows heat, swelling, and tenderness to touch. On passive flexion and extension, there is prominent painful reaction. In chronic cases, the joint shows less swelling, but incomplete flexion and extension are obvious. There is nodding when animal walks. Steps are short in posterior.	Horse	Blood needle	45
	Mule	Electroacupuncture	18
	Cattle	Small wide	125
The affected limb touches ground with the tip of hoof. Lifting the affected limb forward is difficult. Flexion, extension, abduction, and adduction of the affected limb shows painful reaction. Steps are short anteriorly. Lameness becomes severe during exercise. Backward movement is difficult.	Horse	Conventional method and blood needle	36
	Cattle	Small wide	54

TABLE 2.

Point		Internal of application	Acupuncture treatment	Result		
Main	Accessory			Cured	Better	Ineffective
Qiangfeng	Chongtian, Tianzong, Jianzhen, Jianwaiyong	Once every other day	2–4	100	10	5
Qiangfeng	Chongtian, Bojian, Bolan, Hengzhong	Once daily	3–5	62		3
Zhongwan	Yuntou, Bolan	Once every other day	2–4	102	2	3
Zhongwan	Bojian, Cunsi, Sanyanglao	Once daily or every other day	2–5	196	7	20
Chanwan	Titou	Once every other day	1–2	39	4	2
Qiangfeng	Sanyanglao, Cunsi	Once daily or every other day	1–4	17	1	
Cunzi		Once every other day	1–3	120		5
Baihui or Shentang, Huantiao	Jumuo, Huantiao, Dakua or Huanzhong, Huanhan, Hangou	Once daily or every other day Every 5 days	2–5	29	4	3
Huangjin		Once daily	1–4	48	2	4

SOME EXPERIENCES ON THE TREATMENT OF SPRAINED LIMBS

- Acupuncture treatment for the sprain is the same as in other disease treatment. The curative effect is dependent on the proper chosen points and diagnostic accuracy. Acupuncture treatment is only suitable for the soft-tissue sprain. If any hard tissues are involved, acupuncture treatment must be used in combination with other treatment.
- Strong or high frequency electrical stimulation is an important factor in the treatment for satisfactory results. In the conventional methods, the stimulation can be intensified by pushing and pulling the needle or by retaining the needle in the point and walking for 20 min to 30 min.

REFERENCES

Electroacupuncture treatment of domestic animal sprains. 1974. Dept. of Anim. Husbandry and Vet. Med., Huazhong Agricultural College. Hubei Agr. Sci. 10. p 34.

Yu Chuan et al. 1979. Chinese Traditional Veterinary Medicine. (Textbook of Agricultural College). Agricultural Press.

Study on acupuncture treatment of horse and mule limb disease. Gansu Vet. Sci. Res. Instit. In: Selected Papers of Chinese Traditional Veterinary Medicine. p 49. Agricultural Press.

Wang Lao-Tung. 1977. Small wide needle acupuncture for treatment of cattle limb sprain. In: Selected Papers of Chinese Traditional Veterinary Medicine Experience Vol. III. p 8. Scientific and Technical Press.

Wang Shu-tang. 1977. Small wide needle acupuncture for treatment of horse and mule shoulder sprain. In: Selected Papers of Chinese Traditional Veterinary Medicine Experience. Vol. III. p 441. Scientific and Technical Press.

New developments of electroacupuncture. 1973. In: New Treatments of Veterinary Medicine. Vol. II. p 8. Sanxi Anim. Husbandry and Vet. Med. Res. Instit.

Song Da-Lu et al. 1980. Electroacupuncture and Acupuncture Anesthesia in Domestic Animals. Shanghai Scientific and Technical Press.

Yang Jin-Zhong et al. Acupuncture to Mule and Horse Limb Sprain. (Unpublished.)

43

EQUINE IMMUNIZATION

Robert Ball

INTRODUCTION

A properly conceived and administered vaccination program is one of the most important aspects of a good preventive health care program. Any vaccination program for horses should be designed according to the geographical location, the particular use of the animal, and the specific problem encountered on the farm. A vaccination program, if it is to be successful, should be incorporated with sound management practices including proper nutrition, sanitation, frequent deworming, segregation of new arrivals, and limitation of horse movement. The animal's health status will directly affect the animal's immune response. Therefore, only healthy animals should receive vaccination.

STRANGLES

Strangles, or distemper, is an infectious condition of horses caused by the bacteria *Streptococcus equi*. It is characterized by an extremely high morbidity (approaching 100%) with low mortality in most uncomplicated cases. Generally speaking, the disease involves the upper respiratory tract of horses of all ages. The incubation period is 4 days to 8 days. Typical clinical signs of the disease include fever, lack of appetite, purulent nasal discharge, and difficulty in swallowing and breathing. Inflammation of the mandibular and pharyngeal lymph nodes, with abcessation and later rupture, is a common clinical sign. The organism is spread by direct contact with infected horses, or indirectly by contaminated premises. Contaminated premises can harbor the organism for several weeks. A carrier state may exist for as long as 10 mo following infection. Possible severe complications of strangles include secondary pneumonia, purpura hemorrhagica, and "bastard strangles." Bastard strangles is a systemic form of *Streptococcus equi*. Different strains of *Streptococcus equi* have been substantiated; however, the existence of more than one antigenic type is still questionable.

An M-protein extract vaccine is now available, as well as a killed whole-cell bacteria. The M-protein extract vaccine has the advantage of creating minimal local tissue response and lessened likelihood of injection-site abcessation due to the elimination of reactive bacterial components found in killed, whole-cell bacteria. Horses showing clinical signs of disease should not be vaccinated with the M-protein extract vaccine. Otherwise, an initial series of three immunizations is recommended given 3 wk apart and followed by a single yearly booster.

INFLUENZA

Equine influenza, or "flu," is a viral disease of horses that attacks the entire respiratory system. Horses of all ages may be affected, but it is a major problem in 2-yr and 3-yr-old horses. Common clinical symptoms include fever, anorexia, a dry nonproductive cough, and muscle soreness. The incubation period is from 1 day to 3 days. The infected horse can shed virus for as long as 8 days. Transmission of the virus is by inhalation. The influenza virus invades and multiplies within the respiratory airways. The virus essentially denudes the lining of the respiratory tract; therefore, this protective blanket is lost. This greatly increases the susceptibility to secondary bacterial infection, particularly the *Streptococcus* species. In uncomplicated cases of flu, regeneration of the respiratory epithelium will take approximately 3 wk.

There are two subtypes of the influenza virus that have separate and distinct antigenicity. These two subtypes are referred to as A-Equi-1 and A-Equi-2. A horse that is infected with one subtype is still susceptible to infection with the other subtype because there is no cross immunity. Vaccines produced in the U.S. incorporate both subtypes. Another factor to consider is the capability of the influenza virus to alter its antigenic structure. Changes in antigenicity are referred to as "drifts" or "shifts."

Primary immunization includes a series of two IM injections followed by a single annual booster. Some data and clinical observations indicate that horses in high-risk populations should be vaccinated every 3 mo to 6 mo. Occasionally, a horse will react to influenza immunization with a mild fever, general depression, or soreness at the injection site.

RHINOPNEUMONITIS

Rhinopneumonitis, more correctly called equine herpesvirus-1, is responsible for four clinical manifestations of disease in the horse: upper respiratory disease of young horses, abortion in mares, neonatal foal disease, and neurological disease.

The upper respiratory disease of young horses is characterized by fever, aqueous nasal discharge, and mild cough. This form of the disease closely resembles the common cold of humans. The incubation period is from 2 days to 10 days and transmission is by inhalation or contact with infected premises. This form of the disease is a particular problem anywhere large numbers of young horses assemble, such as sales, horse shows, racetracks, etc. It has been estimated that one-third of the upper respiratory tract infections of horses at racetracks in the U.S. are caused by this virus. Recovery from uncomplicated cases takes about 10 days. Foals should receive their initial vaccination at 2 mo to 3 mo of age, with a booster dose 4 wk later. Young performance horses should be vaccinated at least twice a year, or inapparent reinfections may recur as early as 5 mo following natural infection.

This same virus also causes abortion in mares from the fifth month of gestation to full term. Mares that abort due to this viral infection do not have to have prior or previous clinical signs of the upper respiratory disease. Many infections in mares are not apparent prior to abortion and positive confirmation of equine herpesvirus-1. No long-lasting protective immunity exists following natural infection or vaccination. Therefore, mares are susceptible to reinfection and abortion year after year. A killed virus vaccine (Pneumabort-K from Ft. Dodge Laboratory) is available for immunization and protection against the upper respiratory tract and abortion forms of the disease. Mares should receive immunization during the fifth, seventh, and ninth months of pregnancy.

Another form of the disease is characterized by the birth of weak foals, infected by the virus in utero. They usually die within a few days of birth because of respiratory distress or pneumonia. For these foals, usually, even intensive medical support fails.

The last form of this disease complex is the neurological variety with paralysis. Typical clinical symptoms are incoordination of the rear limbs, paralysis of the tail, and urinary incontinence. The neurological form may be seen along with the upper respiratory tract form or may precede the abortion form in mares. Horses that become recumbent are associated with a poor prognosis. Euthanasia may be recommended for humane reasons. Some horses will stabilize and improve 48 hr after initial sign of the disease. No vaccine marketed in the U.S. claims any effectiveness against the neurological form of the disease.

TETANUS

Tetanus is a highly fatal, infectious disease of horses, caused by *Clostridum tetani*. This bacteria produces a toxin that attacks the nervous system of horses and causes muscle spasms and paralysis. The organism is present in large numbers in the intestinal tract and feces of normal horses. Though puncture wounds are especially dangerous, any break in the skin or mucous membrane is a potential portal of entry for this organism. Horses of all ages may be affected and death occurs in approximately 70% of the cases. Clinical signs of infection include erect ear and tail head, muscular rigidity, difficulty in swallowing and breathing, prolapse of the third eyelid, and a typical "saw-horse stance." Affected horses overreact to any external stimuli. Death occurs due to asphyxiation or aspiration pneumonia.

Tetanus toxoid vaccines are available for active immunization. In a previously unvaccinated horse, immunization consists of an initial dose followed by a booster dose in 4 wk. Foals should receive their initial dose at 2 mo to 3 mo of age. Those horses with an unknown vaccination history, after sustaining a wound, should be given tetanus antitoxin and tetanus toxid at separate injection sites, followed by a booster of tetanus toxoid in 4 wk. Tetanus antitoxin of equine origin provides passive immunization and protection for only 2 wk to 3 wk. All horses should be vaccinated annually. Pregnant mares vaccinated at least 30 days prior to foaling will provide immunity to their foals via the antibodies in the colostrum. Newborn foals whose dams have not been vaccinated prior to foaling should receive 1500 units of tetanus antitoxin at birth.

EQUINE ENCEPHALOMYELITIS

Equine encephalomyelitis, commonly called sleeping sickness, is a viral disease of horses that causes central nervous system disturbance and generally a high death rate. Three distinct viruses present in the U.S. are responsible for the sleeping sickness syndrome: the Eastern, Western, and Venezuelan viruses. All three are present in Texas. Along with horses, other mammals, including man, are susceptible to infection. However, horses infected with the Eastern and Western viruses are not sources of infection in humans. Horses infected with the Venezuelan virus can be a source of human infection. Equine encephalomyelitis in humans results in a flu-like syndrome that may progress into a central nervous system disturbance and could cause death.

Encephalomyelitis usually occurs during the spring and summer, coinciding with the mosquito season. The Eastern and Western viruses are maintained in nature by reservoir hosts that include birds, rodents, and reptiles. Outbreaks of the disease occur when the virus spills out of the reservoir hosts and into the general bird population. Mosquitoes feeding on infected birds can then transmit the disease to other mammals, including horses and man. Horses infected with either Eastern or Western encephalomyelitis are considered dead-end hosts and therefore incapable of spreading the disease further.

The Venezuelan virus is maintained in nature in the rodent population, not in the bird population. As with the other two viruses, the Venezuelan virus is transmitted to horses by mosquitoes. This virus differs in that the horse is not a dead-end host but is actually an amplifying host. Therefore, an infected horse is a possible source of infection for other horses and man due to the viremia associated with this particular virus.

Horses infected with equine encephalomyelitis may show a wide range of clinical signs. Initially, there is a febrile response with an accompanying anorexia and depression. Diarrhea may be present in those horses infected with the Venezuelan virus. Classical clinical signs of infection are those referable to the central nervous system. These clinical signs can be quite variable from horse to horse. They include a change in behavior, headpressing, circling, blindness, paralysis, and ataxia. Mortality rates range from 75% to 90% for the Eastern virus; from 19% to 50% for the Western virus; and from 40% to 90% for the Venezuelan virus. Death caused by the encephalomyelitis virus is usually preceded by recumbency, convulsions, and a comatose nature. Surviving horses usually have a diminished learning capacity and have been referred to as "dummies."

Treatment of infected horses is supportive only. Control of equine encephalomyelitis is provided by well-planned vaccination programs. Since all three viruses are present in Texas, an inactivated trivalent vaccine incorporating all three viruses should be used for immunization. Horses should be vaccinated at least 1 mo prior to the mosquito season. A modified live vaccine for Venezuelan encephalomyelitis is available but is not approved for use in pregnant mares. The modified live vaccine for the Venezuelan virus provides protection within 3 days while the inactivated vaccine requires 7 days. Horses infected with either the Eastern or Western virus do not have to be quarantined; those infected with the Venezuelan virus must be quarantined because of their ability to act as hosts.

RABIES

Rabies is a viral disease that attacks the central nervous system. All warm-blooded animals are susceptible to rabies. Transmission of the disease occurs from the bite of infected animals. It is primarily a problem in carnivorous animals, but horses can also be infected by the bite of a rabid animal. In Texas, skunks are the greatest carrier of rabies. At the present time, no approved vaccine for horses is available. Clinical signs of rabies in horses are extremely variable and positive diagnosis is confirmed only at necropsy.

CONCLUSION

Having discussed the infectious diseases common to horses, it is apparent that a well-planned vaccination program is of utmost importance. When the cost of treatment of a sick animal, plus the debilitating effects of disease and the resulting loss of time are considered, the expense of an adequate vaccination program is minimal.

Part 11

CONFORMATION AND TRAINING

44

HANDSOME IS . . . FORM AND FUNCTION IN THE PERFORMANCE HORSE

Matthew Mackay-Smith

A formula for the construction of the ideal horse has been the performance horsemen's goal for thousands of years. The wish is fond and the goal is ever receding for two reasons: 1) performance is extremely complex and the mixture is never the same in two performance horses, and 2) various performances require different sorts of horses so one man's perfection is another man's liability. There is, however, a relative relationship between construction and performance which bears a good working relationship to fairly simple mechanical principles. These can be discussed in some detail.

When we speak of idealizing the horse, we have to choose between an average horse for the widest variety of sports and a specialized horse for a particular sport. Considering the role that temperament, coordination, and education play, the discussion becomes effective only in comparing two horses of very similar performance expectations. Consequently, in choosing a performance horse, you should know the mechanical requirements of the sport, the temperament of the horse that you are looking at, and the kind of education and training that he will receive in order to evaluate his physical attributes.

If we divide all sports into three categories based on the type of exercise required, we can make some generalizations about the physical requirements, at least in terms of overall type, frame, muscling, and fat cover. Looking at the endurance horse, we would prefer him to be wiry with a high surface-to-weight ratio so that he would be an efficient shedder of heat. His frame should be light without sacrificing strength. The muscling should be flat, giving an angularity to the frame and outline because of the high proportion of slow twitch muscle fibers. The fat cover should be slight.

For horses in strength sports, such as pulling contests, roping, and the like, the type should be relatively massive; the frame heavier; the muscling bulky with intermediate fast twitch fibers and the fat covering moderate. The strength horse can use the additional weight to improve his leverage. The speed horse such as the sprinter, cutting horse, etc., should be muscular without being massive, have a moderate frame with bulging muscles exemplary of the fast twitch fiber type, and have a minimal fat covering.

It is the validity of these general observations about conformation that gives both an impetus to and a context for a more specific look at conformation. We understand that the motion of the horse is a mechanical event. We also know that efficient machines are designed and engineered for the job. The design deals with proper proportions and arrangements of parts. The engineering specifies what materials, the appropriate dimensions, and the actual construction of the parts. If we look at a catalog of parts, we can make some generalizations about how changes in their design or engineering would give them an advantage or disadvantage for a particular activity.

Beginning with the hoof, we can see that a narrow upright roof is good for hard ground and for turning quickly. A wide sloping or shallow foot goes well on soft or boggy ground. Draft horses are usually better off with a wide foot than a narrow one. Large, well-made feet will take a lot of pounding and are consequently good for endurance and for traction under a wide range of conditions. Small feet break over more easily in every direction and can lend agility to the horse if they are also strong and properly made.

The pastern comes in various lengths and attitudes. A low or easy pastern gives the horse a fluid gait and makes the ride smoother. An upright pastern is an advantage for power because it enables great application of force to the leg without overbending it. It is also an advantage for endurance because as fatigue sets in, the upright pastern will not sink so low to the ground. A long pastern increases the elasticity of the gait and lengthens the stride of the middle-distance racehorse. A short pastern is consequently an advantage for power or acceleration.

The cannon bone should be short for agility, but longer for staying power in a racehorse. Knees are best conformed neither over nor back; but when jumping is the specialization, a knee which is a little over will bend more easily and enable a better folding of the front legs over the fence.

The shoulderblade should lie in an angle which is moderate; extremely laid-back shoulders are an advantage for an elevated or round gait but are limited in the forward reach of the gait. A very long shoulderblade is an advantage for jumping because it improves the leverage for the lifting of the forehand as the jump begins.

A long back is an advantage for sprinting. A short back is exemplary of the stayer. The croup when flat gives greater elevation to the hind action and consequently is desirable in gaited horses. A steep croup gives greater power and acceleration, particularly when it is accompanied the way by a gluteal hump over the top of the croup, perhaps best remembered in the phenomenal racehorse Secretariat.

If the whole leg is relatively short, it improves the horse's power and acceleration. If it is relatively long, it improves his staying power. If the hock is bent more than average, it can be an advantage for power. If it is straighter than average, it can be a benefit in endurance or staying capacity.

It is most important, however, to remember that extremes of any of these variations are distinct weaknesses and decrease the horse's overall capability. Consequently, when we come to design a horse even for a specific sport, as well as particularly for general athleticism, we must look at the measurements as relative and avoid trying to develop a formula of absolutes. If we continue with our design for such a horse, we would want him to have front pasterns which were of medium length and moderate slope so that a straight foot axis is easy to obtain and maintain. The front cannon bones should be relatively short in proportion to the fore-arm. The knees should not be calf-kneed or back-at-the-knee and not excessively over or bucked. The forearm should conversely be long in proportion to the cannon bone. The arm or humerus should have its center over the center of the coffin joint. Standing at the side of the horse and visualizing the location of both the shoulder joint and the elbow joint, or even marking them with chalk, can enable you to estimate the angulation and position of the arm bone. The arm should describe an angle with the shoulder blade of approximately 110 degrees and an angle with the forearm of 120 to 130 degrees. The shoulder blade should be long and well defined, leading up to shapely narrow withers. Unless you need very high action in front, the shoulder should be only moderately laid back. Extreme lay back of the shoulder, as mentioned, can limit the forward extension and elasticity of the gait. The length of the shoulder blade, the way it fits into the withers, and its position on the side of the chest are more important than its angulation.

The hind pastern should be a little steeper than the front, perhaps 8 to 10 degrees. When the foot is under the middle of the gaskin and the hip joint, the cannon bone should be perpendicular to the ground. The hock should have moderate angulation--too straight or postlegged hocks are often seen with very low or easy pasterns which is a very unsound conformation. Too bent a hock leads to curbiness and weakness. The gaskin should be long in proportion to the cannon bone, but not so markedly as the relationship between the forearm and the cannon in the front leg. The thigh bone should be much shorter than the gaskin and angulated on the gaskin so that the hip joint can be connected to the standing coffin joint with a line which runs through the middle of the gaskin.

The athletic horse's head is fine in inverse proportion to the length of the neck. A short-necked horse needs a heavier head to achieve the same cantilevered balancing action as a very fine head on a long neck. Again, moderation tending toward length and fineness are representative of the most athletic horses. The back should be relatively short, although not grotesquely so and broad with good spring of rib, but over a chest cavity which is deep and relatively narrow. The croup should be long with an angulation to the horizontal of between 20 and 30 degrees.

We all know that it is possible to build a clumsy and unattractive horse with every property described in this list. Something of the artist's eye, together with the gift of grace, is needed to finish the picture both visually and physically. We recognize that crookedness of construction or extremes in the properties we discussed are sources of unsoundness. There are described here a few absolute advantages for soundness, which were detailed exhaustively by Gouboux and Barrier (1800s). First among these is probably a large, strong hoof. The description "no foot, no horse" is absolutely reliable. Many people paint themselves into a corner by getting a lovely athlete on second-class hooves. This makes a nightmare for the farrier who often gets blamed for inadequacies that are not his.

Next in order of importance would be large, deep joints. "Deep" is used to describe the dimension from front to back. If the joints are wide, they have good resistance to twisting forces. If they are deep, the tendons and ligament responsible for motion have good leverage at the joints so that they resist overbending of the joints during support.

Another absolute advantage is so-called "flat bone", which refers not just to the bones themselves, but also to the arrangement of the tendons in the cannon area behind the bone. If the tendons stand well out from behind the cannon bone, the horse is said to have flat bone. This again indicates good leverage for these tendons to act through. The horse that is tied in under the knee is certainly more given to unsoundnesses of the entire spring apparatus of tendons and ligaments than is the horse whose tendons stand well away from the back of the cannon bone. A short cannon bone with a long forearm and a similar relationship of the hind cannon bone to the gaskin decreases the likelihood of wear and tear on the knees and hocks. It also improves the horse's agility and general athleticism.

Another clear cut advantage in the athlete is straight leg axis as viewed both from the front and back and also from the side to the foreleg. Likewise, a straight foot axis running from the fetlock's point of rotation through the pastern and hoof indicates a proportionate balance between tendons and suspensory apparatus which is likely to be the most durable.

In summary, then, if we wanted to look for an overall athletic horse working from a short list of virtues, in addition to a good set of legs, we would want to see him with a moderate-sized head on a long lean neck, a torso which has a deep narrow chest with well-sprung ribs and a strong loin. We would want the back to be equal or shorter in length to the croup and equal or shorter in length to the body depth. We would want the ribs closely coupled to the hips even in a longer back. We would want the arm or foreleg to be placed well forward on the torso with moderate shoulder angulation to give the horse a long underline in proportion to the length of his back.

Competitive achievement is made up not only of physical aptitude but also of temperament and of education. A horse that has any two of these attributes may be competent in his sport, but to be a super horse he will have to have all three.

REFERENCES

Gouboux, A. and G. Barrier. 1892 (2nd Ed.). The Exterior of the Horse. Lippincott.

45

PROGRESSIVE PRECISION TRAINING: THE MOST HORSE

Matthew Mackay-Smith

Performance-oriented horsemen have two major goals: 1) to make the most of each horse, and 2) to develop the super horse. (Maybe we should put those goals the other way around for some.)

Traditional methods of trying to reach these goals have been extremely wasteful. They waste potential never realized, they waste effort on horses who are never going to make it, and they waste time and money on injury and down time which could have been anticipated and prevented. The waste of traditional methods grows out of traditional perceptions. The first of these is "He's bred for it, so keep on trying." The second is "His Daddy won on this program, so stick to it." The third is "His Daddy was sore a lot of the time too, just tough it out." All these acknowledge important aspects of horse training, but tell only a limited part of the story.

Certainly pedigree is important, but you can't make a silk purse out of a sow's ear and you can't make a top performer out of an individual who is physiologically or mechanically unsuited to the work. Yes, work is necessary to make an athlete, but each horse's program must be tailored to his individual mechanical and physiologic needs. And each horse is his own control, determining what the next step in his program should be. Yes, hard work can create discomfort, but by measuring the discomfort before it becomes damage we can anticipate and prevent injury and reduce down time from accidents. **Most athletic injury is in fact not accidental but the result of ill-advised or inappropriate work or performance schedules.**

The serious breeder, (or owner, or trainer) of performance horses therefore has additional needs which are in no wise met by conventional methods. First he needs a system of early recognition of superior performance ability to answer the question, "How good can this horse be?" Second, he needs complete development of that horse's ultimate athletic capability to fulfill the potential he has recognized. Third, he needs a systematic and objective way of reducing the risks of training and performance to improve the horse's overall record and cost efficiency. These needs are met by the application of a multidisciplinary scientific approach to athletics called "sports medicine" or "sports science." The subject is too vast for any comprehensive treatment here, but an outline of the theory and methods will provide a departure point for those interested in a deeper study.

The theoretical approach is grounded in some simple truths about the nature of the athletic animal. First of all, performance is related to genetically-determined capabilities which are modified by nurture and training to deliver performance. Because of the built-in nature of many of these features, the proper scientific measurement of various aspects of the young horse can be used as a predictor of his ultimate capability or to determine the existence of significant limitations. Second, when worked, the horse's body undergoes resulting physical and chemical changes which are a precise reflection of the amount and nature of that work. When that work is within tolerable limits, an adaptive response creating higher and higher levels of work-tolerance occurs. When the work is excessive, measurable evidence of its damaging effect can be used to quantify the damage. The tissue involved in the damage has a biological response rate which dictates the time which will be required for recovery from the injury.

What follows is a discussion of the application of scientific measurements using these basic principles to the goals of early assessment, optimal development, and minimal risk in the horse athlete.

We know that heart size related to body weight is an important determinant of athletic ability in strength and in endurance. Heart size determines the amount of blood it can pump per stroke. Since the maximum heart rate for horses is about the same regardless of heart size, and since the number of beats per minute is characteristic of a particular effort level, it follows that the horses with the larger hearts will be able to pump more blood per minute and consequently distribute oxygen more efficiently.

Three measurements of heart size can be used in horses a year old or less to determine this aspect of future capability. First of all, the resting pulse can be used, since the fewer beats per minute a horse uses, the larger the heart must be in order to maintain the minimum resting circulation. Resting pulse should be taken systematically at the same time every day with the horse in truly quiet and restful surroundings in order to be meaningful.

A second method is echocardiography in which an ultrasonic picture of the heart is developed and gives not only its absolute size, but gives a moving picture of its function as well.

A third method is to develop a heart score in which certain measurable aspects of the electrocardiogram are factored with the horse's weight to determine the relationship between heart volume and the horse's size.

A second early assessment technique takes advantage of the difference in fiber type of the muscles characteristic for various speeds of muscle contraction and therefore athletic capability. There are two basic fiber types in muscles. The first has two or three subtypes. This type is called "fast twitch" because of its ability to contract very rapidly. Some of the fast twitch fibers cannot directly use oxygen for contraction and are therefore limited by the energy they contain when the exercise begins. They are exhausted in 8 to 25 seconds, depending on the degree of training. The other type of fast twitch fibers can be taught to use oxygen through appropriate training. These fibers are responsible for high intensity work of between 20 seconds and 3 or 4 min. They can also participate in longer exercises when only some of them are being asked to participate at any given time. These oxygen-using fast twitch fibers represent the majority of fibers in Thoroughbred horses and also a high proportion in Standardbred racehorses. The non-oxygen using fast twitch fibers are most representative of the all-out sprinting horse such as the Quarter racing horse.

The other main type of fiber, the slow twitch fiber, can operate only in the presence of sufficient oxygen. Therefore, it is characteristic of exercises which the horse can sustain for long periods. The capacity of the slow twitch fibers to sustain exercise at a more rapid rate can be increased through intensive long-distance training which improves the distribution of blood capillaries on the fiber surface for the most efficient oxygen delivery.

The fiber type distribution achieved by a horse by the time he is a year old is characteristic of that horse for the rest of his life. Therefore, sprinters can be distinguished from middle-distance horses and from endurance horses at this early age. This does not mean that the classification is precise. It does mean that horses with a high proportion of slow twitch fibers will be intrinsically slower than comparable horses with a predominance of fast twitch fibers. It also implies that horses with the largest number of intermediate or oxygen-using fast twitch fibers will have the greatest versatility in training for various sports.

The neurology of gait production, or the nervous control over the pattern of the gait, is determined genetically and changes only slightly with training. This change is significant but is not determinant of absolute capability. Computer reading of successive high-speed photographic picture frames can analyze the neurologic control or timing of the gait to detect both those with superior control and those with inferior or defective gaits. Animals who are foul-gaited will be severely handicapped in any training program pointing them toward high performance. Likewise, horses with superior gait characteristics will be able to handle increments of work more readily and to get on with the business of adaptation to high performance.

A fourth approach to early selection is in body measurement. By measuring the segments and body proportions of superior performers and inferior performers, and by computer analysis of the data, some sports now have available analytical services which relate the probability or capability of high performance to specific physical proportions and constructions. This, again, is largely genetically determined, and consequently tends not only to hold up throughout the horse's performance but to be a benchmark for selection of animals for breeding stock.

The optimal preparation of horses for high performance is achieved through **progressive precision training.** The important features of this approach include an early start in the training process. The animal's body is most capable of responding to work when the animal is immature and still growing. The junior athletic programs for human athletes are witness to both the benefits and the perils of an early start. Young athletes are capable of rapid development in a sport, but excessive or ill-advised exercises can quickly snuff out the prospects of early promise through injury. Consequently, the Little League program has certain prohibitions about the length of the game and the type of pitch which young pitchers are allowed to use. Nevertheless, there is no baseball major leaguer who began to play the game after the age of 18, which approximates the 2- to 3-yr-old age of the horse. Our best human athletes have been in some kind of progressive exercise of a systematic nature since the age of 10 or earlier. Consequently, with newer and better methods of monitoring the effects of exercise we can now look forward to starting young horses as early yearlings or shortly after they have been weaned in order to obtain the maximum athletic development.

The second feature of this program is progressive loading with adaptive plateaus. That means that the work level is increased by measured stages and that the work is kept at that level until evidence of the body's adaptation to the new load is seen. Each increase in the work will produce visible and measurable evidences of stress which are spelled out elsewhere in text and articles. Briefly, these include signs of slight fatigue with a prolongation of pulse recovery time, shortening of the gait, and elevation of blood-chemistry features characteristic of increased muscle exertion. Over the next 2 to 5 days, as the adaptive process occurs, the horse shows a return to the attitude and appearance of the unstressed athlete ready for additional work increment.

Still another feature of progressive precision training is sport specificity. This means that, as early as is practical, you should start using the gait and maneuvers (but not the speed) which are characteristic of the sport. The learning of high performance by a horse involves a great deal of neurologic conditioning and patterning as well as development of bones, tendons, and muscles. The imprinting of neurologic patterns or "engrams" from thousands and thousands of repetitions of a particular maneuver gives ease to the performance which is sparing of the horse's effort and protective from injury. Once these neurologic patterns are in place, the strength and speed levels of the performance can be picked up much quicker and with greater effect.

The last major feature of this approach to preparation is that the work should progress and proceed from long slow distance to strength, to speed work. Long slow distance implies that level of effort which can be done entirely with the oxygen energy pathway for periods of 20 min to 30 min or more, at speeds which will increase from introductory levels of 5 or 6 miles per hour for 1/2 mile, up to 8 or 10 miles at 10 or 12 miles an hour, and then 5 to 7 miles at 15 to 18 miles per hour.

When a level of long slow distance achievement that is either appropriate to the horse's eventual sport or nearing his physical tolerance is achieved, the distances are shortened again, and the speed is further increased. This begins to use and develop the higher proportion of the intermediate fast twitch fibers capable of using oxygen to increase their chemical efficiency and blood vessel availability. Only in horses whose careers demand the ultimate sprinting capability should all-out speed be included as part of the training program--and then only as the last part. Speed is so destructive that its promiscuous use in horses whose careers will never require it merely flirts with injury rather than adding to performance capability.

This slow distance-to-strength-to-speed progression uses pulse rates as the most important criterion of work level achieved. Progressive loading in long slow distance will bring the horse's steady state pulse rate from about 125 up to as much as 175 or 180. Work which stimulates the intermediate fiber types uses pulse rates in the 190 to 200 bracket. The pulse rate characteristic for a given strenuous speed will slowly decline as the animal becomes more fit. Finally for all-out speed, pulse rates as high as 240 or more are typical.

If during a given work the pulse rate for a given speed goes up, that is an indication of increased loading or stress levels having been achieved and signals the end of that work effort regardless of what was scheduled for the day. If the pulse rate for a given speed is constantly dropping from day to day, this indicates that no further training effect is being achieved at that speed and that a higher level of effort should be laid on.

Finally, to reduce the risk of a progressive program and to carefully monitor the result, a number of progressively more sophisticated methods are available.

First of all you should rely heavily on your subjective analysis, letting the horse tell you how the work is going. The signs of a stressful work include a slight decrease in appetite or the speed of eating up, a barely perceptible dulling of the horse's attitude, and perhaps a little drooping of his appearance overall. He may show a little puffiness around some of his lower leg joints or a little shortening of the gait from slight overall soreness.

A scale or weight tape can be used to follow the horse's body weight closely. Increments of work tend to bring a valley in the daily weight graph. The horse should recover his standard weight in 2 days to 5 days after each work increment. More prolonged weight loss indicates failure to increase the energy level of the diet or some deep-seated adverse affect of the work increase.

Numerical monitoring can also be achieved through following heart recoveries. The recovery after a standard work is kept on a graph daily, or two or three times a week, on a systematic basis. The faster the recoveries get, the greater the adaptation to the work; the more prolonged the recoveries are, the greater the stress effect or progressive load which the work imposed. Likewise, there will be a small but graphically visible change in the resting pulse upward for a day or 2 after a hard work; progressively downward as adaptation takes place and his fitness increases over the months and years.

Another numerical measuring device just becoming available is a soreness meter. This is a forceplate on which the horse stands with one leg of a pair and which measures the balancing effort the horse has to make to hold up one whole end on one of his front or back legs. Since pain in the leg increases the vibration or balancing effort in the leg, a computer analysis and numerical readout of that effort gives a soreness number which can be followed from day to day and week to week through training to anticipate leg pain before it becomes visible as lameness.

Another simple device for checking the stress effects on the limbs and muscles is an infrared thermometer. With some practice, this measurement of heat radiating from specific body parts gives an idea of their recent activity and of the biological response to exercise in recent days and weeks. The persistence of heat in or around a joint, tendon, or muscle suggests an adaptive effort which is bordering on injury or frank injury needing rest and attention.

At still another level, the progress and effect of training can be followed in the laboratory by measuring blood levels of specific muscle enzymes and the by-products of muscle activity such as lactic acid, and by measuring various properties of the muscle itself through a needle biopsy technique.

This overview of a progressive precision training program beginning with early assessment and supported by objective monitoring and risk reduction gives the horseman with a performance connection a whole battery of methods and tools to get much closer to the super horse and to reduce the wastage and injury in horses of less extraordinary ability.

46

JUDGING HUNTERS, JUMPERS, AND HUNT SEAT EQUITATION

Don Burt

RULES AND PROCEDURES (RULE BOOK)

Horse show judges' responsibilities include making sure 1) that rules and class procedures are followed according to the rule book; 2) that a bookkeeping system adequate for writing of results is used; and 3) that knowledge and experience are the basis used in judging. A "hunter" is a show-ring approximation of a horse galloping and jumping cross-country, originally behind a pack of hounds chasing a fox. A "jumper" is a test of the horse's athletic ability: his fitness, his agility, his power and his intelligence. "Equitation" is that position of the rider that produces the maximum effectiveness from the horse with a minimum of visible aids. Factors in judging these three divisions are 1) hunters: pace, stride, manners, style of fencing, way of going, performance, equipment, fitness, soundness--each relating to the horse galloping (in a group) across country far away from home base; 2) jumpers: judged on faults and sometimes time; and 3) hunter seat equitation: judged on rider's position and effectiveness. The horse is not to be considered as long as it can do the required work. However, a rider will often be able to keep his aids more invisible and appear to be more effective on a smoother, more experienced horse. The following guidelines should prove useful in planning and judging these divisions and skills.

Course Description and Requirements

Hunters

a) No spread over 4-ft wide
b) No square oxers allowed
c) Natural fences preferred
d) On-stride distances between fences
e) No triple bars or hog's backs or swedish oxers
f) No triple combinations
g) No illegal heights

Jumpers

a) About 50% spreads
b) Some square oxers
c) Striped poles and brightly painted fences
d) Some tricky distances
e) Triple bars, hog's backs, or swedish oxers permited
f) Triple combination desired
g) Fences usually higher than hunter fences

Some ingredients of a good hunter course

a) A flowing, smooth path--no sharp turns
b) At least one change of direction (lead)
c) No scary fences
d) One or two easy fences at the beginning of the course
e) At least one combination (in and out) towards the middle or end of the round
f) At least eight separate jumps required of the horse
g) A different number of strides in each line of fences
h) No distance problems

Hunter seat equitation

The easier equitation classes stress the position factor more than the effectiveness factor. These courses resemble hunter courses more than jumper courses. The harder equitation classes, such as medal classes, stress effectiveness because most of the good riders have their positions down pat, and a course that was too easy would not test them adequately. Courses for these equitation classes more closely resemble jumper courses.

BOOKKEEPING

Over-Fences Classes

General

a) Judge's obligation: The obligation of a judge is to have a system that accurately places each horse in position relative to the entire class.
b) Giving reasons for discussion: Although whatever system used is personal and the judge is giving his subjective opinion as to the relative placings, it is important that the judge be able to give specific reasons to support his opinion. With permission of the

judge and steward, an exhibitor has the right to question a judge regarding the performance of his horse. (Decision is unprotestable.)

c) Bookkeeping system: It is only by having a good bookkeeping system (which accurately describes everything a horse does from the time judging starts until judging stops and the card is turned in) that a judge can fulfill his obligation to an exhibitor when questioned.

d) Scoring system: The only time the type of scoring system should be used in which each fence is given a numerical score and the scores are added to produce a final score is when judging jumpers. In judging hunters and hunter seat equitation, a shorthand should be developed that describes each fence and what happens between fences.

Jumpers

The scoring of jumpers is largely mechanical. That is, the points or faults are automatically given when a horse hits a fence, knocks one down, stops, or commits some other error. Judging jumper classes is not easy, however. Factors that complicate jumper judging are: the clock, the technicalities of the rules, and the judgment calls. Because of these factors, an assistant should be provided so that a judge can keep his eyes on the horse during the whole performance.

Judging difficulties

a) The clock: In classes in which time is a factor, the horses usually travel rather fast and turn abruptly. Sometimes a horse will stop and disturb the fence at the same time so that it must be reset before the horse can start again. The judge must be up on all the possible situations regarding the stopping and starting of time and make sure that the timekeepers are briefed properly on the signals.

b) The technicalities (subject to local conditions):
- Illegal or inadequate courses:
  - No poles over brush jumps
  - No starting and finishing markers
  - No oxers and(or) spread fences
  - Two or more faults occurring at one obstacle
  - Bottom poles falling down with top one standing

Course set in such a way that a judge cannot adequately see all of the fences
Incorrect distance between fences

- Problems with equipment and personnel:
  No stop watches
  No knowledgeable people to run the timing devices
  No whistle
  No recorder or assistant provided
  Lack of communication among the judge, timekeepers, and announcer

c) Judgment calls:
  Loss of forward motion
  Crossing one's track
  Circling
  Stopping and knockdown vs knockdown only
  Addressing a fence

Hunters

The score card basically consists of two parts--the numbered boxes that describe each fence with the notes or comments section and the numerical score part.

a) Develop symbols that describe the horse's performance (figure 1 and 2).

b) Develop three categories for your comments: 1) one for general impressions; 2) one for way of moving; 3) one for style of fencing:
  (tns) Tense horse
  (scp) No scope
  (str) Strong
  (ch) Choppy strided
  (gm) Good mover
  (fm) Fair mover
  (bm) Bad mover
  (gj) Good jumper
  (fj) Fair jumper
  (bj) Bad jumper

c) Develop an eye for judging hunters: When you first start judging hunters, your eye is not usually fast enough to catch multiple problems that occur at a fence, and so a judge will usually just put one or two major things down in each box. At first, because of this, a judge need not have very many symbols that he uses. But as he judges more, he will develop his eye to see more faults at a fence and will need more symbols to describe them quickly.

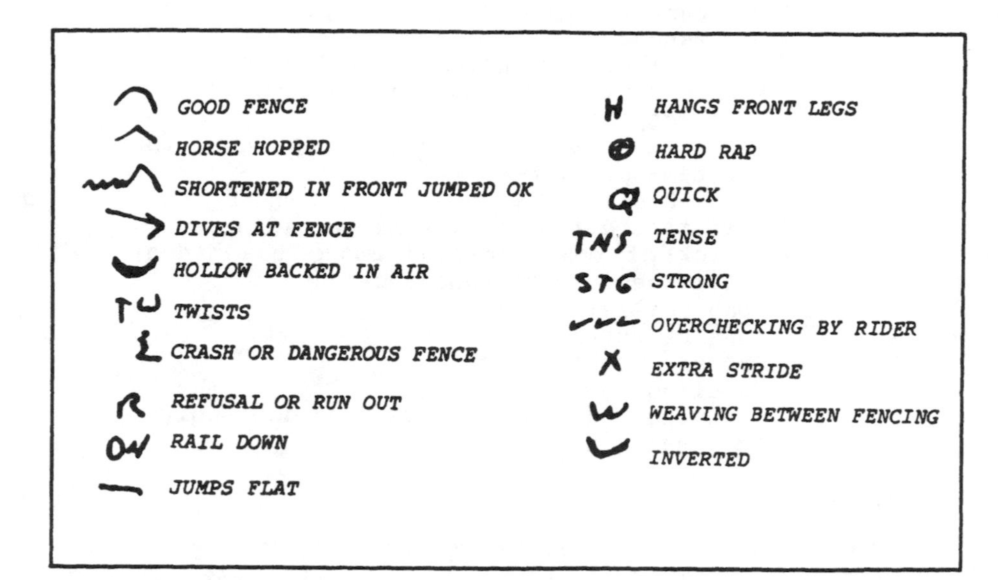

Figure 1. Scoring symbols.

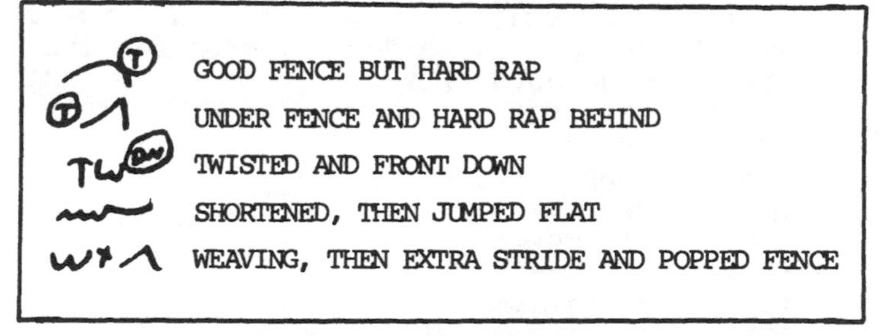

Figure 2. Symbols used in combination.

d) Grade the performance: Once you have written down the description of the horse's performance at each fence and between fences, list general comments and rate him as to moving and jumping style, then give the total round a numerical "grade." Below are suggestions for a system for the numerical score.

0 = not completed or eliminated
40 = two or more major faults (rail down or stop)
50 = one major fault
60-69 = poor performance (D)
70-79 = average performance (C)
80-89 = good performance (B)
90-100 = excellent (A)

e) Place horses relative to each other: Although you should try to have an objective standard of what you think a particular round should be, realize that it is necessary to place the horses relative to each other. Therefore, if you think that a particular round deserves an 82 after comparing it with the other rounds that you already have given an 82, you might decide to give it an 81 or 83. This is where the written symbols and comments become invaluable, since you do not always remember each horse's performance.

f) Develop efficient scorekeeping procedures: There are two other things that one can do as a procedural matter to facilitate the book-keeping (scorekeeping) process. One is to write the type of fence above the boxes containing the fence numbers. This makes it easier to describe to an exhibitor at which fence his horse made a mistake. It also is a better memory jogger when comparing rounds of two close horses. The second thing is to keep a running order of the high-scoring rounds that you feel might be in the ribbons at the top of the card. Then you will be able to avoid delays in time from the time the last horse leaves the ring until you turn in your card. Exhibitors and management both appreciate an efficient judge who does not delay the show with his bookkeeping procedures.

Hunter seat equitation--over fences

Regarding a bookkeeping system for equitation over fences, the procedure for marking a card is the same as for hunters but with a number of exceptions:

- The main exception is that you are now judging the rider, not the horse. Although the horse's performance is irrelevant, it goes without saying that a smoother, more seasoned horse will cover up many of the equitation problems that a rider might have. We must then look at the rider and pay careful attention to his or her position and his or her effectiveness with position, and particularly how obvious the rider's aids are to obtain that effectiveness.
- We do not need the column for the horse's way of moving because the horse isn't being judged.

- We do not need the column for the horse's way of jumping (style of fencing) because we are not concerned about it in equitation classes.
- We need to use a set of symbols that are rider oriented, not horse oriented. Possibly some of the hunter symbols would apply, but definitely the rider symbols would be more important than the horse symbols, in general, with the exception of the horse symbols that reflect a serious omission or commission of the rider.
- An example of some of the equitation symbols that can be used are as follows (figure 3):

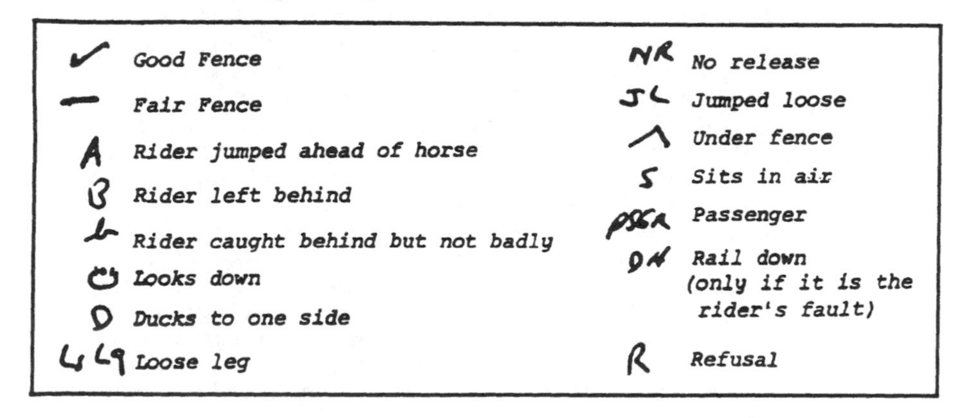

Figure 3. Equitation symbols.

Scoring comments:

- Some general scoring comments that can be used are as follows: loose, bad hands, bad eyes, elbows out, hands flat, hands too high or too low, lost stirrup, ducks, looks down, stiff, no release, rounds back, throws body.

47

GROUND TRAINING AND BREAKING THE HORSE TO RIDE AND(OR) DRIVE

Joe Staheli

An initial phase of training is called ground training. It includes: 1) a patient and gentle method of teaching the horse to lead, 2) teaching the horse to be tied so that he is quiet and safe when tied, 3) teaching the horse and the horseman principles of safety on the ground, 4) teaching the horse and horseman how the horse's feet may be picked up for inspection and care without danger to either (figure 1), 5) teaching the horse and horseman the proper method of using the lungeline as part of the training, and to walk, trot, canter (figure 2), and halt, and 6) teaching the horse and horseman to use long driving lines to help learn obedience and coordination. "Sacking the horse out" is a process used in gentling and training in which the horse is slapped with a sack or saddle blanket. "Saddleing for the first time" requires time and patience. The horse must become familiar with the appearance, smell and feel of the saddle. Handle the saddle gently by placing it on the horse slowly and easily without undue noise to avoid exciting him. Let the stirrups and cinch down gently. Cinch the saddle just snug enough to keep it in place. "Ground driving with long lines" is usually done when a horse wears an open bridle and a surcingle; it helps teach the horse obedience and coordination. After these first experiences the young horse is gradually advanced through the 1) first ride, 2) harnessing, 3) driving on the ground with shafts, drags, and carts, and 4) using side checks and overchecks (to keep horse's head in proper position).

To Pick Up a Front Foot

Stand on the left side of your horse, facing his rear. Place your left hand on the horse's shoulder. Bending over, run your right hand gently but firmly down the back of the leg until the hand is just above the fetlock. Press against the horse's shoulder with your left hand, thus forcing his weight onto the opposite foreleg. Grasp the fetlock with the fingers.

When the horse picks up his foot, place it between your legs, and support it on your left knee.

To Pick Up a Hind Foot

Work from left side. Stand well forward of hind-quarters, facing rear. Gently stroke back as far as point of hip with your left hand. Stroke leg gently but firmly with right hand down as far as middle of cannon. Press against horse's hip, forcing his weight onto the opposite hind leg. Grasp the cannon just above the fetlock with your right hand, lifting foot directly toward you so leg is bent at hock.

Then move to the rear, keeping the hind leg next to your thigh. Avoid holding the foot out to one side. The discomfort of this position will make him resist. Swing your left leg underneath the fetlock to support the leg firmly.

Figure 1. Proper method of picking up a horse's feet (Mississippi Agricultural Extension Service Publication 450).

Gaits

A gait is a manner of walking, running or moving. The three natural gaits of any horse except the Tennessee Walking Horse are the walk, trot and canter. The natural gaits of the Tennessee Walker are the walk, the running walk and the canter. Two other gaits--the slow gait and the rack--are artificial and must be learned.

Walk

A 4-beat gait with the feet striking the ground in the following order: right front, left rear, left front, right rear. The feet should be lifted from the ground and placed down flat-footed.

Trot

A 2-beat gait in which the left front and right rear feet and right front and left rear feet strike the ground together. The horse's body remains in perfect balance. The trot should be balanced and springy.

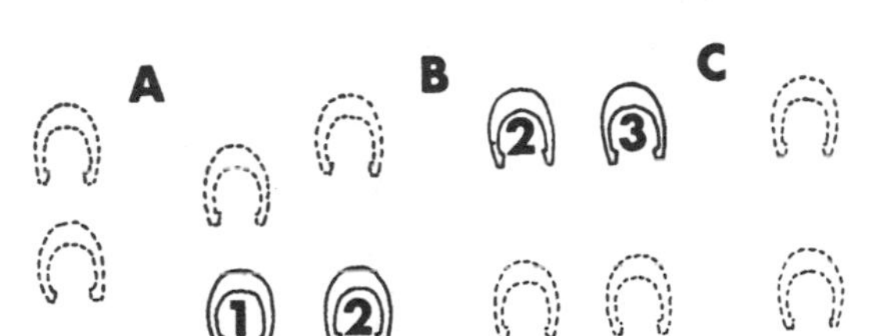

Canter

A 3-beat gait that should be slow. The canter is actually a restrained gallop in which the horse may lead with either of his front feet. The lead foot will be the first to leave and the last to strike the ground. When a horse leads with his left foot, the feet will strike the ground in the following order: Right rear; left rear and right front; left front.

Pace

(stepping pace modified)

Fast, 2-beat gait in which lateral fore and hind feet hit ground simultaneously. Faster than trot but slower than run or gallop. Pace popular gait in England. Objectionable in soft underfooting. Makes way for quick bursts of speed. Speed gait instead of road gait.

Slow gait — an artificial 4-beat gait of the 5-gaited horse. It is smooth-riding gait, but very tiresome to the horse. Each foot is moving more or less separately in this order: left rear, left front, right rear, right front.

Rack — the only difference between the slow gait and the rack is speed; the rack is faster.

Running walk — the natural gait of the Tennessee Walking Horse. It is a very smooth gait with a gliding rhythm that is quite pleasant for the rider. The horse will average about 7 miles per hour in this gait, and his head will have a definite up-and-down movement. This is a 4-beat gait, with the feet striking the ground in the following order: left front, right rear, right front, left rear.

Figure 2. Source: Mississippi Agricultural Extension Service Publication 450.

48

HOOF CARE AND EVALUATING THE FARRIER

Joe Staheli

The adage "no foot, no horse" points out the importance of a horse's foot. The foot is contact point for the horse and the ground. The foot and its various components are important weight-bearing, and shock-absorbing mechanisms of the horse. A horseman should understand the structure and function of his horse's feet and work with a farrier (horseshoer) to properly care for them. The structure of the hoof includes three main parts--the horny wall, the horny sole, and the frog (figure 1).

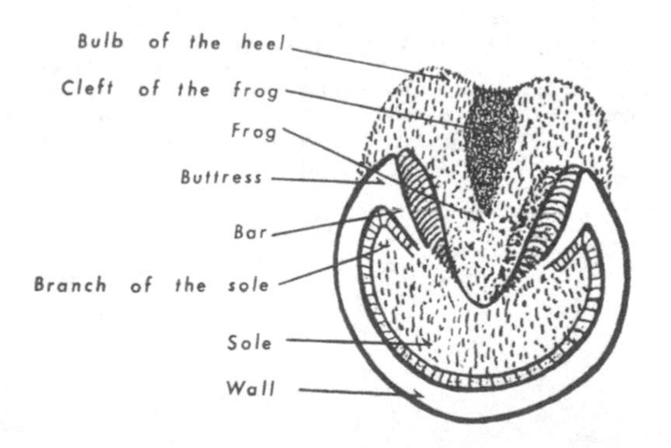

Figure 1. Parts of the hoof (Mississippi Agricultural Extension Service Publication 450).

The horny sole grows out from the live fleshy sole and forms the bottom of the foot. The horny wall and sole are tough protective surfaces that cover the sensitive inner tissue with its complex series of blood vessels and nerves. The frog is a semisoft, elastic, triangular-shaped structure that grows on the sole between the bars. The frog acts as a cushion, absorbing shock when the foot strikes the ground. It is the horseman's responsibility to inspect and care for the horse's feet daily and clean mud or manure from around the bars and frog as needed. This can be done with a hoof hook. An experienced horseman can keep the bottom wall of the unshod horse's hooves level by proper and careful use of the rasp. The horseman depends on the farrier to design, make, and fit the proper shoes for the horse when shoes are needed. Horses should be shod when they are to be used on hard surfaces or in other situations that will damage the feet. Also, shoes may be used to correct hoof structure or growth, to protect the hoof from abnormalities, and to make changes in gaits and action (figure 2).

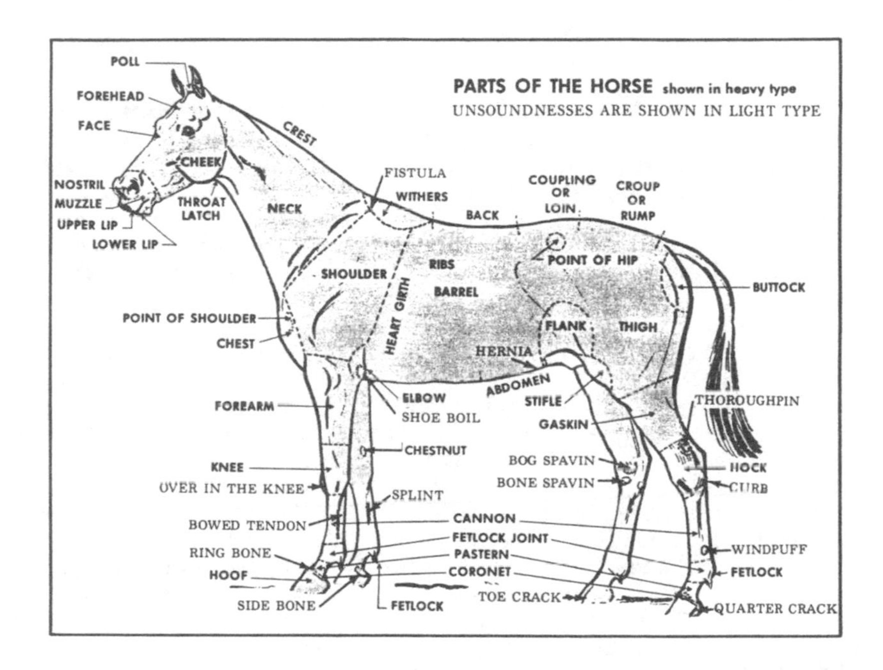

Figure 2. Source: Iowa State University Cooperative Extension Service Publication C-512.

The horseman needs to be able to recognize when he is receiving proper advice on these matters. The farrier must be sufficiently knowledgeable, skilled, and experienced to perform his duties properly to maintain the horse's feet in a healthy condition.

49
TRAINING THE WESTERN HORSE AND THE ENGLISH HORSE

Joe Staheli

The "Western" horse is considered to be a horse of any type, breed or cross identified with western U.S. It is generally associated with ranching and the ability, skills, and disposition needed to perform as a ranch horse. The western saddle and tack are also identified with this type of horse and the horsemen who use it. The western seat or stock seat describes the position of the rider in the saddle. He sits in the deep part of the saddle, with shoulders square, head erect, knees slightly bent and with the weight directly over the balls of his feet. One hand holds the reins, the other is comfortable at the side, and his center of gravity is directly over the horse's center of gravity. The horse requires training for this type of riding. He must learn to go forward and to know and use the gaits, the walk, the jog (trot), and the lope (gallop) on the proper commands by the rider. The horse and rider must be trained to function as a unit in good form or a collected manner. Both the horse and rider must be trained to be ready to act or respond to a command or a stimulus. The horse must be trained for movement and action in proper form with the right degrees of impulsion and flexion. Special manuevers for the horses such as pole bending and special physical features of the horses require attention of the trainer and the rider in developing the athletic ability of the horse. The "English" horse may be of any type, breed, or cross identified with the style of equitation and activities that originated in England. It is associated with the style of horsemanship used in riding in the hunt, riding gaited horses, and performing in events typical of English-type horse shows. The English seat (sometimes called the "saddle seat" or the English hunting seat) is comfortable and efficient for riding any gait for any length of time. The rider conveys the impression of effective, easy control of the horse. The saddle and the tack are specialized for the English rider. The saddle has a flat seat, a slightly raised pommel without a horn, and is generally very light in weight. To be ridden for English events, the horse requires basic training similar to other horses on going forward and in the gaits (walk, trot, canter). The horse and riders need specific training to be able to function as a coordinated unit with good form and a collected manner in English events. The training helps the horse and rider understand the principles and terms of equitation and their implication--impulsion, horse elevation (front), flexion and suppleness, and breaking at the pole.

Part 12

SHOWING AND SELLING

50

GROOMING SHOW AND SALE HORSES

Don Burt

Keeping our horses healthy and happy in what is for them "unnatural" circumstances requires daily concern and care. We must feed them properly and supply them with a constant source of good water. We must trim their feet, brush them, clip them, and exercise them. We must help our horses survive our world, which is alien to them.

Horses in the wild went ungroomed. A million years of evolution fitted them with unique traits to help them live healthily and happily, roaming vast grasslands. Their hooves were kept trimmed from constant travel over rocky terrain. Their hair grew to protect them in the winter and was shed in the spring to allow them to keep cool in summer.

Paleontologists tell us that many varieties of horses have come and vanished over the course of the creatures's existence. Horses on dry, upland ranges developed small, hard hooves, and horses living in damp meadows came to have large, round hooves. Horses in some environments were small and robust. Others living in cold forests became large, hairy creatures. As climates and topography changed, the adaptable equine changed, or perished.

Charles Darwin called it survival of the fittest. Only horses that were fitted with the most suitable traits for the environment survived. Today, we find ourselves with animals who can supply their own needs on a vast, open range. But how many of them are at home on the range? Not many.

We put our horses in damp, closed-in stalls or dry, dusty corrals. The lucky ones roam a 5- or 10-acre paddock. We have changed our horse's environment rather drastically. He is bound to perish if we don't care for him.

The daily care needed by our equines is known as grooming. It has always surprised me to find that many horse owners think of grooming as something done only for the show horse. Actually, every horse needs grooming to stay fit and healthy. As a veterinarian, I see many horses every year who die or are permanently disabled as a direct or indirect result of neglect. This is not careless neglect, but is usually a result of ignorance.

CROWDING

We crowd our horses into small areas and greatly increase the danger of infectious diseases. Many of the equine diseases we have today probably did not exist 10,000 yr ago when horses roamed in small numbers over large areas. Today we must worry about encephalitis, flu, rhinotracheitis, and other problems. It is our responsibility to vaccinate against diseases whenever practical. The first thought of a good groom is immunization. Even the healthy horse should see the vet 2 or 3 times a year for shots.

Very few horses today are raised on terrain that will keep hoof growth from becoming excessive. Periodic trimming is therefore essential. Otherwise, hooves will crack and split and lameness may result. Horses that are ridden heavily, especially on hard or rocky surfaces, need to be shod.

Diet, weather, and the type of ground will determine how often hooves will need trimming. Under some conditions, trimming is necessary every month. Often, every other month is sufficient. The average interval is about 6 wk.

A groom should clean the horse's feet every time he brushes him. This can be done with a simple hoof pick. The mud and caked dirt should be removed from each groove between the frog and the hoof bars. These form a V-shaped crevice that starts near the heels and points toward the toe. The frog should not be trimmed, but all dirt scraped from the sole. With the sole of the foot clean, a groom can check for thrush, a degenerative condition of the sulci (grooves) or frog. A black necrotic, foul-smelling material will be found along the sulcus or on the frog if thrush is present.

PARASITES AND TREATMENTS

One of the biggest problems created by crowding horses into small acreages is the proliferation of internal parasites. These terrible creatures come in many sizes and shapes and have various methods of infestation.

Their methods are made possible because horses live year-round on the same plot of ground. It is no longer possible for the horse to gallop over the hill and graze on fresh pasture that hasn't felt an equine hoof in years. As a result, vast numbers of worm eggs can now be found wherever horses are kept.

Their numbers tend to build up year after year. Worm eggs are swallowed with feed or water. They mature in the horse, produce millions more eggs that pass out of the horse and increase the supply of eggs on the ground. It becomes a vicious cycle, and the horse is caught in the middle.

Once inside the horse, these worms cause severe problems. Adult worms rob the gut of vital nutrition that would otherwise go to the tissues of the body. The worm larvae bore holes through lungs, liver, kidneys, and sometimes gain entrance to the brain or spinal cord, causing permanent damage. Some larvae set up housekeeping in the blood vessels that supply the intestines. Over a period of time, lesions in vessels become so large from larvae that the blood flow is blocked. This kills the tissue of the intestine, which then breaks open. This allows the intestinal contents to spill into the abdominal cavity and creates an overwhelming peritonitis that causes severe colic and results in death.

Only regular worming with an effective anthelmintic will help. There is no way to completely eliminate all worms--worm eggs are too common and worms are too resistant to drugs, disinfectants and other control measures.

Today we have a variety of drugs that are effective against most internal parasites. The groom who switches from one drug to another will be more effective in eliminating worms.

Younger horses are more susceptible to parasites. Most horses build up some sort of an immunity against them. This means that foals should be wormed every 2 or 3 mo, yearlings about 4 times a year, 2 yr olds at least twice a year, and older horses once a year.

Great progress has been made in developing anthelmintic drugs that are easy to administer. They used to have a bad taste, and most horses would refuse to eat grain that was doctored with them. Now most horses will consume the new anthelmintics, and for those who still refuse, there are a variety of paste-wormers. The pastes are squirted onto the back of the tongue with a disposable plastic syringe. The drug adheres to the tongue and inner surface of the mouth and usually every drop is swallowed. For the few horses who manage to avoid swallowing it, there is the veterinarian with his handy stomach tube. The horse does not like having a tube down its nose, but it gets the job done and saves the horse from more unpleasant conditions.

EXERCISE

The horse is built to be extremely mobile. His huge muscles need exercise; his tendons need continual stretching to stay strong. It is criminal to confine a horse to a box stall for days with no chance for exercise. Exercise increases circulation and helps keep all the body tissues supplied with nutrients and drained of wastes. A confined horse without proper exercise is unhealthy.

Even a horse in a paddock may need more exercise than that of running up and down the fence. This is especially true with potential performance horses. Exercise builds muscle tone and size and makes a horse look better. Even a horse being shown at halter should have a comprehensive exercise program.

Exercise can be in the form of walking or jogging in hand, on a hot walker, lungeline, or under saddle. Ponying is a common form of exercise since most horse owners do not have a hot walker. Ponying allows a groom to exercise two horses at once. A stock saddle is used, and the lead shank of the horse being led is wrapped around the saddle horn.

Lunging is popular because it allows a horse to be trained during exercise. The horse can be worked in a bitting rig while the groom stands in the center with the horse traveling around him or her in a large circle. Proper lunging should afford equal time in both directions. Frequent rest stops keep the horse from becoming sore and resentful. The horse should be made to move out in long, energetic strides between rest stops.

GROOMING

The wild horse has a coat that is somewhat water repellent and good protection from the cold. The human brings him into a barn in bad weather, cinches a saddle on his back, and proceeds to ride him. When the horse sweats, lather is formed, which dries on the surface, interfering with body cooling in the summer and good hair insulation in the winter. Saddle and girth pressure often cause sores on the ungroomed, dirty horse.

Good grooming involves the removal of grease and scaly skin with a curry comb and brush. Occasional bathing is helpful. Once the coat is adequately cleaned, daily brushing will keep down dandruff.

The skin continually produces a natural oil that gives the hair a desirable gloss, and daily grooming distributes this oil over the hair and removes dirt.

Horses kept in stalls and small paddocks should be groomed every day whether they are ridden or not. A good grooming massages the skin, which increases circulation and promotes healthy hair growth. A daily grooming gives a person an opportunity to check a horse for cuts, bruises, sores, and general healthiness. People who toss their poor critter a flake of hay and run will often overlook serious problems for several days.

It is best to groom systematically, using the same equipment every day in the same order, starting on the same area of the horse each time and ending on the same area. This will put the horse at ease and allow him to anticipate what is happening.

A rubber curry comb should be used first to loosen dried dirt and manure. A metal curry comb is too sharp for comfort; a flexible curry comb can be used vigorously over the muscles. Some grooms rub in a circular motion.

A dandy brush is used next. It should be used with brisk strokes except on the face and lower legs. Strokes should be in the direction of hair growth. When a horse is shedding or is extremely dirty, the brush will have to be cleaned frequently. This can be done with a metal curry comb, on a fence or stall. A thorough groom may like to go to a short-bristled body brush next to get down between the hair and push out the fine dirt.

A human hair brush works well on the mane and tail. The daily use of a comb on the mane and tail will eventually thin and shorten them too much. Periodic attention should be given to the hair roots of the mane and tail. A buildup of dirt and dandruff in this area can cause such irritation that the horse will rub against objects until the hair is short.

Grooming is especially important after a long strenuous workout, otherwise a horse can end up with cramps, muscle aches and chills. A wet horse should be walked until his breathing returns to normal. Some horse owners like to walk a horse until he is dry, while others prefer to rinse the horse with water to get rid of the sweat.

A large body sponge works well for this. After washing, a towel can be used to help dry the hair, but walking after washing is the best method of drying. An excessively hot horse should not be allowed to drink too much water or colic may occur. Small sips of water over an hour or so should be given. It is best to blanket a horse after cooling out in cold weather.

Racehorse trainers rub their horse's legs with a brace and wrap them after a stiff workout. This assures that the legs are carefully examined after each workout for heat and swelling. The brace is rubbed in from the knees and hocks down to the coronet bands. The rubbing and the action of the linament increase circulation so that excess fluid in the tissue is absorbed. Rubbing and massage usually do as much good as the linament, especially if the massage is long enough. Legs that have a tendency to swell or develop warm spots are usually wrapped with a cotton or elastic leg bandage after being rubbed down.

STALLS AND FACILITIES

When we take a horse to the races or to a show, we are putting him out of his natural element even more. He is completely at the mercy of his caretaker. Unless he is especially well cared for, he can develop colic or injure himself.

It is up to the groom to inspect the stall before the horse is put in it; sometimes there are nails or slivers protruding. The floor should be inspected before bedding is put down. There should be no exposed electrical wires that a horse might chew.

Most grooms bring their own supply of screw eyes that they can put in the walls for buckets and a hay net. The edges of the stall walls should be banked a bit to keep the horse from rolling over on his back and getting stuck against the wall. A bank also helps keep out drafts.

There is always the danger of a fire at a track or show, and a groom should keep this in mind. The horse should not be locked into his stall; no smoking should be allowed near stalls; electrical outlets should not be overloaded and a halter and shank should be left in or near the stall in case a quick evacuation is necessary.

Being a good groom is important if you own a horse; your horse's life depends on it. He must look to you for his every need. We have taken away our horses' freedom; but they can be happy and content if we provide the care and concern necessary to keep them healthy and fit.

51

SALES PREPARATION OF YEARLINGS: EUROPEAN STYLE

Michael Osborne

Management of thoroughbred-horse farms requires understanding of many phases of biology, animal behavior, and human behavior. The author's experiences in the operation of thoroughbred farms in Kentucky and Ireland provide insights to many aspects of horse husbandry, marketing, and related technology. An overview of the author's observations is offered to assist horsemen and horse-farm managers to improve specialized phases of their horse operations.

The basic program for preparation of thoroughbred yearlings for sales is as follows:

- Sixty days prior to date of sales, the yearlings are placed in suitable stabling.
- Have veterinarians check teeth, parasites, feet, heart, and external genitalia.
- The animals are walked in a straight line to observe their gait and to make corrective shoeing.
- Shoe initially with aluminum racing plates. Young immature horses have problems when shod with steel shoes because the extra weight on the foot may cause gait changes with resultant damage from striking.
- Diet should include a top quality protein source of soybean meal and a vegetable oil, which ensure coat bloom. The energy content of the diet is relative to the amount of forced exercise--walking and lunging.
- Walking should start at 15 min/day building up to 1 hr/day for the 2 wk prior to the sales. Walking exercise should be clockwise and counter clockwise and at the fastest possible pace.
- Yearlings should be "sacked" every day from the beginning of the preparation period--rubbed and smacked with a jute sack all over the body. This makes a young horse accept grooming, handling, and breaking with greatly reduced reactions.

- Yearlings should be taught to stand balanced with all four limbs exposed to the viewer who is standing 6 ft from the horse's shoulder.
- The horse should be taught to stand and walk with its head at normal height and should always be turned counterclockwise when shown at the walk.
- Length of tail creates an optical illusion. The longer tail makes a horse appear shorter. A shorter tail makes a horse appear taller.

52

TYPES OF ADVERTISING IN MARKETING HORSES

Ben A. Scott

Advertising is, of course, one of the most essential items in any marketing program--and is perhaps the most misused by people not familiar with its benefits.

Advertising is divided into two basic categories: institutional and direct.

INSTITUTIONAL ADVERTISING

Institutional advertising is defined as advertising used in selling the name of the "institution," not just in selling the product. In the horse industry, we would refer to institutional advertising as promoting the ranch name, a trainer's name, and(or) promoting a specific entity and not a particular product. The dollars spent on this type of advertising are probably the most difficult for most people to spend because it is difficult to ensure success. Usually, we are geared to advertising only if we have something specific that we are selling. The average individual tends not to see the value of cold, hard dollars being spent strictly on a name. In reality, it is often the best advertising dollar spent.

The whole key to institutional advertising is, of course, repetition. Institutional advertising takes planning because to do it only once, twice, or three times a year is a waste of money. Institutional advertising is advertising that is put in front of the public on a monthly, daily, or weekly basis throughout the year. Signs are, in a sense, institutional advertising. Repetition is the key, and sellers must plan for the expense of this type of advertising.

As an example, in a breed magazine, one would have to decide how much money is to be spent annually on institutional advertising, divide it by 12, and then determine the amount of space to be purchased per month. In institutional advertising, it is advisable (a prerequisite) to have some standard logo, name, print, style, or picture so that the repetition is maximized. If you change the farm name, the type, and perhaps the color every other month, you lose the value of the repetition.

Time should be spent analyzing institutional advertising with the desired end result in mind. Are you advertising that your location is extremely convenient to some aspect of the horse industry? Are you advertising a particular aspect of performance of the breed you offer? The emphasis of your advertising should be toward that aspect of your ranch that is your specialty. This aspect should be reflected in the ad or your logo so that it is appealing and attracts the potential customer.

If possible, you should seek some advice regarding colors and style when putting together a particular logo. It should be easily legible. It should be in colors that do not detract from the information being presented. And, it should be in colors and type that are in keeping with your farm signs and image. Often, something looks very good when you are close to it, but once you step aside and look at it from 10 or more feet away, you find it is not even legible.

If you are unable to secure help from an advertising person, there are some basic rules. Don't be afraid of white space. That is, don't be afraid to leave some areas without any color or type, etc. Keep it simple. Convey your message. For example, it is not uncommon to see an advertisement that reads as follows, "Bunker Hill Ranch--Specializing in halter horses, cutting horses, stock horses, draft horses, donkey, mules," etc. In other words, if you say "specializing in," do not list 10 items--because that is not specializing.

Institutional advertising is not an absolute necessity, but if an individual truly wants to grow, the greatest items to advertise are yourself, your farm, and what you want to do. That is institutional advertising. Definite amounts should be budgeted for advertising every year. One of the advantages of this is that the IRS looks at an advertising program very heavily when determining whether you are in the horse business as a hobby or as a business. You should keep all advertising copy, all information used for it, and all the expenses regarding it. Keep it in a file so that you can instantly refer to it and it also will be readily available in the event of an audit.

DIRECT ADVERTISING

Direct advertising is advertising that is directed at consumers with the specific idea of selling a particular product--in this instance, a horse. A good example of direct advertising is classified advertising in newspapers or the classified advertising section of the breed magazines. This type of advertising offers a particular horse for sale, a breeding stallion standing at stud, etc.

In direct advertising the amount of information is not the secret to success--rather the manner of presentation is most important. Oftentimes, the more information presented on horses, the more the possibility of presenting negatives. As an example, "Excellent trail horse for sale. Top 10 in the nation. Black, four stockings and a blaze"--at this point, you have presented your product. Most people however, will advertise things like, "ridden by amateur." This statement could be a negative factor for some buyers; "limited showing" also could be a negative. Always present strong points that cannot be taken as negatives in direct advertising.

In preparing an advertising program, the marketing area is a primary aspect; it determines the media to use and the various methods available in the market area. Is the advertising going to be international, national, regional, state, or local, or possibly a combination of these? After you have answered this question, you can begin to analyze the advertisement media that are available for the product. Carefully consider the kind of horse market you are entering. Depending on whether you are advertising ranch horses, show horses, halter horses, stallion service weanling, cutting stock, etc., you must analyze which of the media (i.e., TV, radio, printed matter) etc., will reach a particular person in that particular market area.

EXAMPLES OF MARKETS AND SURVEYING THE MARKET

Two prime examples of market studies were done in the motorcycle and instant coffee industries. For many years the market for motorcycles in the U.S. was a very limited one. Use of these vehicles was stigmatized by the actions of the people associated with them. Motorcycle owners were not seen as family men or young lawyers; they were thought of as lower class, leather-jacketed gang members. However, when the Japanese decided to market motorcycles in the U.S., they did a market study noting that one of the largest hurdles to overcome would be the motorcycle image. The results were that the entire marketing of the Honda motorcycle was based on convincing people that very successful individuals and family members rode motorcycles. Consequently, their entire theme for their advertising was, "You meet the nicest people on a Honda", and their television advertising was geared toward those "nice people" riding the motorcycles. They overcame the once-limited market and actually out sold two very large motorcycle companies in the U.S. It is important to realize that in marketing horses (depending on the type of horse and where you are located) you will have to determine your potential customers and how to sell to those people. There is a possibility of enlarging the market in your particular area, on your own, by promoting the aspect of the horse industry that you want to develop.

Another successful market study dealt with instant coffee. When instant coffee originally came on the market, it was a failure. Findings of a market survey and a market study, indicated that the reason the product was a failure was that housewives (who were buying coffee) associated instant coffee with being lazy. Also, it was not a "fitting" drink for coffee drinkers because people still thought they must have ground coffee. The instant coffee people turned this image around by promoting instant coffee through use of well-known "personalities" who were shown serving instant coffee. For example, actor Robert Taylor's wife was seen serving instant coffee to him as he came into the backyard on horseback. Instant coffee became a big seller after overcoming the negative reactions.

We in the horse industry, are faced with many stereotyped images dealing with horses, depending on where we are and the exposure that we have to potential buyers. For example, is there the "cowboy" attitude? Is that a different attitude from those who have Arabians? Is there a "rugged" or "elegant" image with regard to riding and showing horses. As an individual who markets horses, as in all product lines, it is important that you recognize these attitudes so that you can capitalize on that attitude or, possibly, attempt to change that attitude with potential clients.

BUDGETING FOR ADVERTISING

Advertising is an integral part of any business operation and often is the sole source of customers. That being the case, one must give it the attention it deserves, because income will be directly related to advertising programs. Nothing sells by reputation alone. Reputations are gained only by advertising the product. A good rule of thumb is that 5% of net income should be devoted to advertising--but the advertiser must be willing to spend part of that money prior to obtaining the income, understanding that it creates the income. Your advertising program should become the most dominant aspect of your horse marketing.

53

EFFECTIVE MARKETING METHODS FOR HORSES

Ben A. Scott

Horses usually are marketed by three methods: livestock auctions, ranch sales, and horse brokering that involves listing agents, trainers, or professional horse dealers. If horses are to be marketed, production should be geared specifically toward a marketing goal. Horses create special marketing problems sometimes because of the very personal involvement of people.

LIVESTOCK AUCTION

The livestock auction is the first and foremost marketing method for horses; it is the backbone of the horse industry, the cattle industry, and any other type of industry related to livestock. Granted, there are many people who (because they may not have known what they were doing) have had a bad experience at an auction. And some auctions are probably not as reputable as they should be. But, for the most part, most livestock auctions are accepted, ethical, and they do a tremendous job for the industry.

It is usually the inexperienced individual at an auction who has a bad experience because he does not understand the basic rules of the auction. For example, some auctions have "no hold"; that is, you put an animal up for sale and it must sell. Other auctions allow the owner to bid. If that bid purchases the animal back, a commission is paid on that bid. Other auctions allow a minimum bid to be established and the auctioneer does his best to get it. Some auctions do not allow you, as an owner, to bid at all. Consequently, an owner may get someone to bid on the animal and run the price up. The important thing when using the livestock auction is to find out the rules and regulations of the auction you are considering. Don't put a peformance horse in a yearling race-type auction. Try to look for the best auction that suits your animal--not only for what it is and does, but for the best prices that have been received in past auctions. It is better to use an auction that has been in existence for awhile and has established a reputation. This is not to say that a newly established auction group or sale should not be used; however the first one or two sales are not as likely to be as good as the following ones. (Of course, this is not all bad; during the following year, there should be more buyers there who have heard or seen what went through the auction and the prices that were brought by the horses.)

One area of concentration should be on conditioning. As a matter of fact, many sales have an award for the best-conditioned horse. Yet, many people figure it is just a game of luck and they put an unfit animal through a sale. Here again, sellers should time the sale so that their horses are at the peak of condition. This can be very difficult if your horse is from the Northwest and you are gearing for a January sale; unless, of course, you have the facilities to keep the horse fit and in peak condition without its winter coat. So, you must analyze your market area and the time of year when your horses are best fit; then determine the auction that will do the most good for you and your animal.

Those people who do not have the facilities, or the location, or do not really have the personal ability or desire to do marketing on their own, need sales media for successful marketing. Those who want to be in the horse business, but don't have marketing abilities, should gear to an annual sale so that: 1) the sale expects the animals and wants them and 2) the farm program and budget are geared to the cash flow at that particular date.

As a seller, you should be cautious about putting horses in a sale and expecting too much money for them. If the horse does not sell and has to be held for another year, or must be marketed in a different way, your program could be set back. Over the years, it is important to stick with a particular sale and--win, lose, or draw--sell your product. You are going to have some years when you may not do so well, but you may hit that year when the market is hot, or the buyers are hot, and you will get far more for your animal than you expected.

To help in choosing the right sale, talk to people who have done well and who consistently use a particular sale. Go to the sale in advance and talk with them; get some help. Most people are willing to give you the information you need, but you have to ask for it. You should use these recommended livestock auctions, providing that, as an individual, you investigate it and do it on a very educated and intelligent basis. In the event a sale does not work out, don't give up. Shop for some other auctions in other areas and constantly try to improve your use of the livestock auction.

RANCH SALES

People involved in the horse industry usually start with ranch sales and often continue with them as an additional marketing method. The use of ranch sales must begin with the acknowledgment that the ranch is your "store" and you must proceed from there--that is, you will need signs, stock, display area, and sales people. The seller needs to know the basics of selling and follow with good sales practices.

READY TO SHOW

Through advertising, signs, or other media, a customer may show up unannounced at the ranch to look at a horse. The seller is so thrilled to have a customer, he immediately heads for the barn to begin showing what he has for sale. Whether the customer is there by an appointment or unannounced, you should be very careful about how you begin to show the animal for sale. The first thing you should do is have a plan in mind when a person comes without an appointment. Keep the customer occupied at the house or somewhere on the ranch while the horse is being groomed to show. The horse should be brushed, hoof blacked, and shined. If you have a halter horse to show a customer, the horse should be as completely fit and ready as if you are entering a halter class (including using a silver halter). Too often, sellers forget that first impressions are lasting impressions and get thrown off balance by the customer who always says he can see beneath the dust, etc.

Whenever possible, show a prospective customer only a horse that is completely fit and ready to go into the show ring. If you are selling a performance horse that doesn't look particularly good unsaddled, keep the buyer occupied and away from the barn until the horse is groomed and saddled, complete with show tack. Plan ahead to have someone available to do this for you. Do not use a training saddle and bit. Sellers often think that because they know the horse so well, everyone else does, too. That is not the case. Whether they buy the horse or not, they will go away with only one picture in their mind--the picture of the horse as it was presented to them. Don't show the horse in the stall with straw in his tail, unclipped, etc. Show the horse as it would show at the horse show. Wear proper attire (not old work clothes) when showing a performance horse to a prospective purchaser. The difference in a presentation can mean the difference in a sale or no sale. The odds are that, if the customer has seen perhaps three horses that particular day, yours would be the only one shown in this manner. You must assume at all times that every individual is a buyer and wants to see the best. It is an extremely good rule around the ranch that no prospective customer see the horse for sale until that horse is ready to show. Stallions should be shown as to purpose; brood mares should not be ridden; weanlings, not broken to halter, should be shown loose.

Your ranch, stable, farm, or ranchette projects your feeling about the animal that you are selling. Many people infer that it takes money to have a nice place. However, there are many small 2 1/2 acre ranches with two or three box stalls and an older home where everything is very well kept, orderly, and clean. The horse industry is no different from any other business. There are those who are successful at it and those who are not. There are those who like a clean store and those who have a store that is so unkempt that it is impossible to find anything in it. There are those who keep good records and those who keep poor records. Through the years the most successful people are the ones who keep their places up and their horses in good condition. But this does not necessarily reflect dollars spent or volume turned; sometimes the interest in horses is not an entire family's interest; perhaps it is only that of the wife, the husband, or a child.

I am speaking directly to those people who are looking at horses as contributing to a portion of their income, whether it be merely to offset the expense of showing or to generate an actual profit. In any case, the merchandising of a horse is essential to what they are trying to achieve. Some people take themselves out of the market of expensive horses. Their product is good, the horse is worth the money, but it is just not presented in an environment that is comfortable for people who are dealing in large amounts of money. People should spend the same amount of time cleaning their barns and fields as they do cleaning up their horses. In the business world, many of us are controlled by various county, city, and federal laws telling us how to keep our business. There is no type of control exercised over how an individual keeps his horse barn. Each and every individual who is truly interested in doing well and improving himself in the horse business must constantly look at the place where he plans to merchandise his product; this includes the barns and corrals. All of these areas should be kept in good repair and in a good state of cleanliness, and this can be done without spending large amounts of money. A person is better off with three box stalls that are clean, correctly done, and safe than 10 box stalls that are half-way neat and clean. The upkeep of a ranch should be directly related to your horses; you should give as much care to upkeep as you give to your product, the horse. By presenting your horses in the proper fashion and proper environment, you can expect to get top dollar and achieve what you want to achieve. The end result will be pride of ownership and profit from horses.

Probably the most damaging thing done by some people involved in the horse industry (which can hurt sales more than anything) is to show horses that are not for sale to a prospective client or customer. Because these people are proud of their animals and what they have accomplished, they often insist that a prospective client look at all their horses that are not for sale prior to showing the one that is for sale. It is imperative that sellers do not take a horse out of a stall and show it to someone if it is not for sale. I don't care how proud of the animal the seller may be; he merely confuses a prospective buyer. The buyer may question why you are selling the animal that is for sale when you have all those good ones that are not for sale. If one is asked about a horse that is not for sale, merely say that it is a horse you are riding. Don't say anything else. Soft pedal it, and don't offer information such as, "That's my husband's super stock horse that is undefeated this year," and then proceed to elaborate on how great that other horse is that isn't for sale while showing the stock horse that is for sale. Show only the one that you have for sale and show it properly. In the event that the people say to you, "Do you have anything else for sale?" ask them their feelings about the horse that you are selling that you have just showed them. Ask them if the horse fits their needs and if they are interested in the horse. In other words, do not distract your buyers from what he came to see until he absolutely refuses to do anything about it. At that time, you might try to move him to another horse. To bring a buyer into your barn and show him 15 horses, of which 12 are not for sale, is suicide. In fact, a rule of thumb is that every time you show a prospective buyer another horse, your odds for a sale go down 20%--merely because the buyer becomes confused. You also must instruct whoever works for you or with you that this is the way it will be done and that they are not to discuss other horses or show them. Zero in on the horse for sale and stick with it until you know you don't have a sale or that you have made one. Then, in the event the sale fails and you feel you want to show another horse, go back to the barn and only show the one horse that you want to sell. This will save you time and money--and many wasted hours, as well.

HORSE BROKERAGE

Agents who broker horses are trainers, dealers, and livestock listing agencies; all usually operate on an individual basis with many variables in the commission structure.

Trainers are used most commonly when a seller has a horse in training and has agreed to sell at a given price and pay the trainer a commission. That commission usually is 10%. It is almost always a verbal agreement and involves no expense to the trainer, because many times the horse is in training. For this type of selling, the one important ingredient is that the price be the same, even if the seller also is working on the sale. The seller should not quote a lower price even though he does not have to pay a commission if he sells it.

The seller should also avoid quoting net prices and allowing the trainer to "stack on" any price he wants just because there is no set price on the horse.

Many expensive horses are sold through trainers and it is a valuable way to sell, but the seller should be very specific in the details and avoid problems by requiring a written agreement.

Dealers

As his name implies, the horse dealer makes his living buying and selling horses and he constitutes a valuable portion of the marketing area. In using a dealer, sellers must realize that the dealer must purchase at a price at which he can resell and make a profit. Also, the seller is usually not involved in the representation of the horse. This is not necessarily bad, but sometimes the credibility of the seller could be damaged if the dealer makes false statements regarding the horse. When consigning a horse to a dealer, the agreement should be in writing and the seller should take the time to investigate the credibility of the dealer if not known personally.

Listing Agencies

More listing agencies have been appearing on the horse marketing scene in the last few years because sales have been increased. Any horse-listing agency has written contracts, similar to those in real estate. Anyone using a listing agent should be sure to read the agreement thoroughly to make sure of his commission, of the length of the agreement, and of the stipulations if someone else should sell the horse or if the owner sells it. The good listing agencies will be very specific in their guarantees of advertising brochures, etc. A seller, when marketing through a listing agency, should determine (in writing) what the agency will do for the seller; he also should be familiar with the cancellation provision of the agreement.

Regardless of the marketing method used, it is imperative that the marketing of horses be given careful attention because it is the marketing program and its success that determines the profitability of the horse program.

54

U.S. IMPORTATION AND EXPORTATION OF HORSES: METHODS AND REQUIREMENTS

Ralph C. Knowles

INTRODUCTION

The advent of the "Jet Age" has enabled people and horses to travel internationally with great speed and flexibility. During recent years, from 21,000 to 24,000 horses and other equidae have been imported into the U.S., annually.

Relative to the rest of the world, the U.S. has a healthy horse population. This great health status is not by accident but by design. The eradication of diseases such as dourine, glanders, Venezuelan equine encephalitis, and contagious equine metritis from the U.S. required uncommon persistence and several herculean thrusts by the U.S. horse industry, the veterinary community, and state and federal governments.

A healthy U.S. horse population represents a population that is highly susceptible to invasion by foreign animal diseases. For example, the 1971 invasion of Venezuelan equine encephalitis (VEE) into South Texas from Mexico was a real threat to the U.S. equine population. The tremendous force exerted against VEE, through vaccination, quarantine, and mosquito control, limited this disease to 26 South Texas counties and a loss of only 1,500 horses at a cost of 19.5 million dollars.

When horses are offered for importation into the U.S., the U.S. Department of Agriculture (USDA) has the prime responsibility for inspecting for infectious diseases and parasites such as ticks.

IMPORTATION

Horse diseases of primary concern are: African horse sickness, dourine, glanders, equine infectious anemia, equine piroplasmosis, Venezuelan equine encephalitis, and contagious equine metritis. Freedom from infestation with ticks also is highly important.

Horses offered for importation into the U.S. are scrutinized by physical examination, serological testing

TABLE 1. SUMMARY OF U.S. EQUINE IMPORTATION ELEMENTS

Disease	Physical examination	Type of serological test	U.S. quarantine holding period[a]	Other methods used	Special considerations
Dourine	Yes	CF[b]	3 to 4 days	None	None
Glanders	yes	CF	3 to 4 days	None	None
Equine infectious anemia	Yes	AGID[c]	3 to 4 days	None	None
Equine piroplasmosis	Yes	CF	3 to 4 days	Courtesy test[d]	Tick infestations
Venezuelan equine encephalitis	Yes	Not used	7 days	None	Origin - South and Central Ameria and Caribbean Islands
Contagious equine metritis (CEM)	Yes	Not used	3 to 4 days[e]	Overseas certification and treatment	Origin - CEM-listed countries
African horse sickness	Yes	Not used	60 days	None	Origin or transit through African continent
Tick infestation	Yes	Not used	3 to 4 days	None	

[a]General time period, if otherwise found healthy.
[b]Complement fixation.
[c]Agar gel immunodiffusion (Coggins test).
[d]An overseas CF test used to preclude shipment of infected horses to U.S. quarantine stations.
[e]Breeding animals are released only to designated states for postentry quarantine and further qualifying tests and treatments. This generally takes 45 days to 60 days.

and, in the case of African horse sickness, a prolonged quarantine period. Table 1 summarizes the elements of this scrutiny.

Because U.S. animal import requirements are subject to change, it is advisable to seek information on current requirements from the USDA federal veterinarian in charge in your state. This person can be located through your state veterinarian or in the telephone directory (this office is usually in the state capital) listed under U.S. Government, USDA, Animal and Plant Health Inspection Service, and Veterinary Services.

Unless you are experienced in the international shipment of livestock, it is usually advisable to employ an international livestock broker or freight forewarder to handle details of your horse importations into the U.S.

EXPORTATION

Generally, the USDA expects horses exported to foreign countries to comply with the health requirements of the receiving country. Such requirements vary with the receiving country and are subject to change. It is advisable to obtain specific information from the USDA veterinarian in charge.

Importation and exportation of horses regarding the U.S. are complex operations. Proper planning and allowing time for the necessary steps can simplify this operation and preclude undue delays of shipments.

Part 13

FACILITIES AND BEHAVIOR

55

SMALL-SCALE FARM DESIGN FOR THE HORSE BREEDER

Mark M. Miller

Not all of us are fortunate enough to start from the beginning in the design of our farms; more often, we are constrained to live with someone else's mistakes in the overall farm layout.

The ideal farm would be square, rather than oblong, and would be laid out like the spokes on a wheel with barn, feed, and storage centrally located and pastures radiating out to the property line.

Driveways should make a straight entrance to the barn area and should end either in a large parking area or should circle the barn to facilitate turnaround for horse trailers. Signs directing traffic to appropriate places create a more businesslike atmosphere on the farm.

Large pastures seem to survive heavy grazing better than do smaller pastures. The latter tend to become dirt lots within a few months of over populating.

However, cross-fencing is the best investment you can make in your fields. By rotating stock, the pastures can be rested and grazed on a regular basis and thus can be managed to support more horses than the same area left unmanaged. Generally, your pasture is both the cheapest and the most sizable element in your horse's diet; pasture rotation can provide substantial additional nutrition for only the cost of cross-fencing.

Cross-fencing also is especially valuable as a means of separating horses according to nutritional requirements. Thus, weanlings and yearlings can be fed higher-protein diets, whereas broodmares, lactating mares, and other stock can be grouped for specific diets. Each large pasture should have a holding pen or feedlot built into it. The holding pen should be large enough to keep all the horses comfortably for up to several weeks so that a wet or newly seeded pasture can be rested.

Individual feeding pens ensure each horse its proper diet, but they are not always economically feasible. Having one or two holding pens can serve the same purpose for the underweight or overweight horse on a special diet, or for the pregnant mare who is near term. Holding pens also come in handy for injured horses and for horses that require isolation for disease, weaning, etc.

Each feedlot can be facilitated with a catch pen, preferably a long narrow dead-end run that is helpful in catching horses that have not received regular handling.

Feedlots that will house mares and foals should be equipped with a fenced creep area where sucklings can have access to their extra ration.

In colder climates, pastured horses need access to a dry windbreak, usually in the form of an open-sided shed. Such sheds should be large enough that the horses can be accommodated without crowding.

In hot climates, shade should be provided. Nurseries have a porous roofing material that blocks sun but allows water to pass through that works quite well for shade sheds. Horses will be more inclined to use shade if a small supply of hay is kept underneath.

Gates between holding pens and pastures should be large enough to drive a tractor through. Designing a series of gates so that stock can be moved easily can save labor. Catch pens can be arranged as lanes between feedlots, offering a savings in building material as well as providing isolation.

Fence can be built from a wide range of materials--all of them, as you will discover, are guaranteed to be able to injure a horse. The factors of cost, maintenance, effectiveness, and safety must all be taken into account. If you live on a major highway or near any kind of hazard, then security must take precedence. High-tension wire is cost effective for fencing areas but cannot be guaranteed to hold a horse. If you live in the desert, rubber is cheaper than wood, but you may lose your best horse to impaction if he finds the strands are a delectable treat. Field wire is very effective, but by the time you put a board on the top to keep your horses from mashing it down, you could have put up a board fence and had a farm as fancy as your neighbor's. Of course, if your neighbor has a three-board fence, you might notice how many boards are down at his place or how many weekends he has spent patching his fence.

If you really want to go first class, there are the new latex-covered, man-made boards that bolt together and come with a 5-yr guarantee. But at $4.50 a foot, they are generally not practical for farms larger than a postage stamp.

Barbed wire can only be justified out on the range. While there is no such thing as the perfect fence, barbed wire is a crippler of horses. I do not advocate it in any way, shape, or form--not as a top strand nor as any part of horse fence. Chain link holds a horse very well but must be over 5 ft to keep a horse from attempting to jump over it. If a horse got hung on the edge of chain link there would be no way to pull him off. The only thing that seems certain is that your horse will find a way to get hurt on his fence or he will get out trying.

The barn is the most significant building you will plan for your farm. There are various barn designs available;

for example, single row, double row, and middle aisle, etc. In northern regions, the barn should be aligned with prevailing winds. A southern exposure can also be beneficial. Extra insulation should be considered if the barn is to house show horses or if heating is to be installed.

In hotter climates, a northern exposure is preferable, and ventilation becomes a priority. A wind funnel or removable siding can keep barn temperatures down when the thermometer rises.

Drainage can make the difference in whether the barn functions or fails. When building in a low spot, a foundation of clay or soil can raise the stalls higher than the surrounding area. This not only facilitates draining of stalls but prevents rain from draining in.

Roofing materials range from the utilitarian to the elite. As a rule of thumb, the less you spend on your roof, the more you should spend on roof sealer.

Skylights help to brighten the barn area, but they will require maintenance caulking to prevent leakage.

Flooring ranges from clay and sand to asphalt and concrete. Horses housed on hard ground are more prone to puffiness and swelling in their legs.

No matter what you stand them on, some horses are going to dig--but indestructible surfaces are so much harder on horses that I think you are better off to keep them on softer ground and repair the stalls on a regular basis.

A wide, safe aisle is a must in a large barn where horses must pass; a grooming area that is separate from such an aisle will keep traffic moving in a busy barn. A wash rack with central drainage will keep everybody happy and will help your boots to last longer too.

Tack should be stored in a separate area. Plan cabinet space for extra supplies and closet space for rain gear, hats, and storage of blankets.

If the barn is more than 200 yd from the house, you will pat yourself on the back a thousand times for being smart enough to include a bathroom in the barn plan.

If your barn is as large as it can be and still be workable, plan an adjacent area for a pole barn. Pole barns are cost effective for hay storage and for parking tractors, mowers, and farm vehicles. A tool shed with a carpenter's workbench should have a place in the scheme of things.

Once construction begins, you will be making decisions every day. A few days of planning before you begin can simplify the decision-making process by giving you a clear picture of the goals you are trying to reach.

56

CONTROLLED GRAZING AND POWER FENCE®

Arthur L. Snell

A double-barreled revolution has swept the cattle industry in the last 5 yr: controlled grazing and the introduction of Power Fence® have both dramatically affected the future of the industry in a positive and productive manner.

CONTROLLED GRAZING

Controlled grazing can be described best as a grazing pattern designed to increase carrying capacity, eliminate overgrazing and overrest, and utilize herd effect. (Some definitions of these terms and others are included at the end of this article.) In a controlled grazing operation, a grazing unit is divided into subdivisions or paddocks with all livestock concentrated in one paddock and then rotated according to plant growth. The number of subdivisions or paddocks may vary from 8 to 40, depending on conditions.

Controlled grazing on our dry, brittle lands of the western half of the U.S. was brought to this country by Allan Savory from Rhodesia and by Stan Parsons from South Africa. (A great many ranchers, soil conservation personnel, and government agencies are aware of the Savory Grazing Method and its long-range effect on both man and the land.) Concurrently, a great deal of information about controlled grazing practices started pouring into the country from New Zealand, the United Kingdom, France, and Argentina. Although the system was designed for more stable environments with higher rainfall, the fundamental principles of controlled grazing are essentially the same in all environments. However, the dry, brittle areas must be more carefully managed.

Overgrazing in many parts of our native rangeland and improved pastures is one of the major problems in today's livestock grazing management systems. Overgrazing generally occurs in lightly stocked areas where cattle are left in the same pasture for extended periods of time. Livestock selectively graze palatable species and then regraze them until they are damaged or dead. Overrest often occurs in the same pasture at the same time. Bunchgrass species become dry, fibrous, and are neglected by livestock. These plants, as they mature and begin to die, are rendered useless as forage. It is ironic that understocking, overgrazing, and overrest occur all at the same time in the same grazing area. In grazing areas that are free of livestock for extended periods, overrest causes severe loss of forage when the neglected bunchgrasses mature and die. According to Savory, the current management system of most of our rangelands is causing desertification at an alarming rate. The only real solution to this problem is controlled grazing.

Under controlled grazing, all animals are bunched in smaller grazing areas called paddocks. The herd effect then becomes important. Breaking up the soil cap and preparing the seed bed is the benefit of concentrated hoof action. Concentrated recycling of manure and urine adds to and develops soil quality. In a controlled grazing program, cattle are moved frequently; thus bite damage is reduced significantly, allowing roots to develop through leafy grass structures.

In New Zealand, paddock rest is the key to success in controlled grazing in their stable environment. Figure 1 shows the effect of various numbers of paddocks on the rest factor. Note that a 12-paddock system rests 335 days out of

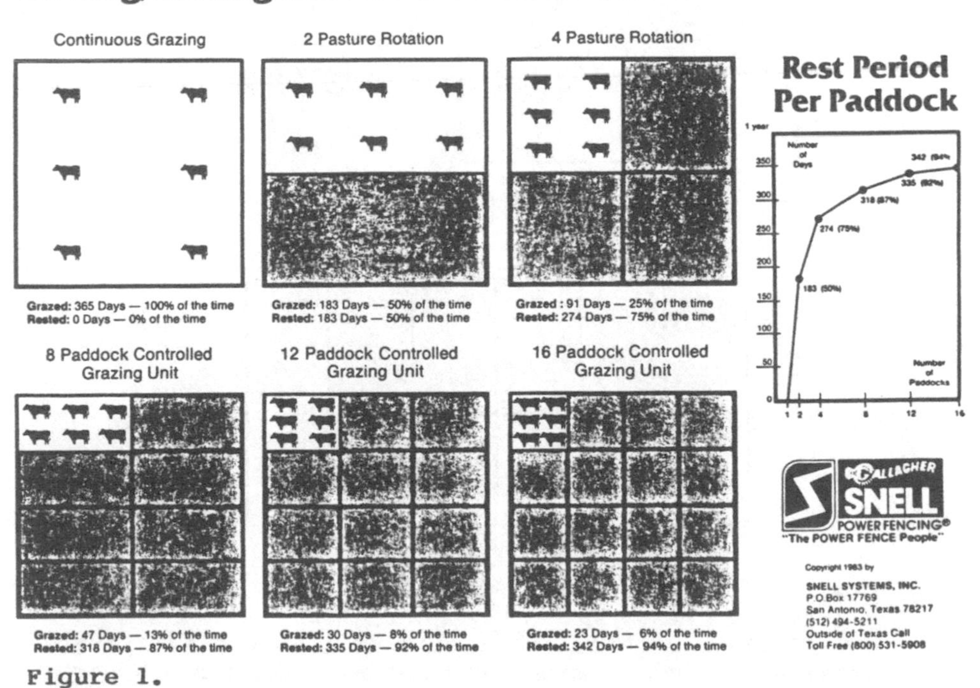

Figure 1.

the year or 92% of the time. This allows grass species to establish adequate root systems, and overgrazing and over-rest are eliminated because livestock are moved through the grazing unit on a time-controlled basis. In a stable, high-rainfall area, a 20-day rest period would be adequate. In drier, more brittle climates, or during the nongrowing season, rest can extend from 40 days to 90 days/paddock. The sun provides 90% of a plant's nutrients. Because harvesting of sunlight is the basic job of grazers, controlled grazing offers great opportunities. As shown in figure 2, controlled grazing, with time off and time on, allows development of the leafy structure and thus growth and development of the root system.

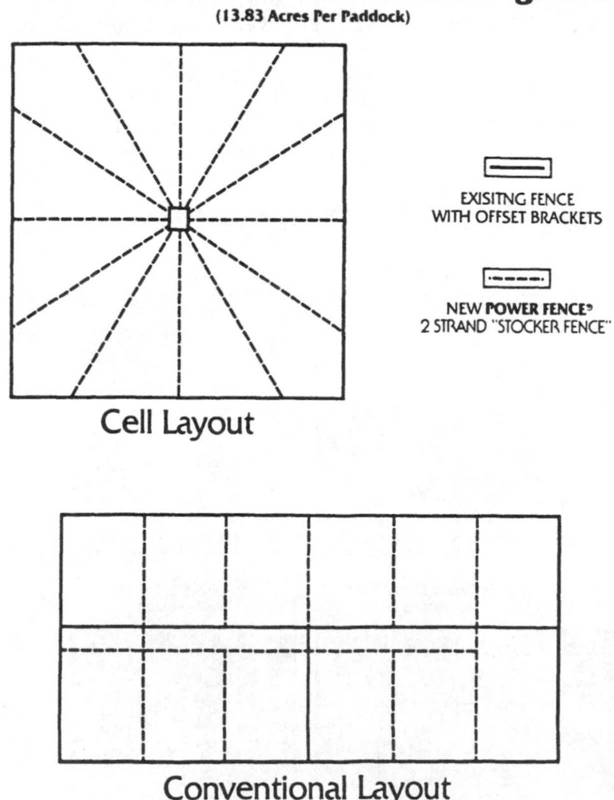

Figure 2.

An effective controlled grazing program can extend the grazing season. In my own experience on a small ranch north of San Antonio, our animals obtained 40 days to 60 days of extra grazing per year when we applied this grazing pattern. There are many beneficial aspects of extending the grazing season--from getting more productivity out of winter grasses to extending the dry-matter utilization of native grasses.

Our experience with controlled grazing has permitted us to double and even triple our current stocking rate. Doubling your current stocking rate is generally accepted as being a practical and achievable goal. Obviously, you can't double stocking rates in the middle of January, but because of increased forage production, adding to stocking rates is an accepted practice and very low in risk.

Improved forage production is certainly one of the great benefits of controlled grazing. With the benefits of herd effect and control of rest periods and grazing periods, roots develop, native species reestablish, and total forage production increases dramatically. This extra production, of course, permits the increased stocking rates.

In New Zealand, recycling of manure and urine back into the soil through high stock density provides as much as 600 lb of nitrogen per acre. It's a terrible waste to allow livestock to bed down in the shade and drop their manure; these nutrients remain under the trees instead of on the pasture. Careful planning will assure that this doesn't happen. Increased herd density also helps break down fibrous matter, weeds are tromped out and brush generally begins to regress. Soils and plants appear more healthful.

Obviously, controlled grazing can add considerably to management skills. Simply laying out paddocks or subdivisions and calculating cattle moves is only the beginning of increased management participation. While management is intensified by the routine movement of cattle, the benefits include more frequent observations of the livestock. Improved overall management becomes more efficient, from ranch planning to marketing.

Controlled grazing offers great flexibility--for example, at my ranch, we often run replacement heifers one paddock ahead of the main herd so that they get the best forage. Their numbers are not so great that they significantly diminish forage available to the following cow herd. In times of heavy grass growth, when cattle cannot keep up with production, certain paddocks can be eliminated from the grazing unit and be used for hay or silage--again adding flexibility to the concept.

There are disadvantages to controlled grazing. One disadvantage is increased management involvement. Many ranchers have an established way of life and prefer not to get involved in the extra management activity required by a proper controlled grazing program. Another disadvantage is the possibility of decreased individual animal performance during the first year or two. This is more than offset by the increased stocking rate. Animal performance tends to stop decreasing as the system develops and gets on stream, even increasing as forage improves.

In setting up a controlled grazing unit, you should decide whether to add to your existing fencing system to make more subdivisions, or to set up what is called a Classic Cell System. Figure 2 shows some options on paddock layout. Once this layout has been decided upon, you should carefully plan the location of water points and make certain that sufficient water is available for concentrated numbers of livestock. Careful consideration of terrain and cattle working facilities is also very important.

Establish the standard stocking rate for your area based on Soil Conservation Service figures. Decide what your new stocking rate will be and start your financial planning and projections from that point.

The minimum number of paddocks that we recommend for any controlled grazing system is 8, but preferably 12, 16, and even up to 42. The more paddocks there are, the more flexibility the grazer has. Fewer than eight paddocks does not offer enough paddock rest under all conditions. In paddock layout, livestock-handling facilities should be considered and located in the most efficient area. When planning a program, it is most important to decide the rest period per paddock based on the time of year and plant growth conditions. A rule of thumb: fast growth, fast rotation; slow growth, slow rotation.

The time of year, type of forage available, and financial planning should all be part of your controlled grazing program. Failure to know what you are doing, particularly in the dry brittle areas of the U.S. can lead to disappointment. The author can provide information on in-depth training on controlled grazing programs.

It is ironic that both of the revolutions we mentioned earlier, Power Fencing and controlled grazing, hit the U.S. at the same time--they were meant for each other. Certainly, in large controlled grazing programs in the western U.S., Power Fence made the program possible because of its significantly reduced cost over that of barbed wire and its extraordinary ability to manage livestock.

POWER FENCES®

Most of you have heard of Power Fences. The concept originated in New Zealand and Australia and most of the technology we have today came from these two countries. A Power Fence is electric fencing with emphasis on quality, higher-priced energizers, and proper engineering design on the fence itself. The new high-powered, low-impedance energizers most commonly used on the market today are imported from New Zealand. These low impedance energizers are short resistant--that is they won't short out when weeds and grasses contact the fence line.

Traditional electric fencing has had a bad image because electric fence lines have required extensive maintenance and cutting of weeds and grasses. Shorts and fence failure were constant concerns. The new high-powered energizer solved some of these problems in that the design of the circuitry allows us to establish a Power Fence system that is effective as well as reliable.

Bull control is one application for Power Fence--either with three- or four-wire fences, or by simply offsetting a hot wire on an established barbed or net wire fence. Properly installed, a Power Fence can totally control bulls--in fact, there are two very large AI laboratories near San Antonio that have bull runs of Power Fence.

Because Power Fence affects the animal's nervous system, they can be trained to its use relatively easily, but such training is necessary (figure 3). Stallion control is easily achieved with the properly installed three-wire Power Fence. On our own breeding farm, north of San Antonio, our mature stallions are put in paddocks each day with total confidence in the Power Fence control. Sheep fencing/ predator fencing with Power Fence to keep coyotes out of sheep flocks is a widely used practice. In fact, the development of the modern-day Power Fence occurred because Australian sheepmen needed a fence that would control sheep and yet be inexpensive. Elephant control is now a standard procedure with Power Fence on plantations in Malaysia. The elephants are excluded from cash-crop areas with a two-wire fence (figure 4). Controlling deer, elk, and other game with Power Fence is a standard procedure in many states and countries throughout the world today.

A properly installed Power Fence uses 12 1/2 gauge hi-tensile wire. Under no circumstances should soft, low-tensile wire be used. Soft wire is commonly available and is cheaper than the hi-tensile but will cause a great many problems if installed as a permanent fence. Fiberglass posts are used in Power Fencing systems along with insul-timber posts, a self-insulating, high-density wood that has become the product of choice for the Power Fence. Ratchet-type line strainers, tension springs, cut-off switches, and many miscellaneous accessories make the Power Fence a modern engineering fete that has saved the ranching and farming communities millions of dollars during the past several years.

CONCLUSIONS

Why consider a controlled grazing project? There are a multitude of reasons, but Iowa Western Community College, in a carefully conducted controlled grazing program, came up with a reason that justifies the effort: "You like the cattle business and want to increase your income."

Doubling your stocking rate, increasing forage production, and extending the grazing season are but a few of the economic rewards derived by combining the powerful management tools of controlled grazing and Power Fence.

Figure 3. Control through an animal's nervous system eliminates the need for physical and painful barriers such as barbed wire. Power Fence is used to control all types of livestock and predators.

Figure 4. Two-strand Power Fence keeps elephants out of the palm tree plantations in Malaysia. Note "hot" wire running over top of post to keep elephants from pulling out with trunk.

DEFINITIONS

Controlled grazing - a grazing pattern designed to increase carrying capacity, eliminate overgrazing and overrest, and utilize herd effect. A controlled grazing unit is divided into eight or more "paddocks" with all livestock concentrated in one paddock and rotated according to plant growth.

Overgrazing - consists of livestock selectively grazing and constantly regrazing desirable grasses; generally occurs in lightly stocked pastures where cattle are left for extended periods of time. Most pastures under current livestock management conditions are understocked and overgrazed.

Overrest - grass plants (particularly bunchgrasses) when rested for long periods, will mature, become fibrous, and die back from the centers. This condition occurs in totally rested pastures and also occurs in pastures where livestock are lightly stocked. Overgrazing and overrest may occur simultaneously.

Herd effect - is the impact on soil and vegetation from large numbers of livestock concentrated in small areas. Herd effect includes bite damage, hoof action, recycling of manure and urine, and animal behavior.

Nutrient cycle - describes the recycling of manure and urine through high-density livestock populations to increase nitrogen in the soil. Hoof action aids the nutrient cycle by breaking down dry matter and breaking up soil cap.

Brittle environments - generally defined where rainfall is not satisfactory for plant growth during part or all of the growing season.

Stable environments - generally areas with rainfall exceeding 24 in./yr; the more rainfall, the more stable the environment.

Strip grazing - exposing cattle on grass to fresh feed on a daily basis by portable or movable fences or even stationary fences; generally provides enough forage per strip for one day's feeding.

57

LIVESTOCK BEHAVIOR AND PSYCHOLOGY AS RELATED TO HANDLING AND WELFARE

Temple Grandin

Reducing handling stresses can help improve livestock productivity. For example, research indicates that agitation and excitement during handling for artificial insemination can lower conception rates. Excitement during handling will raise body temperature. Stott and Wiersma (1975) reported that an elevated body temperature at the time of insemination of a cow can "affect the mortality of the embryo 30 to 40 days later." Excitement prior to insemination depresses secretion of hormones that stimulate contractions of the reproductive tract that move the sperm to the site of ovum fertilization. Handling stress can lower conception rates when Synchromate B is used to synchronize estrus. Stress caused by collecting blood samples 24 to 36 hr after implant removal drastically reduced conception rates,but the cows still displayed estrus behavior (Hixon et al., 1981).

A separate chute should be used for AI and "doctoring" so that cows will not associate breeding with nose tongs and needles. Cows can be easily restrained in a dark box chute that has no headgate or squeeze (Parsons and Helphinstine, 1969; Swan, 1975). The wildest cow can be inseminated with a minimum of excitement. The dark box chute is easily constructed from plywood or steel; it has solid sides, top, and front. A small window can be made in the front gate to entice the cattle to enter. When the cow is inside the box, she is in a snug dark enclosure. A chain is latched behind her rump to keep her in (figure 1). After insemination, the cow is released through a front or side gate. If wild cows are handled, an extra long dark box can be constructed. A tame cow that is not in heat is placed in the box in front of the cow to be bred. The wildest cow will stand quietly with her head on the rump of the "pacifier" cow.

ISOLATION AND INDIVIDUAL DIFFERENCES

A cow isolated alone in a breeding pen can become highly agitated and stressed, but a pacifier cow can help to

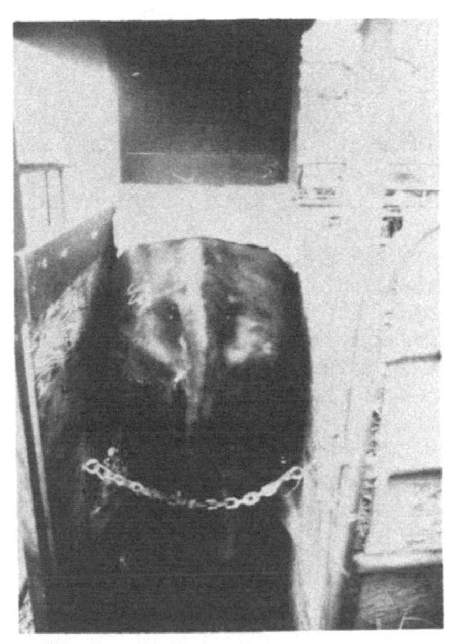

Figure 1. A cow will stand quietly in a dark box AI chute for insemination or pregnancy testing. This chute has no headgate or squeeze.

keep her calm. Body contact with herdmates is calming to cattle (Ewbank, 1968) and this principle is utilized in both herringbone milking parlors and a herringbone AI facility. (A dairy cow left by herself in her stanchion without her herdmates showed elevated cell counts in her milk [Lynch and Alexander, 1973]).

A single steer or cow separated from its herdmates during handling can become highly agitated and is likely to injure itself trying to jump the fence to rejoin its herdmates. An animal may become separated from the group when all its herdmates have walked up the chute and it is left alone in the crowding pen. If the animal refuses to enter the chute, let it out of the crowding pen, and bring it up with another group of cattle. Many handler injuries have occurred because a person got in the crowding pen with a lone, excited animal. Isolation and the sounds of anxious bleating is stressful to sheep (Lankin and Naumenka, 1979).

Robert Dantzer (1983), an innovative animal behaviorist from France, reported that when animals (laboratory rats) were confronted with an unpleasant stimulus such as a shock, the corticoid (stress hormone) levels in the blood were highest when a lone animal was shocked. When another rat was placed in the cage, the secretion of corticoid was reduced in the previously lone animal; with "company" it was less stressed by the shock.

There are large individual differences in an animal's reaction to a stressful situation. Livestock with similar genetic backgrounds will vary greatly in their reaction to stress. Ray et al. (1972) and Willet and Erb (1972) found that there were large individual differences in the stress reaction to restraint in a squeeze chute. Some animals had high corticoid levels, even though they had no visible signs of agitation.

The animal's reaction to a handling procedure is affected by an interaction between genetic background and previous experiences. The way an animal is reared will affect its behavior when it becomes an adult. Brahman-cross cattle are more excitable than Hereford cattle. Guernsey cows are more stress susceptible than are other breeds (Moreton, 1976). There are also large individual differences within a breed. Up to 30% of Merino sheep became so disturbed when they were separated from the flock that they could not be used in physiological experiments (Kilgour, 1971). Other individual sheep were more tolerant of being separated.

NOVELTY AND STRESS

Cattle and sheep are creatures of habit and they become stressed when they experience a novel or painful situation. Novelty can be a strong stressor if the animal perceives it as being threatening. Dantzer notes that the degree of stress imposed by a novel situation depends on how the animal perceives the situation and this perception depends on the animal's prior experiences. The more novel a situation, the more likely it will be stressful.

Not all novelty is stressful and only a dead animal is totally free from stress. Feedlot cattle will readily approach strange objects in their pen such as a manure spreader. They will move away when the spreader or loader first enters the pen, and then approach and sniff the object (figure 2). Rearing environment will affect an animal's reaction to novelty (Moberg and Wood, 1982). Lambs raised in isolation will withdraw from a novel stimulus such as a toy horse. Lambs raised with their mothers will approach the object. Blood corticoid levels were the same in both groups. Rearing environment has a similar effect on pigs. Pigs reared in a group in an environment containing toys would approach a novel object more quickly than would pigs raised in pairs in small pens (Grandin et al., 1983).

Dantzer (1983) designed a clever experiment to test the effect of novelty on the stress response of calves. Stress was indexed by measuring corticoid levels in the blood. The

Figure 2. When a novel object enters a pen, most cattle will retreat and then return and investigate. The animal's reaction to novelty in its environment depends on its prior experiences.

calves were provided with new experiences that had different degrees of novelty. Veal calves were raised either inside a building in stalls or outside in group pens. When the calves reached market weight, they were subjected to an open-field test in both an indoor and an outdoor arena. The open-field test was conducted by placing each calf alone in a strange arena and observing its behavior. A blood sample was collected after the calf had been in the arena for a few minutes. Being alone in the arena is stressful to herd animals such as cattle and sheep.

Calves raised indoors had higher corticoid levels when they were placed in the outdoor arena. Calves raised outdoors had higher corticoid levels when they were placed in the indoor arena. Both the indoor and the outdoor arena were stressful to all calves, but the arena that was most novel was the most stressful. The calf's reaction was determined by its experiences during rearing.

REDUCING HANDLING STRESSES

Cattle and sheep are less stressed and shrink less when they are handled in familiar corrals. Livestock will shrink less the second time they are transported because the truck is less novel the second time. Corticoid responses that occur during handling are reduced when animals become familiar with procedures. Kilgour (1976) suggests that animals could be preconditioned to handling stresses. Sheep that are accustomed to people have a lower output of glucocorticoids when transported than do sheep that have been put out on pasture (Reid and Mills, 1962). Handling feeder calves prior to shipment from the ranch may help reduce stress (Phillips, 1982). To reduce stress during AI, cows could be walked through the chutes prior to insemination.

Livestock that are handled every day become accustomed to handling procedures and there is little or no stress. Restraint in a squeeze chute is stressful to most cattle; however, cattle that are placed in a squeeze chute every day will get accustomed to it. Heifers used in an educational farm demonstration for children at the MSPCA Macomber Farm in Massachusetts became so accustomed to the squeeze chute that they would walk in and wait for the headgate to catch their heads.

PASTURE LAYOUT AND STRESS

Many ranchers are using short-duration grazing systems. In these systems, cattle are moved to a new pasture every few days (Heitschmidt et al., 1982; Savory and Parsons, 1980). Heitschmidt et al. (1982) report that the success of the Savory system and similar systems in New Zealand is because they allow cattle to be moved to the next pasture with little or no stress. The wagon-wheel layout with corrals and watering facilities in the hub makes switching pastures easy. The cattle or sheep learn to move to the next pasture when the gate to that pasture is left open. They do not have to be driven.

Figures 3 and 4 illustrate pasture-rotation layouts that can be used with conventional, short-duration, or Savory pasture-rotation systems. By adding or subtracting gates, these designs can be used with 4 to 16 pastures. Cattle waiting to be worked can be kept separated from worked cattle. The wide 7.62 m x 25 m alley has double block gates for keeping cattle groups separated. The layout shown in figure 3 will handle 450 pairs, with all the animals contained within it. Sorted calves, worked cattle, and unworked cattle can be kept separate.

The design of the corral layout inside the octagonal alley allows all of the handling procedures to be done in the curved lane and diagonal sorting pens. The cattle are gathered in the wide octagonal lane and in the gathering pen of the corral. The 3.5 m (12 ft) wide curved sorting reservoir lane serves two functions. It holds cattle that will be sorted back into the diagonal pens and it also holds cattle that are waiting to go to the squeeze chute, AI chute, or calf table. When cows and calves are being separated, the calves are held in the diagonal pens and the

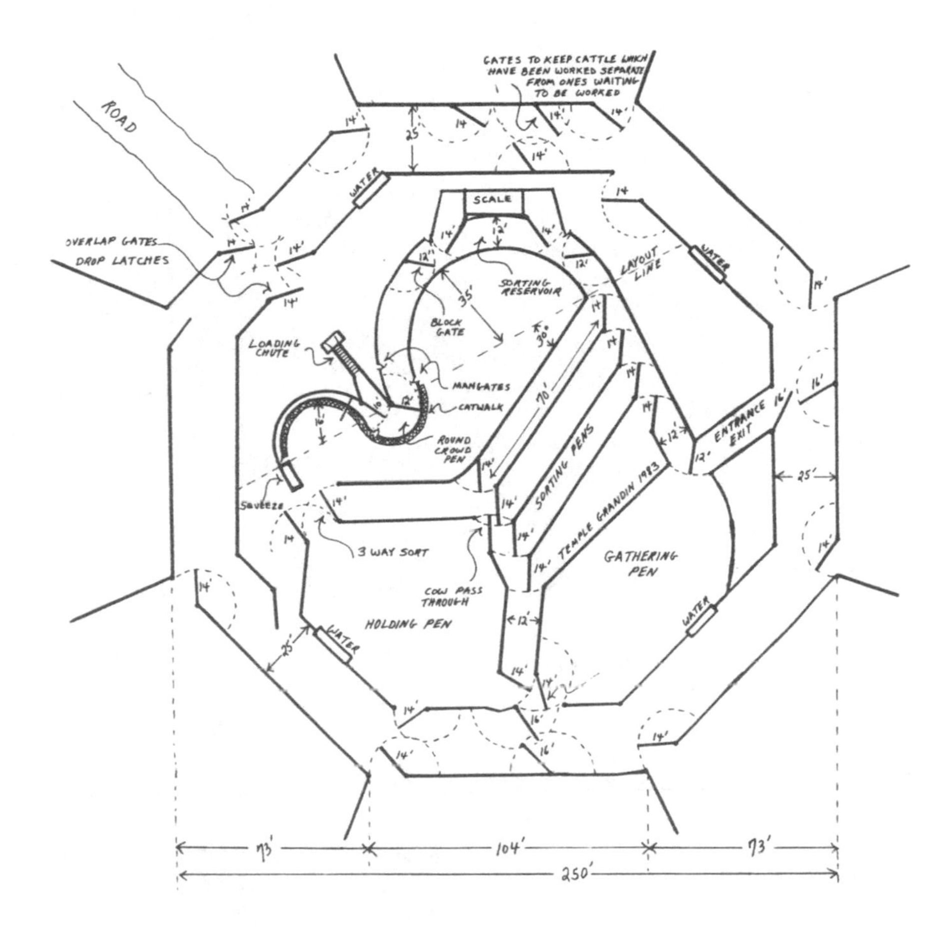

Figure 3. Pasture rotation layout which can handle 450 head. It contains a curved corral system for more efficient handling.

cows are allowed to pass out of the diagonal pens into the large post-working pen. (For a more detailed description of the corrals, refer to Grandin [1983a] in Vol. 19 of the Beef Cattle Science Handbook.)

HANDLING FACILITY DESIGN TIPS

Install solid fences in single-file chutes, crowding pens, and loading chutes to prevent the cattle from being "spooked" by people and other moving objects outside the

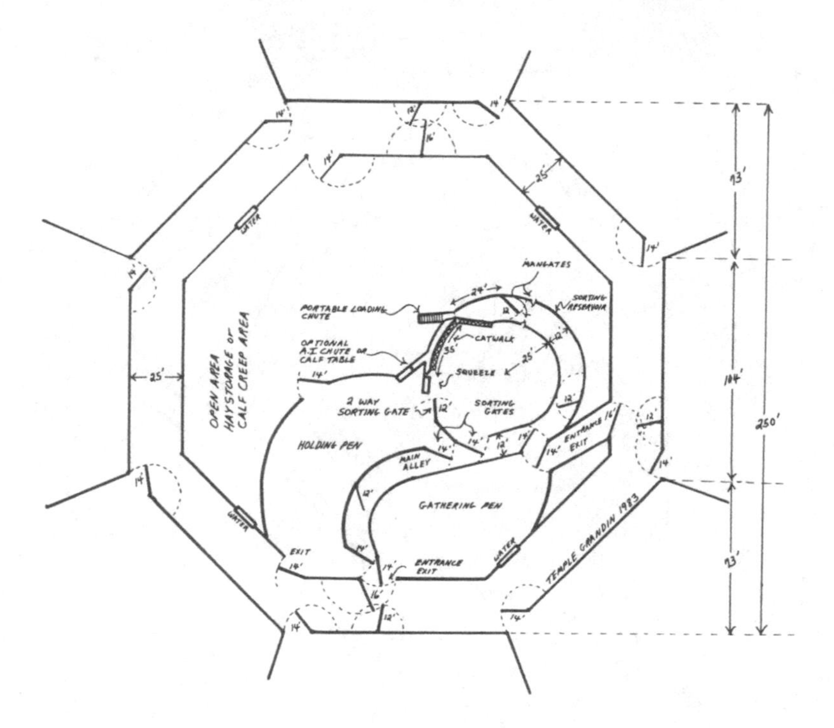

Figure 4. Pasture rotation layout with an economical corral system.

fence (Grandin, 1983b). The crowding-pen gate should also be solid, otherwise the animals will face the gate instead of facing the entrance to the single-file chute. However, sliding gates and one-way gates in the single-file chute and loading ramp should be constructed so that animals can see through them. This design promotes following behavior. For more information on these basic principles refer to "Sheep Handling and Facilities" (Grandin, 1984).

Man-gates should be installed in cattle facilities with solid fences for handler safety. In alleys and other areas where cattle are crowded, the fences should be constructed from substantial materials. If cable or thin rods are used, install a wide belly-rail that the animals can see.

A curved chute works better than does a straight chute for two reasons. First of all, it prevents the animals from seeing the truck or squeeze chute until they are part-way up the chute. A curved chute also takes advantage of the natural tendency of cattle to circle around the handler. The handler should work from a catwalk that runs alongside the inner radius of the curved chute (Grandin, 1980).

EFFECT OF NOISE AND EXCITEMENT ON HANDLING

High-pitched sounds such as cracking whips are stressful to cattle (McFarlane, 1976), which are more sensitive to high-pitched noises than are humans. Ames (1974) found that the auditory sensitivity of cattle was greatest at 8,000 hz. The human ear is most sensitive at 1,000 to 3,000 hz. New research by Kilgour et al. (1983) indicates that cattle are sensitive to sounds up to 18,000 cycles. High-pitched noise from motors and other equipment in milking parlours may be irritating to cattle. When steel handling facilities are used, banging and clanging can be reduced by installing rubber stops on gates. When a hydraulic squeeze is used, the motor should be removed from the top of the squeeze and placed to one side. Noise from a hydraulic pump will increase balking (Grandin, 1983c). Falconer and Hetzel (1964) found that barking dogs and the sound of exploding fire crackers increased thyroid hormone levels in sheep. Sheep slaughtered in a noisy commercial abattoir had higher corticoid levels compared to sheep slaughtered in a small research abattoir (Kilgour and Delangen, 1970).

Cattle and sheep become more difficult to handle if they become excited. If an attempt to restrain an animal is badly handled, subsequent attempts to restrain the animal will become more difficult (Ewbank, 1968). Excited cattle and sheep should be allowed to settle down before handling. Dogs can cause animals to become very excited. Allowing a dog to bite sheep doubled the glucocorticoid levels compared to trucking for 90 min, shearing, or dipping (Kilgour and Delangen, 1970).

FLIGHT ZONE

When a person or dog penetrates the flight zone of either cattle or sheep, the animals may become highly agitated if they are unable to move away. This problem is especially difficult in confined areas such as crowding pens and the drip pens near a dipping vat. The size of the flight zone depends on the tameness or wildness of the animal; this zone is 1.52 m (5 ft) to 7.61 m (25 ft) for feedlot cattle and up to 30 m (100 ft) for cattle on mountain ranges (Grandin, 1978). The best place for the handler to work is on the edge of the flight zone (C. Williams, personal communication). When the flight zone is penetrated the animal will move away. The animal will stop moving when the person is no longer within the flight zone. If the handler penetrates the flight zone too deeply, the animal will either break and run away or attempt to run back past the handler. When cattle or sheep are being moved down an alley, the handler should **retreat** or **back up** if an animal attempts to turn back; he must move outside the animal's flight zone.

Yelling and noise can increase the size of the flight zone. A handler of cattle who is positioned on the edge of the flight zone may find himself deep inside the flight zone if he yells or if the cattle approach noisy equipment.

When livestock are being handled in a single-file chute or other confined area, the handlers should refrain from leaning over the chute directly above the animals. Cattle will often become excited and rear up when a handler leans over them, because he has deeply penetrated their flight zone (figure 5). The animals respond by leaping and rearing in an attempt to increase the distance between themselves and the handler (Grandin, 1978). Handling is most efficient when handlers work from a catwalk that runs alongside a chute. Overhead catwalks should not be used.

Figure 5. Cattle will rear up in a single-file chute when a handler penetrates their flight zone. They do this in an attempt to increase the distance between themselves and the handler. The handler should back up and get out of the animal's flight zone.

MAN/ANIMAL INTERACTION

Seabrook (1972) found that the personality of the dairy herdsman affected milk yield. A quiet confident person was the best herdsman for a small dairy (J. Albright, personal communication). In pigs, there is a significant correlation between "the behavioral response of pregnant sows towards human beings and recent reproductive performance on the farm" (Hemsworth et al., 1981). Sows with low reproductive performance were wary of people and were less likely to approach a strange man.

The way a person handles animals can also affect animal weight gain. Hemsworth and his colleagues subjected growing pigs to three different treatments. The treatments were 1) pleasant handling when a pig approached, 2) unhandled control, and 3) slapping or shocking a pig when it approached. Pigs in the pleasant handling treatment gained the most weight.

REFERENCES

Ames, D. R. 1974. Sound stress and meat animals. Proceedings of the International Livestock Environment Symposium, Amer. Soc. Agr. Eng., SP-0174. p 324.

Dantzer, R. and P. Mormede. 1983. Stress in farm animals: A need for reevaluation. J. Anim. Sci. 34:103.

Ewbank. 1968. The behavior of animals in restraint. In: M. W. Fox (Ed.) Abnormal Behavior in Animals. Saunders, Philadelphia. p 159.

Falconer, I. R. and B. S. Hetzel. 1964. Effect of emotional stress on TSH on thyroid vein hormone level in sheep with exteriorized thyroids. Endocrinology 75:42.

Grandin, T. 1978. Observations of the spatial relationships between people and cattle during handling. Proceedings, Western Section, Amer. Soc. Anim. Sci. 29:76.

Grandin, T. 1980. Observations of cattle behavior applied to the design of cattle handling facilities. Appl. Anim. Ethol. 6:19.

Grandin, T. 1983a. Design of corrals, squeeze chutes, and dip vats. In: F. H. Baker (Ed.) Beef Cattle Science Handbook, Vol. 19. Winrock International, Morrilton, AR. pp 1148-1163.

Grandin, T. 1983b. Livestock psychology and handling of facility design. In: F. H. Baker (Ed.) Beef Cattle Science Handbook, Vol. 19. Winrock International, Morrilton, AR. pp 1133-1147.

Grandin, T. 1983c. Welfare requirements of handling facilities. Commission of European Communities Seminar on Housing and Welfare. Aberdeen, Scotland. July 28-30, 1983.

Grandin, T. 1984. Sheep handling and facilities: In: F. H. Baker (Ed.) Sheep and Goat Handbook, Vol. 4. Winrock International, Morrilton, AR.

Grandin, T., S. E. Curtis and W. T. Greenough. 1983. Effects of rearing environment on the behavior of young pigs. Paper presented at the 75th Diamond Jubilee Meeting of the Amer. Soc. Anim. Sci. Washington State University, Pullman. July 26-29.

Heitschmidt, R. K., J. R. Frasure, D. L. Price and L. R. Rittenhouse. 1982. Short duration grazing in the Texas experimental ranch: weight gains of growing heifers. J. Range Manage. 35:375.

Hemsworth, P. H., A. Brand and P. Williams. 1981. The behavioral response of sows to the presence of human beings and its relation to productivity. Livestock Prod. Sci. 8:67.

Hixon, D. L., D. J. Kesler, T. R. Troxel, D. L. Vincent and B. S. Wiseman. 1981. Theriogenology 16:219.

Kilgour, R. 1971. Behavioral problems associated with intensification of sheep. Proc. N.Z.V.A. Sheep Section, 1st Symposium. Massey University. pp 144-154.

Kilgour, R. 1976. Sheep behavior: Its importance in farming systems, handling, transport, and preslaughter treatment. West Australian Dept. Agr. Perth, Australia.

Kilgour, R. and H. Delangen. 1970. Stress in sheep resulting from management practices. New Zealand Soc. Anim. Prod. Proc. 30:65.

Kilgour, R., L. R. Matthews, W. Temple and M. T. Foster. 1983. Using operant test results for decisions on cattle welfare. The Conference on the Human Animal Bond. Minneapolis, MN. June 13-14.

Lankin, V. S. and E. V. Naumenka. 1979. Emotional stress in sheep elicited by species specific acoustic signals of alarm. Degat I.P. Pavlova 28:994.

Lynch, J. J. and G. Alexander. 1973. The Pastoral Industries of Australia. Sydney University Press. Sydney, Australia. pp 371-400.

McFarlane, I. 1976. Rationale in the design of housing and handling facilities. In: M. E. Engsminger (Ed.) Beef Cattle Science Handbook, Vol. 13. Agriservices Foundation. Clovis, CA. pp 223-227.

Moberg, G. P. and V. Wood. 1982. Effect of differential rearing on the behavioral and adrenocortical response of lambs to a novel environment. Appl. Anim. Ethol. 8:269.

Moreton, H. E. (Duchesne). 1976. Management and behavioral factors affecting the incidence of dark cutting beef. Paper presented at Brit. Soc. Anim. Prod., University of Bristol, Langford. Bristol, England.

Parsons, R. A. and W. N. Helphinstine. 1969. Rambo A.I. breeding chute for beef cattle. Plan No. C, Univ. of Calif., Dept. Agr. Eng., Davis, CA.

Phillips, W. A. 1982. Factors associated with stress in cattle. Symposium on Management of Food Producing Animals. Purdue University. West Lafayette, IN.

Ray, D. E., W. J. Hansen, B. Theurer and G. H. Stott. 1972. Physical stress and corticoid levels of steers. Proc. West. Sec. Amer. Soc. Anim. Sci. 23:255.

Reid, R. L. and S. C. Mills. 1962. Studies of carbohydrate metabolism of sheep. Aust. J. Agr. Res. 13:282.

Savory, A. and S. Parsons. 1980. The Savory grazing method. In: M. E. Engsminger (Ed.) Beef Cattle Science Handbook, Vol. 17. Agriservices Foundation. Clovis, CA. pp 215-221.

Seabrook, M. F. 1972. A study to determine the influence of the herdsman's personality on milk yield. J. Agr. Labour Sci. 1:1.

Stott, G. H. and F. Wiersma. 1975. Embryonic mortality. Western Dairy J. April. p 26.

Swan, R. 1975. About A.I. facilities. New Mexico Stockman. February. pp 24-25.

Willet, L. B. and R. E. Erb. 1972. Short-term changes in corticoids in dairy cattle. J. Anim. Sci. 34:103.

58

HORSE SENSE: WHAT KEEPS HORSES FROM BETTING ON PEOPLE?

Michael Osborne

Management of thoroughbred-horse farms requires understanding of many phases of biology, animal behavior, and human behavior. The author's experiences in the operation of thoroughbred farms in Kentucky and Ireland provide insights to many aspects of horse husbandry, marketing, and related technology. An overview of the author's observations is offered to assist horsemen and horse-farm managers to improve specialized phases of their horse operations.

Man imposes, by lack of appreciation of the problems, a stress situation on horses. These situations arise when:

- Man forgets that horses have a pecking order which remains static. Place 10 mares in a paddock and within 3 days the pecking order of dominance is established. Each time a new mare is introduced, the sorting out process must occur. The more dominant the newcomer is, the more violent the sorting out process will be. To avoid or curtail this "peer pressure," find out if your mare is a dominant or a passive mare. Place the dominant mare alone or in a small group but not with other dominant or aggressive mares. Young foals subjected to aggression among their parents risk bodily harm.
- Horses are less stressed when stabled in open-plan housing. Tie stalls on an open-plan basis will avoid box walking, weaving, and other stable vices associated with solitary confinement.
- Horses have a keen appreciation of sound and respond well to soft music as opposed to heavy metal rock. Teasing mares by using the recorded sound of the stallion mating call is most successful.
- Horses need a high-fiber ration for good health and contentment. Cutting paddock grasses short--topping to 4 in.--causes severe digestive

problems if associated with lush grass growth. The most common cause of colic in these circumstances is an overly full alimentary tract with too little fiber.

Let the paddock grasses grow to the stem stage and, if the paddocks are topped like a golf course, feed hay, ad lib, in the paddock.

Part 14

COMPUTER TECHNOLOGY AND OTHER TOPICS

59

MICROCOMPUTER USAGE IN AGRICULTURE

Alan E. Baquet

Much has been written about microcomputers in agriculture. Farm and ranch magazines, extension newsletters, and computer trade magazines have all carried such articles. Thus, it may seem that discussion of the past usage of this relatively new technology would be somewhat ludicrous. But taking a brief look at the past usage should prove helpful as we consider the present and plan for the future.

THE PAST: A HISTORICAL PERSPECTIVE

As we consider the past usage of computers in agriculture, the parallels between computerization and mechanization are amazing. By 1933, row-crop tractors had been around for a good 5 yr. Most farmers did not own one; a depression was delaying their purchases. Farmers were faced with the decision of when and whether to buy one.

At the end of 1983 microcomputers had been on farms a good 5 yr. But most farmers have not yet purchased one; depressed agricultural prices have delayed their purchase decision.

There are other parallels. Steam tractors had been at work for 30 yr by 1933. And by 1983, farm-management applications of computers also had a 30-yr history, tracing to Fred Waugh of the USDA. The original computing was done in batch mode, with the data (material) taken to the computer or machine. In a similar fashion, grain threshing with steam engines involved taking the grain to the machine. We have moved rapidly from the days of batch computing (stationary threshing) to the day of individual microcomputer applications ("modern" combining).

This rapid transition has caused a great deal of confusion and uncertainty in the minds of farmers and ranchers. What are those things? Do I need one of them on my place? What should I buy? Where should I buy it? All are important questions in the mid 1980s, just as farmers and cattlemen in the 1930s asked, "Do I need one? Should I buy an Ajax? Or a Holt? Would a Case from Manhattan Implement be a good buy?"

As you can see our fathers and grandfathers faced decisions similar to those faced by many today. The decision process regarding new technology is not new, only the type of technology has changed.

THE PRESENT SITUATION: DAWNING OF A NEW ERA

Just as the age of mechanical power brought forth a new productivity in agriculture, the electronic computer era will significantly change the agricultural industry. A computer system is composed of hardware (the machine, screen, printer, etc.) and software (the programs designed to control the machine).

Current applications of microcomputers can be grouped into the following areas: data storage and retrieval, decision aids and analysis tools, terminals to interact with remote data bases, and devices to monitor and control activities.

In each of these uses, the microcomputer system acts as an aid in management. The micro without the owner/operator is of no value. The machine cannot replace or substitute for the manager. It will not do anything that cannot be done in another way. Let us explore some of the applications in each of the areas listed above.

Data Storage and Retrieval

Record keeping is a tedious and time-consuming task. Micros allow the manager to store more data faster and retrieve it quickly. Examples of the type of things that can be stored include production records for livestock, production records for crops, and cost and return information. Storage and retrieval capabilities allow the manager to have access to a wealth of data to be used in analysis and decision-making activities.

Decision Aids and Analysis

This is perhaps the most important area for microcomputer usage. When data and facts are combined or processed to provide new meaning, they provide information to the decision maker. Information supports decision making that leads to action. Measured results produce additional data that can be processed to form new information to complete the information cycle. As the information flows improve, the number of potential sources of action expand for the decision maker.

With suitable software, the computer can perform numerous complex mathematical computations at a very high speed and with such accuracy that it vastly increases a decision maker's analytical power. The computer allows for greater use of sophisticated quantitative tools of analysis and more information in the decision-making process.

Many decisions can be reduced to "what if" type questions. By using appropriate software, the decision maker can test alternatives before taking action. Managers can also do sensitivity analyses to help incorporate production and price uncertainty into decision making.

The computer has the capability to free the decision maker to think more about the data used in decision making and to better analyze the results of the computer's computations. The computer does the computations and can be programmed to remind the user of necessary data requirements. The decision maker's role is elevated to being a thinker and an analyst, rather than a pencil pusher.

Terminals to Interact with Remote Data Bases

Through the use of modems and the telephone lines, microcomputers are being used to access data and information from centralized systems such as AGNET and AGRISTAR. Access to larger systems permits farmers and ranchers to avail themselves of timely information and large data files. However, appropriate usage of this information and data in the overall decision process is still up to the manager.

Devices to Monitor and Control Activities

An important usage for microcomputers will be in the area of monitoring on-going farm activities. As an example, with appropriate peripheral equipment, the micro can be used to monitor the feed intake of dairy cattle. If a particular animal does not eat the normal amount, the computer could provide an appropriate message. Or, if milk production is monitored and a significant change is observed, the computer can alert the manager.

Current application of microcomputers in agriculture is hampered by the lack of readily available application software much as the early adaptation of tractors was hampered by the lack of availability of implements. There are basically three types or levels of software: 1) programs written for a specific application on a specific machine, 2) user-programmable application software, and mass-produced application software.

Initially micro users usually had to write their own software (or have a programmer write it specifically for them). Thus, most early software was type 1 (above). As the general use of micros increased, electronic worksheets and database-management programs developed. These are Type 2 programs that the user can tailor to a specific application. An example would be the use of an electronic worksheet program to calculate adjusted 205 day wt for cattle.

The increased agricultural market for computers has spurred the development of broadly applicable software programs, primarily in the accounting area. As the agricultural market continues to strengthen, we should see increased availability of "off-the-shelf" software.

At present we find microcomputers being used in all application areas with all three types of software.

THE FUTURE: EASYCHAIR FARMING AND RANCHING

When considering the future of computers in agriculture we are tempted to think they will take all the hard work and drudgery out of the business and that all we'll have to do is sit back and push buttons. IT WON'T HAPPEN! Computers will help in ways far more important than that. It used to be that the rancher who got up earliest and worked hardest got ahead. That is no longer the case. Now it is the individual who ranches the "smartest" who gets ahead. Computers will help us "ranch smarter". As the agricultural market continues to strengthen, we will find more and better software available. Changes in hardware are taking place on a continual basis, just as changes in other types of equipment are taking place.

Perhaps the greatest potential for development of computer applications is in the area of monitoring and controlling activities. The dairy industry has been a leader in using computers to monitor and control feeding of cattle. Both the swine and cattle feedlot industries are also making advances in this area.

Development work is under way to use microcomputers to monitor and control the application of irrigation water. This will be a very sophisticated system that measures soil moisture, atmospheric conditions, previous rainfall, etc. These measurements will be combined with information on plant growth to determine the level of water to be applied.

In the area of decision aids, the increased access to virtually instantaneous information through data base sources such as AGNET and AGRISTAR will allow farmers and ranchers to base decisions on the same information available to other sectors in agriculture and business. For example, farmers and ranchers buy inputs at a retail price and sell outputs at a wholesale price. This is not the case for other segments of the industry. The increased availability of timely information to the farmer and rancher will allow him to compete better and to market his products more effectively.

It is not unreasonable to think that in future production, agriculture will be as dependent on the microcomputer as on the tractor. We can survive without tractors, but would we want to?

CONCLUSION

Computers are a part of agriculture. They are here to stay. Their role will be increasing over the next several years. The adoption of this new technology is likely to continue at a very rapid rate. While the potential for using microcomputers is great, we also should be aware of the associated limitations.

Having a microcomputer adds to the number of tools available to the decision maker. The computer and software are only tools and have **no capacity for reasoning** nor more intelligence than a pencil. One must devote time and effort to effectively use the computer and software in decision making.

To conclude that running data through a computer adds reliability and accuracy can be an **erroneous** conclusion. If inputs are not reliable or accurate, the output cannot be accurate or reliable either.

User-oriented software (such as Visicalc or Multiplan) that helps to overcome the programming problem in the use of the micro has been widely publicized. This software, however, **does not eliminate** the need for the proper mathematical procedure and data to solve the problem--a factor sometimes overlooked.

Many analytical tools that previously were used only by researchers with large computer facilities can now be used for decision making on the small business computers. The decision maker's knowledge of the tools, data, and ability to interpret results is now the limitation in effective use of what were once researchers' tools.

If properly utilized, the computer complements the decision maker's experience, judgment, and knowledge. The user must know how to use the information and cannot expect the computer to make decisions. The manager or decision maker must still direct the course of action. The computer can only help lay out the course alternatives and evaluate progress.

Acquisition of a computer probably will not reduce the time spent in management activities. In most situations, computer use will increase the demands on managerial time, but this time will be used in analytical and productive activities.

Standing alone, people are often slow and inaccurate, but brilliant. On the other hand computers are fast, and accurate, but stupid. Teamed together computers and people can provide **management power.** It is management that will determine the future of agriculture.

60
COMPUTERIZED RECORDS AND DECISION MAKING

Alan E. Baquet

The computer is a natural tool to aid the farm or ranch manager in both record keeping and decision making. And record keeping is usually one of the first uses considered for a microcomputer, because it can save a great deal of time storing , sorting, classifying, and summarizing numbers for various types of reports. Keeping complete and accurate records is the first step to making timely decisions. In this paper, we present some important attributes of a good record-keeping system and some decision-making applications with a computer.

MICROCOMPUTER RECORD KEEPING

Two separate but related types of records are important to agriculturalists: 1) financial and 2) physical. Financial records deal with the dollars and cents portion of the business; physical records deal with production aspects such as pounds of gain, hundredweight of milk, and bushels or tons per acre. A good record-keeping system should integrate these two areas. We will focus our attention on the financial aspects.

Keeping accurate financial records assists an operator to know his current position and to set a course toward a goal. Records provide checkpoints along the way to that goal. In addition, a record system helps the operator:

- To measure over a period of time the financial success and progress of the business
- To comply with tax reporting requirements and do tax planning and management
- To establish a factual basis for comparing production of past years with the present and with goals for the future
- To plan for the future by providing data for estimating the effects of operational or economic changes
- To obtain credit

"Shoebox" record keeping no longer has a place in today's sophisticated agricultural industry. As a minimum, managers need a record system that classifies income and expense items and gives enough information for income tax reporting.

Adding crop and livestock production data and separating receipts and expenses according to each enterprise provides the needed data for a total farm record. Such a system will make possible an income statement, a balance sheet, and tax reports. It also offers the potential for computing efficiency and performance ratios. As a result, strong and weak sectors in the business can be located.

In obtaining credit, good farm and ranch records are becoming increasingly important. Credit is based on the ability of the borrower to repay money, and the lenders expect their borrowers to have adequate records to show that their businesses are on sound financial footing and that the operations are producing, or will produce, satisfactory income.

Making decisions is a fact of life. A farm or ranch manager without useful, and reasonably complete, data on past performance has a disadvantage in decision making. For example, a producer who is deciding between growing wheat or barley has a disadvantage if he has no knowledge of his past expenses to produce these commodities. Knowledge of past performance is also important to lenders. A producer who intends to borrow funds to purchase additional land must be able to show the lender that he has the ability to repay the loan. Past records are a valuable aid in justifying the loan. Complete records are equally important to successful credit acquisition for farms and ranches.

Other information that the operator needs is a balance sheet, which is the business' financial picture of assets, liabilities, and net worth over a period of time. One purpose of the balance sheet is to illustrate the solvency of the operation.

Most farm and ranch businesses are still organized as sole proprietorships or family-owned corporations. Financial statements, therefore, generally reflect both agricultural and nonagricultural assets, liabilities, and net worth. The balance sheet,however, may separate personal and business items. This is done so that the financial progress of the agricultural operation can be evaluated properly and separately from nonagricultural interests.

A series of balance sheets, for comparable dates over a period of years, can provide the basis for forming an opinion about the changing financial structure and financial strength of the business. To be most meaningful, these statements should be on the same date each year.

The income statement, or the profit and loss statement, is a second type of report needed by operators. An income statement shows how well the business actually did over a set time period. An income statement includes all expenses and receipts of the business during a specified period (usually one year) and the adjustments for inventory changes. A good record-keeping system should provide this type of information. It is possible to calculate several measures of profitability from the income statement. One measure is the return on investment or the profitability associated with all resources owned by the business. The rate of return realized on the total owned assets is found by dividing the return to capital by the value of assets.

A second measure of profitability is the return on equity capital. This tells the farm owners what rate of return they are getting on net worth. This ratio is found by subtracting interest charges from total return on capital, and then by dividing by net worth.

A cash-flow statement indicates the ability of a business to generate cash inflow through sales, borrowed money, and withdrawals from savings to meet its cash demands (cash expenses, principal, and interest payments on debt, capital purchases, and salaries or family living expenses) during a specified period of time. The cash-flow statement generally projects into the future. This is useful for farm managers or operators.

Too often financial analysis looks backward. However, changes occur too rapidly in modern agriculture to permit survival by hindsight. With larger and larger sums of money riding on "right" decisions, and with profit margins that leave little room for error, the cash-flow statement can be used to plan ahead. A cash flow reflects all the cash transfers that occur in a business.

A cash-flow statement combines and summarizes all the financial affairs in one report: all business income and expenses, nonbusiness income, loans, debt repayment, and even personal withdrawals and household spending in the case of family business.

The most effective cash-flow statement is one that is an integral part of a total record system and provides the benefit of checking actual progress against projected plans.

To illustrate how the three basic financial reports interact, consider the purchase of a piece of capital equipment--a tractor, for example. If we make a down payment and borrow the remainder of the purchase price, we have the following impacts. The balance sheet will show the value of the tractor as an asset and the debt incurred as a liability. The income statement for the year will reflect the depreciation amount as an expense. The cash-flow budget will contain both the cash down payment and future payments on the debt in the year and month incurred.

In addition to the three basic reports, most farmers and ranchers will require additional, more detailed information to adequately analyze their business. Most agriculturalists are engaged in more than one enterprise. For example, on our place in Montana we grow wheat, barley, oats, have livestock, and raise hay to feed to the livestock. For a complete analysis of our operation, we need to know the cost and return information for each of these activities. Our place is typical of most farms and ranches. Enterprise analysis, as it is often called, is very important. Farmers and ranchers need to know which aspect of the business is making money, which is not, etc. To do this, the record-keeping system must have a way to allocate expense items to various enterprises and record income from the various enterprises.

There is yet another complication that a good record-keeping system must be able to handle. When the hay we raise is fed to livestock, a noncash transfer has taken place. Raising hay incurs some expenses. These expenses must be accounted for. Our return on the expenses comes from selling the livestock, not from selling the hay directly. Thus, in a sense, we require the livestock enterprise to "buy" the hay, but no actual cash is transferred. The agricultural accounting system must be able to handle these noncash transfers.

The agricultural production process spans a considerable length of time. For example, winter wheat in Montana is planted in late September and harvested the following August. Livestock production spans a similar time span from breeding through gestation, birth, and weaning before a salable product is available.

The typical farm or ranch operates its financial books on a calendar-year basis. Thus, the production process "spills" over two accounting periods and may go into three. We plant wheat in September, close the books in December, harvest in August of the second accounting year, and we may store the wheat for sale in January or February of the third year. Did we make any money on the wheat crop? Without a reasonably sophisticated accounting program, we don't know, do we? The same accounting "spill over" takes place in livestock production.

For an accounting system to be of greatest value, it must address all of the issues raised above. There are several "off the shelf" accounting programs available on microcomputers that will deal with these issues in a satisfactory manner. It will be up to the individual user to decide which program best fits his or her needs. You should recognize that it will be very unlikely that any program will fit your situation exactly. You will have to adapt your situation to the selected accounting program.

One of the primary objectives of any record-keeping system is to provide information for analysis and decision-making purposes. To provide this information, the system must be used; a sophisticated system that is not used is no good. None of us would have a 200-horsepower tractor pulling a two-bottom plow. The same principle applies to record-keeping systems. Decide on a system that will be right for you and your operation. If you do not plan to use a record-keeping system on a regular basis, at least monthly, then do not get a sophisticated one.

DECISION MAKING WITH A MICROCOMPUTER

Good decision makers are going to be the survivors in agriculture. They will be using the best tools available. One of these tools is the microcomputer. Steps in the decision-making process are:

- Define the problem.
- Collect data related to the problem.
- Analyze the data and determine alternative solutions.
- Project the consequences of alternative solutions.
- Decide on an action.
- Implement the decision.

The computer can assist in some of these steps. While the computer will not define problems, as such, it can point out symptoms of problems. For example, if we are doing a cash-flow projection for the coming year, we might discover that our expenses exceed our cash inflow for the month of July. This will definitely be a concern, but what is the real problem? Are our expenses too high or is our income too low? Or maybe we didn't allocate our operating loan properly.

Once we determine the problem, we can develop some alternative ways to solve it. Let us suppose that in this case our expenses in July cannot be reduced, thus we must increase cash inflow. One alternative would be to increase the size of our operating loan or simply to reallocate it. If we have an excess cash inflow in June or August, reallocation would be an easy solution. If, however, we cannot do that and we must increase our total cash inflow, we need to consider other alternatives. One might be to sell some stored grain. Another might involve selling some livestock. Yet a third might involve borrowing more money for operating expenses.

In the decision-making process, we now need to project the consequences of each of these alternatives. If we consider selling stored grain, we would want to determine the impacts of selling before July vs selling later. To do this, we would need to know our cost of storage and have information on future prices. We should be able to determine the cost of storage from previous records. Expected prices for wheat can be obtained from sources such as AGNET or AGRISTAR. Once this type of information is gathered, we could use an electronic worksheet to develop a program to determine the cost of selling grain in various months. Once the program has been set up, we could do several "what if" evaluations such as what if the cost of storage is $.01 higher? What if the price of grain is lower?

In considering the livestock alternative, we need to know the pounds of gain per day and the associated cost of that gain, as well as the expected price of cattle. We could again use an electronic worksheet to analyze the effects of selling cattle at various times.

The consequence of increasing the operating loan is perhaps the easiest to determine. Finally, we need to look at the impact of each alternative and select the one that has the least detrimental impact. After selecting the best action to take, we implement that plan.

In this simple example, we have illustrated how the computer can help us to identify potential problems before they happen. The computer can then be used to help select and analyze various solutions to the problems. We can use various types of programs in this analysis portion. Some that have been used extensively are referred to as electronic worksheets. Some examples of these are Visicalc, Multiplan, and Lotus 1-2-3. These are powerful yet relatively easy to use.

CONCLUSION

The computer is an excellent tool for storing and retrieving information. It has the ability to store and assimilate volumes of data into meaningful reports for farm managers and operators. The key to successful management, like anything else, is to have the right tools and to know how to use them. The future of agriculture will belong to those managers who use the most modern tools. The microcomputer system, consisting of both hardware and software, is one of those modern tools.

61

AGNET AND OTHER COMPUTER INFORMATION SOURCES

Robert V. Price

The microcomputer revolution has come of age in agriculture. Other speakers on this program will provide invaluable insights into some of the uses for your microcomputer in your farm and ranch business.

However, regardless of the brand of hardware that you decide is the best for your individual operation, a small investment in a computer telephone modem and terminal software (costing somewhere in the neighborhood of $500) will be returned many times over in a very short period of time. The ability to send and receive information over your telephone line literally puts the world at your fingertips, no matter where your microcomputer may be located.

PROBLEM-SOLVING SOFTWARE

Although powerful mainframe computers oftentimes can house problem-solving software that can be of immense benefit in your farming and ranching operation, most producers will opt to acquire as much problem-solving software as possible for their own microcomputer. This will save on the expense involved with telephone bills and computer time on mainframe networks.

However, there may be occasions when it is not economically feasible or technically possible for some desired problem-solving software to reside in the microcomputer. Specialized software that a producer may want to run only once or twice a year may be done more cheaply by connecting to a mainframe computer than by acquiring the software, especially if the software has a high purchase price. Additionally, there may be instances when the matrix needed for the decision software may be too large to reside in the memory of the microcomputer. The powerful mainframe systems can handle such problems with ease because the memory capabilities are virtually unlimited.

I can speak more freely about AGNET, a computer network headquartered in Lincoln, Nebraska, than about most other networks available. This is due only to the fact that I am more familiar with AGNET than the other networks, and does not imply that AGNET is the only one, or even the best one, capable of providing management decision software.

Oftentimes the software housed in a mainframe computer like AGNET may be more sophisticated than off-the-shelf software available for your microcomputer, due to a tremendous amount of Land Grant University research that is supporting the software. For instance, AGNET has a feedlot simulation model that is one of the most complete that I have ever seen. It includes variations in cattle performance based on type of cattle, lot conditions, and area of the country where cattle are being fed. The environmental constraint curves on each of these items are derived from feeding trials involving literally thousands of cattle over a decade or more. It is doubtful whether canned, off-the-shelf software at your computer vendor could begin to approach all these environmental constraint curves found in this particular piece of mainframe software.

This research is not cheap. The value of the software residing in the AGNET computer alone has been appraised by IBM to be worth at least $4 million if all development costs are considered. Appendices A and B list the programs available on AGNET.

INFORMATION NETWORKING

Without a doubt, the main reason a microcomputer owner would be interested in a computer modem would be to access updated information. Information networks are available in this country covering a wide range of topics and interests. The information on these networks is stored in tremendously large data bases that are updated regularly to provide the microcomputer user with the latest possible knowledge.

In Appendix C, I have listed some of the major information networks available in North America, as well as addresses where a producer can follow up, if he has more interest. The two major general-information networks are the SOURCE and COMPUSERVE. Information is available on a wide variety of topics from the latest Dow Jones quotes to the official airline schedule where a subscriber can actually book his own airline tickets. In addition, both networks provide electronic catalog services where a subscriber can actually order goods and services to be delivered at his doorstep simply by typing in the requested information and billing it to a charge card. Also, electronic bulletin boards allow for "classified" advertising. This entails both goods and services that may be needed or offered, help wanted or help available, and other sorts of classified advertising information.

Information networking for agriculture is rapidly becoming the wave of the present. A producer who may want to receive further information on a particular disease that seems to be affecting his cattle herd, no longer will be limited to calling his local veterinarian for such information. A few strokes on the keyboard of a microcomputer may well bring the needed information directly to the kitchen table in a timely manner 24 hr a day, 7 days a week. Much information concerning integrated pest management is also available over such networks.

One of the AGNET programs listed in Appendix A is simply called NEWSRELEASE. The program itself may be a bit misleading as to what information is contained in the program. On any typical day a user of AGNET could call for a menu of the NEWSRELEASE program and find information similar to the following:

REPORT NAME	DATE	DESCRIPTION	# LINES
		BUSINESS & MANAGEMENT	
PIK	Jul 25	USDA--extends loan settlement	27
COTTON	Jul 25	Young cotton leaders sought	8
CROPINSURANC	Jul 22	USDA--county crop insurance programs approved	24
GRIZZLE	Jul 19	USDA--Grizzle named acting deputy assistant secretary	23
NEW	Jul 19	USDA--new agri. personnel named	43
		FOODS	
WORLDHUNGER	Jul 22	USDA--world hunger; what the U.S. is doing	97
FOODPRICES	Jul 22	USDA--June food prices up from year earlier	60
SURPLUSFOOD	Jul 22	USDA on new processing sys. for surplus food	25
		GRAIN & CROPS	
RESERVECORN	Jul 26	USDA--Reserve vs farmer-owned corn is released	43
TOBBACCO	Jul 26	USDA--1983 flue-cured tobacco price support level	126
NEBWEATHCROP	Jul 25	Nclrs--weekly weather crop report ending 7-24-83	54
CROPWTHR	Jul 25	Crop weather 7-18 to 7-24-83 wy crop & livestock rptng svc	44
PIKRICE	Jul 25	USDA--some growers to receive PIK rice from other st	26
SOYBEAN	Jul 25	USDA--U.S.-China soybean germ plasm exchange	26
SALE	Jul 22	USDA--South Africa sale	15
		LIVESTOCK & POULTRY	
BRUCELLOSIS	Jul 25	USDA--Ark. in cattle brucellosis program	43
GOAT	Jul 25	USDA--employees give goat to Calif. girl	42
BOVINEB	Jul 21	tb USDA--Ne. declared free of bovine	44

		RESEARCH & SCIENCES	
INSECTS	Jul 25	USDA--alfalfa fights off insects	76
		SITUATION & OUTLOOK	
ERSWHEAT	Jul 26	Summary of ERS wheat outlook & situation report	114
AGCLIM	Jul 25	Ag. climate situation committee report, 7/25/83	62
ERSOIL	Jul 18	Summary of ERS oil crops outlook & situation report	96
		SOILS & WATER	
WSTATESNEWS	Jul 20	Western states water newsletter	125
		OTHER	
BLOCKTRIP	Jul 26	USDA--Block postpones Canada trip	13
NEBCOOPSUR	Jul 26	Current pest survey data available by Ne. Dept. of Ag.	256
MCCONNEL	Jul 25	USDA--McConnel named ag. counselor to Indonesia	20
GOLDBERG	Jul 22	USDA--Goldberg appoints	25
BLOCKMEETING	Jul 22	USDA--Block meeting with Canadian ag. officials	18
AGCOUNSELOR	Jul 22	USDA--new ag. counselor to France	21

As you can see, a vast array of valuable information can be at your fingertips in only one program on AGNET or other information networks.

A user does not have to be a computer whiz to access such information. Most information networks strive to be "user-friendly" and help the user determine answers to the questions. For instance, there is an AGNET program called MICROPROGRAM that lists microcomputer software that is available. The following information would be given to anyone using the program:

Subject Matter Codes:

A - FARM MANAGEMENT
B - FARM/RECORD ACCOUNTING
C - ANIMAL RECORD KEEPING
D - LIVESTOCK FEEDING
E - CROP MANAGEMENT
F - CONSUMER ECONOMICS
G - DATA BASE MANAGEMENT
H - COMMUNICATIONS
I - GAMES
J - MARKETS
Z - OTHER (Buyer specifies subject)

Computer Brands & Models:

A - RADIO SHACK
    1 - MODEL 1
    2 - MODEL 2
    3 - MODEL 3
B - APPLE
    1 - II
    2 - II PLUS
    3 - III
C - COMMODORE
    1 - PET 2001
    2 - CBM 8000
D - NORTH STAR
E - CROMEMCO
F - ITHACA AUDIO SYSTEM
G - VECTOR GRAPHICS
H - SUPERBRAIN
I - IBM PERSONAL
J - XEROX 820
K - ZENITH H-89
Z - OTHER (Buyer specifies micro brand)

Example	Explanation
d,b,1	Asks for all RATION FORMULATION programs for the APPLE II
a,a	Asks for all FARM MANAGEMENT programs for all RADIO SHACK models
i	Asks for all GAME programs
,c,1	Asks for all programs for the COMMODORE PET 2001

ENTER CODES FOR SUBJECT MATTER, MICRO BRAND, AND MODEL
(ENTER COMMAS BETWEEN ENTRIES)

If a user wanted livestock feeding programs for all microcomputers, he could enter a "d." The output would look something like this:

LIVESTOCK FEEDING

```
PIK-PIK       WE HANDLE A WIDE VARIETY OF AG SOFTWARE
TAPE & DISK   WE SPECIALIZE IN SERVING AG MARKETS
 48 K         OUR PIK-PIK EVALUATES THE NEW FARM PROGRAM
              CALL FOR MORE DETAILS
              Programmed for:  APPLE II
                               APPLE II PLUS
                               APPLE III
              Operating System Used:  3.3DOS
              Price:  CALL US ON WATTS OR 402-375-4331
              Seller's Name:   THE COMPUTER FARM
                               WAYNE NE 68787
                               NE & IA WATTS ALSO
                   Telephone:  NE 800-74::672-8
```

MICRO-MIXER DISK 48K	LEAST-COST RATION PROGRAM. SOLVES, CREATES, STORES, RECALLS, UPDATES LIVESTOCK RATIONS INGREDIENTS, PRICES & REQUIREMENTS Programmed for: RADIO SHACK MODEL 1 RADIO SHACK MODEL 2 RADIO SHACK MODEL 3 APPLE II APPLE II PLUS APPLE III COMMODORE PET 2001 COMMODORE CBM 8000 VECTOR GRAPHICS IBM PERSONAL Operating System Used: TRSDOS CPM MSDOS Price: 250 Seller's Name: FARM ACCOUNTING SERVICE MARILYNNE BERGMAN RT 1 ITHACA, NE 68033 Telephone: (402)623-4354

With rapidly changing market prices and the rising importance of sound marketing for survival in the farm or ranch enterprise, it is quite advantageous to be able to receive updated market information when desired any time of the day or night. Most of the agriculture networks are supplying this type of information. A typical menu of files in AGNET's MARKETS program might look like the following:

REPORT NAME	**DATE**	**DESCRIPTION**	**# LINES**
		DAILY	
COLOGRAIN	Jul 27	Colorado country grain prices	53
PORK	Jul 27	Daily slaughter, dressed pork, int & term hog reports	71
FEDCATTLE	Jul 27	Daily terminal feedlot reviews	83
BEEF	Jul 27	Daily slaughter & wholesale beef values	64
MPLSCASH	Jul 26	Minneapolis cash grain closings	24
CMEFATC	Jul 26	CME Fat cattle	17
CMEFDRC	Jul 26	CME Feeder cattle	16
CMEHOGS	Jul 27	CME Hogs	19
CMEBELLIES	Jul 26	CME Pork bellies	17
CBTCORN	Jul 26	CBT Corn	17
CBTOATS	Jul 26	CBT Oats	16
CBTSOYB	Jul 26	CBT Soybeans	20
CBTSOYM	Jul 26	CBT Soybean meal	20
CBTSOYO	Jul 26	CBT Soybean oil	20
CBTWHEA	Jul 26	CBT Wheat	16
KCBTWHEA	Jul 26	KCBT Wheat	16
MPLSWHEA	Jul 26	Minneapolis wheat	15
CSCESUGAR	Jul 26	CSCE Sugar - World	17
CTNCOTTON	Jul 26	New York cotton	17
CMXGOLD	Jul 26	COMEX - New York gold	19

IMMTBILLS	Jul 26	IMM - Chicago treasury bills	17
CMELUMBER	Jul 26	CME Lumber	17
CBTGNMA	Jul 26	CBT GNMA'S	17
FUTUREC	Jul 26	Prints GRAINC & LIVEC reports plus gold & T-bills	169
LIVEC	Jul 26	Livestock futures closings	56
GRAINC	Jul 26	Grain futures closings	99
SOYC	Jul 26	Soybean complex futures closings	54
WHEATC	Jul 26	Wheat futures closings	41
MTGRAIN	Jul 26	Montana grain prices	75
MTLIVE	Jul 26	Montana livestock prices	143
		WEEKLY	
STONEFRUIT	Jul 27	Pnwfruit shipping point trends	49
PNWAPL	Jul 27	Pnw fruit shipping point trends	16
KANHAY	Jul 26	Kansas weekly hay review	39
WFSHPHOG	Jul 25	West Fargo prices for hogs plus sheep and lambs	51
WFCATTLE	Jul 25	West Fargo fed cattle, cow, bull, and feeder prices	52
WAORLVSTK	Jul 22	Washington-Oregon feedlot & range sales	37
SHEEPREVIEW	Jul 22	Weekly sheep summary put on by wlmip	32
GRAINREVIEW	Jul 22	Week in review: grain futures markets (with comments)	66
HOGFUT	Jul 22	Table & comments on this week's hog markets: Murra	60
CATFUT	Jul 22	Tables & comments on this week's cattle markets: Murra	93
WASHHAY	Jul 21	List	28
ROTTERDAM	Jul 21	USDA-FAS CIF Rotterdam import prices	45
JAPAN	Jul 21	USDA-FAS CIF Japan import prices	16
NEWYORK	Jul 21	USDA-FAS Ex-dock New York import prices	16
FOBWHEAT	Jul 21	USDA-FAS FOB, wheat export prices	20
FOBFEED	Jul 21	USDA-FAS FOB, feedgrains export prices	21
PORT	Jul 21	USDA-FAS delivered port export prices	26
		AS AVAILABLE	
ERSWHEAT	Jul 26	Summary of ERS wheat outlook & situation report	114
MTINFO	Jul 25	Description of mtlive and mtgrain markets programs	39
ERSOIL	Jul 18	Summary of ERS oil crops outlook & situation report	96

Do you want to know what foreign country may be in the market for breeding cattle produced in Texas? A simple telephone call and a few keystrokes puts that information at your disposal, complete with necessary contact people and information on how to complete a sale to a foreign destination.

These are just some small examples of the vast array of information that can be and is available on information networks.

SUMMARY

Microcomputer technology is in its infancy; what the future holds can only be dreamed at this point; but it promises to be an exciting future. The producer who is going to spend $3,000 to $10,000 on his microcomputer because he knows that it will be a sound investment, but neglects to buy the computer modem and associated software to access mainframe networks, will be like a farmer who spends $40,000 for a new tractor but neglects to purchase any equipment to pull behind it.

APPENDIX A
GENERAL AGNET PROGRAMS

PROGRAM NAME	DESCRIPTION
BASIS	Develops historical "basis" patterns for certain crops
BEEF	Simulation and economic analysis of feeder's performance
BEEFBUY	Comparison of alternative methods of purchasing beef
BESTCROP	Provides equal return yield & price analysis between crops
BINDRY	Predicts results of natural air & low temp. corn drying
BROILER	Simulation and economic analysis of broiler's performance
CALFWINTER	Analyzes costs and returns associated with wintering calves
CARCASS	Scoring & tabulation of beef or lamb carcass judging contest
CARCOST	Calculates costs of owning & operating a car or light truck
CASHPLOT	Prints a plot of selected cash prices
CODLMOTH	Assists with timing of insecticides against codling moth
CONFERENCE	A continuing dialogue among users on a specific topic
CONFINEMENT	Ventilation requirements & heater size for swine confinement
CORNPROJECT	Projects ave U.S. corn price for various marketing years
COSTRECOVERY	Calculates P.V. of income taxes saved over life of depr asset
COWCOST	Examines the costs and returns for beef cow-calf enterprise
COWCULL	Package to help determine which dairy cow to cull and when
COWGAME	Beef genetic selection simulation game
CROPBUDGET	Analyzes the costs of producing a crop
CROPINSURANCE	Analyzes whether to participate in crop insurance program
CROSSBREED	Evaluates beef crossbreeding systems & breed combinations
DAIRYCOST	Analyzes the monthly costs and returns with milk production
DIETCHECK	Food intake analysis
DIETSUMMARY	Summary of analysis saved from DIETCHECK
DRY	Simulation of grain drying systems
DUCTLOCATION	Determines ducts to aerate grain in flat storage bldg
EDPAK	Demo programs illustrating computer assisted instruction
EGGCASHFLOW	Computes financial analysis for 14-month laying cycle
ERS	Prints situation and outlook reports provided by USDA-ERS
EWECOST	Analyzes the costs & returns of sheep production enterprise
EWESALE	Lists sheep for sale
FAIR	Scoring and tabulation of judging contests
FAN	Determination of fan size and power needed for grain drying
FARMPROGRAM	Analyzes USDA Acreage Reduction Program
FAS	Prints trade leads & commodity reports provided by USDA-FAS
FEEDMIX	Least cost feed rations for beef,dairy,sheep,swine,& poultry
FEEDSHEETS	Prints batch weights of rations including scale readings
FILLEDIT	Constructs and modifies files for use in FILLIN
FILLIN	A "fill in the blank" quiz routine
FINANCE	A package of financial programs
ANNUITY	Solves problems involving periodic payments
AGPLAN	5 year agricultural proforma cash budget
AGSPREAD	Simple model for spreading comparative financial statements
BUYORLEASE	Analyses after-tax costs of alternative financing methods
CASHFLOW	12-month agricultural cash budget
DEP3	Depreciation (3 methods solved simultaneously)
EQUITY	Loan analysis with breakdown of payments to equity & int.
FUTVAL	Calculates future value
LOANSCHEDUL	Prints regular or fixed loan schedules with balloon payments

APPENDIX A, continued

LOAN	Single loan analysis
LUMPSUM	One-time investment
MARGIN	Weekly & monthly bank interest income & expense projections
MULTLOAN	Multiple loan analysis
NETDEP	Computes net declining balance depreciation
FIREWOOD	Economic analysis of alternatives available with wood heat
FORESTPAK	Package of forest management programs
DFSIM	Coast Douglas-Fir management
FORESTECON	Analyzes economic attractiveness of forest mgt. regime
RMYLD	Ponderosa/Lodgepole pine & Englemann spruce-subalpine mgt.
FOODPRESERVE	Calculates costs of preserving foods at home
CANNING	Calculates costs of canning foods
FREEZING	Calculates costs of freezing foods
FUELALCOHOL	Estimates production costs of ethanol in small-scale plants
GAMES	Package of game programs
GRASSFAT	Analyze costs and returns associated with pasturing calves
GUIDES	Prints available reports of reference material information
HAYLIST	Lists hay for sale
HELP	Lists available programs & items of interest to general user
HOUSE	Estimates the costs of heating and cooling a house
INPUTFORMS	Prints available input forms
IRRIGATE	Irrigation scheduling
JOBSEARCH	Matches abilities and interests to occupations
LANDPAK	Package of programs to assist in land management decisions
BUYLAND	Estimates maximum price you can afford to pay for land
CASHRENT	Estimates maximum cash rent you can afford to pay for land
LANDSALE	Compares a land contract sale with a cash sale
MINCOME	Calculates minimum net cash income required to make payments
MACHINEPAK	Machinery analysis package
CUSTOM	Calculates breakeven acreage and custom rates
FIXEDCOST	Estimates machinery costs as a percent of new purchase price
GRAINDRILL	Least-cost grain drill analysis
MACHINE	Determination of field machine costs
SEMITRUCK	Estimates cost of operating a tractor-trailor rig
MAILBOX	Used to send and receive mail
MARKETCHART	Prints various charts on selected future and cash prices
MARKETS	Various market reports and specialists' comments
MC	A multiple choice quiz routine
MCEDIT	Constructs and modifies files for use in MC
MICROPROGRAM	Lists programs for microcomputers
MONEYCHECK	Financial budgeting comparison for families
NEWSRELEASE	A program for rapid dissemination of news stories
PATTERN	Helps select a commercial pattern size & type for figure
PESTREPORT	Contains weekly NDSU Extension Plant Science reports
PIPESIZE	Computes most cost-effective size irrigation pipe to install
PLANTAX	Income tax planning/management program
PREMIUM	Compiles and summarizes fair premiums
PRICEDATA	Prints selected historic cash and/or futures prices
PRICEPLOT	Designed to plot market prices in graphic form
PUMP	Determination of irrigation costs
RANGECOND	Calculates the range condition and carrying capacity
SEEDLIST	Lists seed stocks for sale
SOILSALT	Diagnosis salinity & sodicity hazard for crop production
SPRINKLER	Examines feasiblity of installing sprinkler irrigation
STAINS	Tells how to remove certain stains from fabrics

APPENDIX A, continued

STOREGRAIN	Cost analysis of on farm and commercial grain storage
SWINE	Simulation and economic analysis of feeder's performance
TESTPLOT	Standard analysis of variance
TRACTORSELECT	Assists in determining suitability of tractors to enterprise
TREE	Summarization of community forestry inventory
TURKEY	Simulation and economic analysis of turkey's performance
VITAMINCHECK	Checks the level of vitamins & trace minerals in swine diet
WEAN	Performance testing of weaning weight calves
YEARLING	Performance testing of yearling weight calves

Each of these programs can be executed by typing the program name.

APPENDIX B
SPECIALIZED AGNET PROGRAMS

PROGRAMS WHICH ARE AVAILABLE TO THE GENERAL PUBLIC, BUT ARE DESIGNED TO BE USED WITH ADDITIONAL MATERIALS AND/OR TRAINING FROM PROGRAM AUTHOR(S).

PROGRAM NAME	DESCRIPTION
AFFORD	Financial budgeting model
AGBUS	Agribusiness management game
ANIMAL	Analysis of gain & feed consumption of experimental trials
BIGMGT	Big management farm supply game
BUDGEDIT	Builds and modifies files for use in BUDGET
BUDGET	General accounting & bookkeeping system
BULLTEST	Used for Nebraska bulltesting program
BUSPAK	Package of financial analysis programs
BUDGET	Capital budgeting
CASHFLOW	Discounted cash flow
DEP	Depreciation
GROWTH	Rate of growth in equity
IRR	Internal rate of return
RETURN	Return on investment
CROPROD	Teaching tool to calculate crop yields and economic returns
ECON	Package of teaching programs dealing with economic conditions
FARMSUPPLY	Farm supply business management game
FEEDEDIT	Used for building and editing files for the FEEDMIX program
GRADINGPRO	Package of programs used in grading exams and quizes
INSECTCONT	Insect control teaching programs
LIFESTYLE	Lifestyle assessment
LP	Linear programming model
LPEDIT	Used for building and editing files for the LP program
MARKOV	Markov chain analysis - simulating trends of growth of systems
MBO2	Simulation of meat quality in merchandising
NUTRIFIT	Nutritional recommendations
PCA	Management decision model for Production Credit Associations
PLANPAK	Package of programs for financial analysis and planning
PNWSOIL	Estimates sheet & rill erosion, specifically in Pacific NW
SORTANIMAL	Random sorting & assignment of animals to pens in experiments
STATPAK	Package of programs for statistical analysis of data
SUPERMARKET	Supermarket business management game
TRANS	Transportation model for allocation between supply and demand
WILDLIFE	Pgms simulating enviromental effects on undomesticated animals

Each of these programs can be executed by typing the program name.

APPENDIX C

SELECTED MAJOR COMPUTER INFORMATION NETWORKS

General information networks:

Compuserve
5000 Arlington Centre Blvd.
Columbus, OH 43220
(614) 457-8600

Source
Telecomputing Corporation of America
1616 Anderson Road
McLean, VA 22102
(703) 821-6660

Agriculture networks accessible by microcomputers:

AGNET
Al Stark
AGNET-University of Nebraska
105 Miller Hall
Lincoln, NE 68583
(402) 472-1892

Agri Markets Data Service
Dick Henry
Capitol Publications
1300 North 17th Street
Arlington, VA 22209
(703) 528-5400

Agristar
Warren Clark
Agri Data Resources, Inc.
205 W. Highland
Milwaukee, WI 53203
(414) 278-7676

COIN
Robert Routson, Systems Analyst
Technical Information
Program Development, Evaluation
and Management Systems
Extension Service
National Ag Library, 5th Floor
Beltsville, MD
(301) 474-9020

Agriculture networks requiring special terminals:

Extel
Peter Ganguly
Symons Hall
Department of Ag Economics
University of Maryland
College Park, MD 20742
(301) 454-3803

Pro Farmers Instant Update
219 Parkade
Cedar Falls, IA 50613

Grassroots
Infomart - Grassroots
Ste 511 -1661 Portage Avenue
Winnipeg, Manitoba R3J3T7
Toll free (800) 362-3388

This is not really a "computer" network but instead transmits by public television stations (in test stage at current time):

Infodata
Ron Leonard
A M S,USDA, Rm 0096 South Building
Washington, DC 20250
(202) 447-6291

62

COMPUTERS IN THE HORSE INDUSTRY

Mark M. Miller

Computers are expanding our horizons as they expand our ability to accumulate data in a usable form. Computer skills allow us to collect vast quantities of information in a retrievable form, thus enabling us to utilize that information in making our business decisions. The net result is better management of our time, our money, and our business.

The immediate applications of the computer to the horse industry are the same as those in all businesses. Use of computers in billing, accounting, accounts receivable, inventory, and sales can expedite cash flow in any business by keeping the owner apprised of the situation. Easy access to amortization depreciation schedules may expedite sales.

The Thoroughbred industry has been the front-runner in the use of computers as a means of recording and having access to an infinite amount of data on their breed. Computer services can provide race records of an individual animal, complete with times, distances, and handicaps while simultaneously providing the same detailed analyses of the horses within an entire pedigree and of the progeny of the horse being researched. Through the computer's infinite ability to scan volumes of data for significant relationships, it is possible to receive data on the horse's brothers, sisters, aunts, uncles, and so forth. This retrieval system, called a program, enables us to analyze pedigrees for significant individuals in racing history and, more importantly, for significant crosses.

The show-horse market is developing a similar program suitable for matching buyer to seller, using a list of variables expanded to include age, sex, color, training, and show record. A computer match may even be supplemented with a video cassette of the sale horse.

As buyers increase their use of such services, the availability and accessibility of data will expand, leading to programs that can provide show records of horses within a pedigree and show records of progeny.

On the farm, computerized recordkeeping can enable the owner to analyze the makeup of his herd. Through the use of a computer, the information does not change, but the owner's access to it does. For example, data on each animal are entered in the program to include sire, dam, age, breeding record, and record of progeny sold, held, shown, bred, etc. The computer can then sift out the following information: average age of the herd, average age of breeding stock, average number of mares in production, inventory horses, breeding stock, average age of sale horses, average sale price according to age and(or) sire and(or) dam, a list of pregnant mares in order of due date, and a list of all open mares. Analyzing data in this fashion can help the horse breeder to determine when to sell aging broodmares, to spot problem breeders, and to know the strengths and weaknesses of his marketing program.

In setting up a breeding program, data on the mares can be organized alphabetically, by stallion used, and by breeding status (maiden, open, bred, foaled). Information on number of covers and number of pregnancies can be updated daily or weekly so that the stud manager will have easy access to the total picture. A list of cycling mares shows instantly how many have been covered, how many have received more than one cover, and how many are confirmed in foal. This information is critical in early recognition of problems in the stud barn (such as stallion infertility, non-cycling mares, etc.).

Computers also have the capacity to initiate a "tickler file" or reminder system; for example, a complete vaccination, worming, and Coggins testing program will automatically bring up a list of all horses due for a procedure in a given month. Thus, when a new horse arrives, its history is entered in the system and each month the computer reviews those records and prints a list of horses whose Coggins tests expire within 30 days, who are due for tetanus, etc. This type of program can be expanded to the breeding shed with breeding and teasing records. The tickler file can then print out a daily list of mares due in heat or due for a pregnancy check at 21, 30, 45, or 60 days.

Entries in such a system have been simplified immensely by the arrival of the portable unit. These computers in miniature can be carried around the farm and can operate off flashlight batteries. They can store up to 24K memory and transfer all information to a tape cassette or a disc in seconds. Cost of portable units is under $1,000.

Word processor capabilities enhance computers even more. Not only can the two-fingered typist produce flawless letters, the farm owner can easily edit, update, and produce most standard forms, including sale lists, sale contracts, breeding contracts, monthly inventories, and so forth.

Visicalc is a slightly more complex program that is most useful in making projections on a given set of variables. Sales over the past 3 yr can be entered along with increases in inventory to reveal the average increase in value. That increase, when converted to percentage, can project the sales of the coming year, provided other factors such as inventory, advertising, and promotion remain constant. Inserting theoretical changes in the variables can present a projection of the probable effect. These kinds of data can help the horse owner make more educated decisions on how and where to spend money to effectively produce more business.

Computers are truly making inroads into our everyday lives. With the advent of "user friendly" programming, one need not fear that such equipment is too sophisticated for those of us living down on the farm. With the push of a button aptly labeled ENTER, each one of us can indeed enter the new frontier of data storage and utilization that can enhance our skills as stockmen and improve the management of our farms as a business.

63

NONTRADITIONAL "CONDITIONING"

R. D. Scoggins

Traditionalism is a regular component of many phases of horsemanship; certain areas of our country have traditionally been known for specific aspects of horsemanship, and our mobile population has dispersed many techniques throughout the U.S. and Canada. "Western" style training has even spread to central Europe. American-style "cutting" has become extremely popular in Australia. With all of this willingness to look at other forms of training, increased interest in different methods of conditioning is occurring.

Conditioning requires time, which is expensive; the more skilled the personnel, the greater the expense. The high degree of specialization of today's equine athlete precludes the use of unskilled personnel to exercise them. Training problems can too easily be created by an unskilled rider doing the wrong thing, compounding the mistake.

The schedules demanded of today's equine athlete places more stress on them than those of the past, thus horses must be in the best possible physical condition. At the same time, conditioning must be maintained with a minimal chance of causing lameness.

Conditioning may be described as: those exercises that, when repeated and increased over a period of time, allow development of strength and neuromuscular coordination to a level where optimal athletic performance can occur with minimal risk of injury. Conditioning 1) requires regular work; 2) takes time, 90 days to 120 days (minimum); 3) requires concussion to effect bone-ligament-tendon-hoof; 4) lengthens optimum athletic life of the horse; and 5) **does not alter basic conformation.**

Conditioning of the equine athlete has taken on new emphasis; it requires time, consistency, close attention to detail, and bulldog persistence. Because it frequently is viewed as a necessary evil, conditioning may be delegated to someone less than competent to assume the necessary responsibility. Time is money and shortcuts are sometimes sought to reduce unnecessary expenditures. As a result, a number of methods have been introduced over the years. A review of several of these methods follows, including their pros and cons.

The ultimate proposed use of the horse, its age, breed, financial limitations, and the experience of the personnel should be appraised before selecting a method. When changing the method of conditioning, be sure to acquire some expertise in how to do it properly. Any technique has limitations and potential problems, as well as advantages. Learning how to properly handle a new technique will improve its effectiveness and safety. Close monitoring of results is necessary to be sure the new technique does not adversely affect the overall program.

Swimming has been investigated the most; mechanical-walker information is almost nonexistent; and treadmills have been utilized extensively without being well evaluated. Along with these, a few comments on lunging and driving in harness will be included as conditioning alternatives.

All these procedures are being used most extensively for horses being shown at halter and to a lesser degree for performance horses. All methods have limitations, especially for the performance horse. They also may have specific advantages. Intermittent use of these methods may be more effective than continuous use would be; they can be especially valuable when intermixed with regular work.

Exercise clearly improves both muscle tone and definition; it helps reduce excessive fat and improves symmetry and overall appearance. The associated assumption is that horses are, therefore, more physically fit. This, however, is not necessarily so; exercise is more apt to result in systems not being in a state of synchronized condition. While we develop one system, we may fail to properly develop another one. And usually the most important system, structurally, is the one that receives the least conditioning--that is, most of these methods will not sufficiently condition bone, ligament, and tendon. The cardiovascular system will respond to swimming and the skeletal muscle to treadmill work. Mechanical walkers are relatively ineffective, except for toning. Results will depend on the level of stress placed on the horse and how well it responds to the method being used.

SWIMMING

Swimming has received considerable usage for years; ponds, lakes, rivers, and the oceans have all been used to swim horses. Swimming pools for horses were built after World War II in California and Florida. Since then, pools have been built in nearly every state. Horses appear to enjoy swimming once they have been trained to accept it. Training requires competent (preferably experienced) help. Both head and tail ropes are suggested initially to prevent the horse from sinking, floating on its side, or flipping over backward--which could result in a panic situation, drowning the horse or injuring the handler.

Swimming can aid in conditioning of the cardiovascular system and to a lesser degree the respiratory system. It also will increase development of the pectoral and gluteal muscles. This may be desirable in certain horses for show purposes, if development of the chest or croup area is required. This can result in increased extension of the forelimbs and improve the straightness of the forelimb motion due to pectoral muscle development over a period of time. This could be significant in horses when enhanced extension of the forelimbs or increased height of action is important, i.e., English Performance horses or some halter class participants. Croup muscles (gluteal group) also may be enhanced by regular swimming. Again, this may be especially beneficial in halter horses because it may improve the apparent levelness of the croup. Up to 90 days or more may be required for this improvement. Swimming does not significantly condition skeletal muscle used in running-type performance. It does not improve coordination, quickness, nor action without adequate on-the-ground work. A certain amount of concussion, stress, and strain are necessary to stimulate a response within the tissue. Swimming does not supply this stimulus to bone, tendon, ligament, or joint.

Although race horses have performed using swimming as the only means of conditioning, this is not desirable. Conditioning is an all-encompassing term that indicates improved physical well-being, along with increased agility or coordination. It will be necessary for the athlete to practice his or her skills to become adept at them. This is especially important with the young horse that has not yet learned its activity or type of performance--dancers must learn to dance, cutters to cut, reiners to rein, etc.

Swimming can be used advantageously for horses with minor lower-leg injuries that could be aggravated by concussion; for attitude problems towards work, i.e., sour (they are given a choice--sink or swim); and for thermoregulating difficulties (i.e., anhidrosis--failure to sweat). Very nervous or excitable animals that are difficult to physically exercise may become more calm and easier to handle after a period of regular swimming.

Swimming also can serve as an alternate method of exercise during bad weather, when the trainer is occupied by other duties, and when freshening of the attitude is needed. A fairly short time is required; usually maximum swimming time is 10 min, which may be repeated 3 times on training days (i.e., 2 to 3 times/wk).

Disadvantages of swimming include: cost (construction costs are higher); risk (especially initially); failure to condition bone, ligaments, and tendons; dermatitis that may result from improperly maintained pools; seasonal availability (ponds, lakes, etc.); unbalanced usage in horses with injuries (sore limbs are not exercised); submaximal condition of systems used; and muscles used for normal locomotion are not conditioned.

In summary, swimming can be used advantageously in a conditioning program. It is especially useful in rehabilitation and maintenance of cardiovascular conditioning of horses with minor lower-limb problems. It is best included on an intermittent basis to increase cardiovascular reserves without the danger of leg injury.

MECHANICAL TREADMILL

The mechanical treadmill has been used extensively as a research tool. While a considerable amount of data have been accumulated, only anecdotal accounts are available regarding its value for performance. Speeds range from a fast walk at 4 mph to 5 mph to a strong trot, i.e., up to 15 mph; angle of slope has varied from level to 11 1/2°.

Most of the gait and motion studies are from Europe (especially from Sweden by a group at Stockholm). This work was primarily done on Standardbred trotters. In the U.S., the mechanical treadmill has been quite well-accepted by the Standardbred industry. The Quarterhorse segment also has utilized this equipment extensively. Other breeds have shown variable interest, with minimal Thoroughbred interest. The gaits utilized are the walk and trot because they allow better stability and balance (which may account for the lack of Thoroughbred enthusiasm).

The mechanical treadmill provides submaximal conditioning in most instances. Knowing how to load a horse in a trailer is the basic requirement for experience, along with the ability to observe. Short periods of exercise are used--usually 3 min to 5 min, seldom over 15 min, and rarely in excess of 30 min, even at only a walk.

Treadmills are felt to be the most readily controlled as to adjustment of work load--degree or angle of slope, speed, time (and in some instances environment) can all be regulated. Some horsemanship is necessary to teach horses to load and accept the use of this device. Particular groups of muscles are conditioned. The quadriceps or stifle muscles appear to show the greatest response. Rear limb muscles are worked mostly--somewhat like weight training or hill climbing when the treadmill is set at an angle or slope. Horses are reported to have increased muscular development, especially of the rear limbs; and year-round exercising is provided, regardless of weather. It takes very little room (less than a box stall). Horses travel in a straight line, removing the joint strains that accompany constant turning in circles that occurs with some other forms of exercise; this is a particular advantage to young horses.

Even though treadmills were the "in" thing for some time, interest and usage seem to have diminished. Cost can be a factor--they come in a wide price range, from stripped-down versions to those that do everything except load the horse and put it away.

Rheostat-type speed adjustments are recommended for infinite adjustment. These allow better accommodation to the individual horse. Adjustable front-end heights for changing the angle of the floor are advantages. Roller design is important to minimize vibration to the horse's hooves. Several types have minimized rollers and have sliding belts to reduce vibration and subsequent friction to the horse. This is of considerable importance in the hock joint. Increased lameness has been reported following treadmill work by horses that have lower-limb problems. Rear-limb-lameness problems especially seem to be aggravated by forced treadmill work.

Treadmill usage seemed to evolve from both human and research experiences--researchers find treadmills useful because horses can travel up to 15 mph and never be more than an arms-length away. Measurements and samples can be obtained readily. Telemetry equipment can be attached and motion studies conducted in relative comfort; work times range from 5 min to a maximum of 30 min. Submaximal work is usually all that is available to horses, but one researcher has reported maximal stress in conditioned ponies.

It is particularly difficult for non-Standardbred horses to approach and sustain a 15 mph trotting gait without developing gait-related lamenesses. Noise levels attained by most machines, especially when confined, can become deafening. Breakdowns occur, as with any mechanical device, and usually at inopportune times; Murphy's Law applies. Some warm-up and warm-down at a walk are advised before trotting occurs.

MECHANICAL WALKERS

Mechanical walkers are primarily used for exercise to warm up or cool out horses. While deep-footing encourages muscle development, no significant conditioning occurs because speeds that will cause metabolic stress cannot be attained safely.

Arms on walkers should be as long as possible to minimize circular motion. Adequate strength along with proper bracing of the walker is essential for safety and durability. Most walkers are powered by electric motors and adequate grounding is an absolute necessity. Walkers should be equipped with directional reversing capability to prevent constant one-way movement of horses. Direction of the walker should be changed daily.

LUNGING AND DRIVING

Other forms of exercise used to condition show horses include lunging and driving in harness. Driving allows horses to be conditioned while avoiding the weight of a rider. It combines physical conditioning with the mental development associated with any form of training. It requires the expertise, equipment, and space needed to train a horse to drive. Driving can allow the horse to travel to a great degree in a straight line; the pulling effort involved can be adapted to the horse's abilities. This type exercise has a definite muscle-toning effect and is especially beneficial to the back and rear-limb muscle development.

Forcing horses to trot beyond their comfortable abilities can augment locomotor problems. Working the horse within its capabilities and increasing the distance worked is effective. All systems can be enhanced, but bone, tendon, ligament, and skeletal muscles will benefit most from being exercised aerobically.

The speed of work and the distance alter the stress or level of conditioning. Thoroughbreds, Quarter horses, and other nontrotting breeds have been satisfactorily loped or galloped in harness to obtain sufficient speed to provide conditioning. It may look a little strange but can be used quite effectively.

Lunging has long been used for exercise and training. It may be one of the most poorly understood and most often abused forms of exercise conditioning. Constant travel in a relatively small circle produces considerable stress on joints and limbs. It is especially dangerous in the immature individual. From a training aspect, lunging may tend to cause horses to shift their weight heavily to the forehand, further aggravating front-limb problems. Horses that tend to lug out on a line may place additional lateral stress on their forelegs.

In a study involving young horses being free-lunged in a 45-ft circle for up to 30 min/day, 5 days/wk, 60% showed clinically observable lameness within 4 wk. Most of these horses had medial-splint lameness, others were sore in their fetlock joints, knees, and shins, or tendons.

For lunging, it is useful to have large pens (50 ft or more) with level, mixed-sand footing that allows hoof rotation without sticking, moderate gaits, and frequent reversal of direction, with no more than 10 circles allowed in one direction. Routine use of splint boots will prevent splints and other bumps due to interference. Bandages without adequate padding are useless and may even result in cording problems.

Areas of adequate size, protective covering, proper footing, and adequate training can all help in minimizing injury. The lunging area should not resemble a racetrack; excess speed must be avoided; horses should not lug on the line, and cross-cantering must be prevented. Properly used, lunging can be beneficial as an exercise form. However, insufficient stress can be produced to significantly condition an athlete.

SUMMARY

All of the methods described are principally forms of exercise to improve body tone and shape. None will produce adequate physical conditioning by themselves. Most are submaximal in their level of stress. Physical coordination is not improved significantly with any of these methods. Minimal bone, tendon, and ligament conditioning occur. All have value as adjuncts to conditioning programs in specific situations.

Personnel working with the horses are the most important part of any conditioning program. The best horseman, given the proper amount of time, will produce the best-conditioned horses. Horsemanship is the major factor in any program. The persistent person with a feel for the horse, and who will work with it regularly, will obtain the best results.

"Talent without ambition, ambition without direction, persistence without knowledge, none will gain anything."

64

ADVANCES IN EQUINE HEALTH

R. D. Scoggins

The past several years have brought a number of new health aids and techniques for horses and horse owners.

Because relatively little horse meat can legally enter the U.S. food chain, drug approvals for horses are more easily obtained than are those for meat-producing animals. Although drugs are tested for safety and efficacy before being licensed, unexpected side effects can occur when they are used in an unprescribed manner. When expected results do not occur, use of the drug may be criticized. A thorough understanding of a drug or product action is absolutely necessary. It is also prudent to understand any potential side effects (whether the drug is used properly or abused in some manner). Problems could be created by excessive and prolonged use of anabolic steroids, nonsteroidal analgesics, corticosteroids, ultrasound, hormones, vitamins, antibiotics, protein, electrical stimulation, etc.

If we spend time observing and talking with knowledgeable successful horse people, the fact that there are no short cuts should become apparent. Improved pedigrees are not found in a hypodermic syringe. Improved conditioning does not come in a fancy feed sack. Increased soundness does not come in a laser gun, a battery pack, or a magic boot. All of these things occur from a combination of balanced nutrition, proven pedigrees, structurally sound conformation, and training and conditioning programs that provide the opportunity for evaluation.

Science has made tremendous strides in areas that we often take for granted. Because of improved anesthetics and anesthetic machines, plus improved surgical techniques and aftercare, several thousand horses undergo successful abdominal surgery annually. Colic surgeries, tumor removals, cesarean sections, etc., are all performed routinely in university and privately owned equine hospitals. It is not, and is not intended to become, box-stall surgery. Most of us do not want our appendix removed on our kitchen table. Major surgery on horses is most apt to be successful when performed in a properly equipped and staffed hospital facility.

NEW TECHNOLOGY

Although many of you may never have the need for such equipment, tremendous gains in technology are allowing us to perform diagnostic tests routinely today that were not in existence 3 yr ago. Despite the advances in treatments that you may be well aware of, perhaps the most important advances are the ones in diagnosis. Without a diagnosis, treatment must be hit or miss.

Improved procedures include: 1) transtracheal washes in respiratory cases; 2) peritoneal taps in abdominal disease; 3) fiber optic laparoscopy; 4) ultrasound diagnosis of skeletal tissue, soft tissue, pregnancy; 5) biopsy techniques (uterine endometrium, liver tissue, abdominal tissue); 6) thermography (superficial muscle; skeletal inflammatory lesions); 7) more precise radiographic techniques; 8) improved methodology in culturing infections and sensitivity testing; 9) scintigraphy (radio nuclear scans to detect inflammation within tissue); and 10) computers.

New treatment forms (some are fully accepted and tested, others border on the suspect) include: 1) laser (with tremendous variability in equipment); 2) therapeutic ultrasound; 3) low impulse electrical stimulation (unproven except to reduce pain and edema [swelling]); 4) arthroscopic surgery; 5) surgical implants, metal alloys for pins, plates, screws, etc.; and 6) very lightweight cast material.

New drugs (tested and being used): 1) oral progesterone; 2) ivermectin (injectable dewormer); 3) oral wormers (new formulations, benzimidazoles, levamisole); 4) antimicrobials (for treatment of bacterial infections; Tribrissen--sulfa drug: chemical compound, antibiotics); and 5) prostaglandin-related compounds.

Rapid advances are being made in maintaining and improving horse health. I will discuss some of those items with you and hope to place some of these in perspective for the horse owner.

Constant improvement in anesthetics and surgical techniques allows the salvage of a much greater number of horses than was possible just 10 yr to 15 yr ago--easily within the life span of an average horse.

Bone plating and lightweight cast material greatly improve our chances of saving long-bone fracture cases. A broken leg is no longer an automatic death sentence. The individual horse's attitude towards accepting fixation devices can be as important as the surgeon's skills. Pneumatic splints and equine ambulances at some competitive events, and especially at race meets, greatly improve the chances for recovery from competitive injuries.

Arthroscopic surgery has attained increased usage for repairing certain types of joint injuries. Small chips within joints can be removed before abrasiveness of the particle can cause extensive injury.

Newer, more effective drugs are being evaluated and released as soon as effectiveness and safety are proven. Examples are: oral progesterones for aiding in the control of the estrous cycle of mares; newer more potent antibiotics and sulfas or sulfa compounds to aid the control of bacterial infections; and more specific hormones that help to treat reproductive disease more specifically, i.e., prostaglandins.

Ivermectin (Eqvalen) has arrived on the market with wide acceptance and some problems. Injection-site reactions have been related almost universally to a less-than-adequate technique. Sterile disposable equipment and proper injection-site preparation are a must. This is the recommended procedure for all injections, but some products are less forgiving of mistakes than are others.

Parasite-control studies have resulted in numerous changes in treatment regimens. These should "shake out" through time and use; only then will we be able to adequately evaluate their effectiveness.

Pressure from the industry and practicing veterinarians has resulted in widespread increased usage of this modern-day technology. Computers are being used widely for improved business techniques and for maintaining updated inventories of drugs and health records. In some areas, improved diagnostic capabilities are the result of increased computer usage, which appears to be the next big use of computer technology.

Diagnostic techniques have made some dramatic advances within the last few years. Ultrasound is used for pregnancy testing and to map bone, tendon, and ligament structure. Abnormalities in these structures indicate potential areas of lameness and(or) injury. Other technological improvements include increased use of nuclear medicine scans to locate areas of inflammation. A preliminary examination locates the general site of the problem. A specifically labeled compound is injected intravenously. After a time, the material is attracted in abnormally high concentration to the site of inflammation. A gamma camera picks up low level radioactive emissions; these are focused on a screen and also transferred to film. A person trained in radionuclear medicine can then evaluate the degree and site of inflammation, thus increasing his/her diagnostic ability. This technique also provides an improved method of monitoring some lesions, especially hairline fractures, and allows better evaluation of the healing of an internal disease, such as a lung abscess. We have found radionuclear medicine to be especially helpful in treating long-bone stress fractures, pneumonias, joint problems, and other inflammatory lesions in animals at the University of Illinois Large Animal Clinic.

Thermography shows some useful application in diagnostic areas, especially in detecting inflammatory changes. Horsemen are most familiar with thermovision for detection of sores in Tennessee Walking Horses. Experience with this equipment indicates further study is warranted and may be especially useful where training programs desire regular monitoring to detect early signs of superficial inflammation.

Regular radiographic equipment, both x-ray machines and processing equipment, has become more portable, which allows for more frequent use in stables with adequate electrical power. These improvements also have improved the quality of radiographies, which result in more diagnostic films and fewer "blurograms." Quality control has also reduced the hazard of extraneous x-ray exposure.

A group of veterinary surgeons discussing techniques with students commented that videotapes made from 1970 to 1975 were so outdated in some areas that they were no longer usable for teaching purposes. It is not that the techniques themselves did not have value. They were, and are, useful, but modern inquiry has resulted in newer methods being developed.

This is the way advances occur, usually as a response to a perceived need, not always as rapidly as we may like.

SPECIFIC RESEARCH

Several university research groups are working in specific areas; the University of Georgia is doing colic research; Colorado is known for its reproductive research capabilities; the University of Missouri has done extensive work in founder; the University of Illinois is doing considerable investigation into respiratory disease, reproductive disease, and parasite control programs; and surgical techniques in the area of orthopedics have been developed at Ohio State, Colorado State, and Texas A&M. The Morris Animal Foundation, the Grayson Foundation, and many private horse organizations do a great deal towards funding further equine research. The horse industry will have to pay its own way. Direct donations for further studies can be made to the university of your choice or the Morris Animal Foundation.

An area that has attracted considerable emphasis in recent years is equine sports medicine. It has become a catch-all for every program designed to get a horse into competitive physical shape. Some are unbelievably simple, others are equally unbelievable in their level of complication. The effective level seems to be somewhere in between.

To be effective, conditioning must be consistent (daily), persistent (day after day), and must include a strong dose of common-sense horsemanship. The long, slow, distance work seems to be the biggest cause of "drop-outs." Sixty to 90 days of jogging 10 mi to 12 mi/day is not a fun idea for most of us. Once a horse has been properly conditioned, maintained, and placed in competition, you will probably not accept anything less. Only those interested in serious competition and in performing with consistency will be interested in this conditioning regimen.

A new strangles vaccine became available 3 yr ago. The Chinese claim to have developed a vaccination for equine infectious anemia. Improvements in other vaccines and the development of new ones occur regularly.

Today's veterinarian graduates can become seriously out-of-date within 5 yr of graduation if no attempt is made to keep up. Within 10 yr, they can be hopelessly out of date. Just keeping up with today's technology can be overwhelming.

New ideas are stimulating. Many will prove useful, some will not. Allow your veterinarian the use of his or her professional judgment in the area of adapting changes. Some procedures have stood the test of time so well that new procedures would not be of any additional benefit. Being conservative in some situations can be very prudent. The capability of correcting problems does not relieve the breeder of the challenge to produce sound, useful horses. The basic choice of the foundation for genetic material still lies with the breeder.

65

A FARMER'S OBSERVATIONS OF WASHINGTON, D.C.

Michael L. Campbell

When President Lyndon B. Johnson announced the establishment of the White House Fellows Program in 1964, he stated that "a genuinely free society cannot be a spectator society." For one year, commencing September 1, 1982, I was given the opportunity to participate in the program developed by President Johnson's Administration to draw individuals of high promise to Washington for personal involvement in the process of government.

The White House Fellowship is a highly competitive opportunity to participate in and learn about the federal government from a unique perspective. For one year, the 14 to 20 persons who are chosen as White House Fellows are full-time Schedule A employees of the federal government. They work in a cabinet-level agency, in the Executive Office of the President, or with the Vice-President. Rather than fit the Fellows to their pre-Fellowship specialties, the program aims at utilizing their abilities and developing their skills in the broadest sense possible. In most cases, a Fellow serves as a special assistant, performing tasks for a Cabinet Secretary, the Vice-President, an Assistant to the President, or for appropriate under or deputy secretaries. In this sense, the White House Fellow's year is a high-level internship in government--but it is also much more.

The White House Fellowship program is not a direct federal recruitment program and is not designed to attract people into the federal service in the immediate sense. It is a sabbatical or leave of absence (without salary) from the individual's previous employment. Some Fellows have stayed on for a short while after their Fellowship year and some returned to government (state, local, or federal) in later years. Most Fellows, however, return to their geograhic or professional communities where they can share their new knowledge and contribute to society more ably and productively through a fuller understanding of the federal government. The program is an opportunity for intensive service, with the goal of improving each participant's ability to serve more fully for years to come.

The White House Fellow's activities in Washington represent a dual (work and education) experience. The work assignment provides the Fellow the opportunity to observe closely the process of public-policy development and to come away with a sense of having participated in the governmental process as well as having made an actual contribution to the business of government.

As noted, the program's aims are to tap the resources of the Fellows and to develop their abilities in the broadest sense rather than fitting the Fellows into assignments directly related to their pre-Fellowship specialties.

The educational program is a distinguishing feature of the White House Fellowship. The Fellows participate as a class in a series of off-the-record meetings, usually held two or three times a week throughout the Fellowship year with prominent representatives from both the public and private sectors.

The meetings in the Washington area are supplemented with occasional travel for the Fellows to experience, observe, and examine major issues confronting our society on a firsthand basis. In addition to the domestic focus, Fellows examine international affairs and U.S. foreign policy and develop an understanding of the philosophies and points of view of other governments through overseas travels.

The educational program is typically developed around several broad themes reflecting the interest of the fellowship class and typical policy issues facing the nation as a whole. This thematic approach to the educational component of the Fellowship is designed to provide the Fellows with a comprehensive understanding of exceedingly complex national issues.

The federal government plays a huge role in the operations of all agriculture. In our farming operation, I considered the government to be my partner, and yet few of us in agriculture really know how the government functions or how to make it function more effectively for agriculture. One of the primary reasons I applied for the Fellowship was to acquire just such an understanding.

When my White House Fellow year started, I had no idea how comprehensive this educational process would be. Assigned to the Office of the Chief of Staff, I was able to observe and participate in the functioning of the White House on a daily basis. I also worked with the White House Office of Policy Development, developing agricultural policy for the White House, and with key personnel at the USDA. This overall exposure to the White House and to the USDA, combined with over 200 off-the-record meetings with national and international decisionmakers in Washington, around the U.S., and in the Orient and Southeast Asia, has dramatically increased my understanding of the federal government and agriculture's role in it.

As the second farmer to have participated in the White House Fellows Program, I have returned to agriculture with many firsthand observations on the enormous size, influence, and functioning of the federal government. It has been said that Washington, D.C., is the only U.S. city where the local news is also the national news. Hundreds of thousands of people pour into federal buildings in the District of Columbia each day to make decisions that affect each of our lives.

Few would argue that there is room for improvement in the functioning of the federal government. However, people that differ ideologically can rarely agree on a common solution to any given problem. The Reagan Administration has attempted to use the commission approach to reach solutions for some major problems and in turn to generate a consensus of opinions. A few examples of this approach are the Commission on Social Security Reform, the Grace Commission, and the Commission on Central America.

The relationship between the President and Congress has changed since the Viet Nam War and Watergate. Prior to Viet Nam, the President was the chief formulator of foreign policy. However, since Congress became so involved in foreign affairs during the Viet Nam War, it is very reluctant to relinquish its power, and bipartisan support for foreign policy continues to be a rarity. The Watergate episode and the attempted impeachment of the President diminished the respect that Congress held for the Office of the President. The combined effect and results of Viet Nam and Watergate on the Presidency are hard to quantify, but the adverse consequences are real.

The news media have a great influence and control on the activities of the federal government. Media support or disfavor, for example, with proposed legislation, a political appointment, or budgetary matters can often mean the difference between success and failure. Top-level presidential appointees were unanimous in their feelings that the most difficult adjustment in Washington was that of dealing with the distortions and untruths in the press. Employee dissatisfaction can translate into leaks to the news media. I see no diminishment in the power and influence of the news media as they continue in their roles as communicators and monitors of government activities.

The Department of Agriculture is one of the largest departments in the Executive Branch. As with most other departments and agencies, its responsibilities and functions are the cumulative results of many different administrations and sessions of Congress. Unfortunately, programs that are initiated to solve problems do not necessarily cease when the problem does. I wonder how much longer agriculture will allow itself to be so strongly controlled by the federal government and how much longer the American taxpayer will allow agriculture to be supported at its current level.

Americans may not always view with pride the activities that transpire in Washington, but they can be proud of the city. The capitol of our country, with its beautiful museums, monuments, memorials, and buildings, is a tribute to the people that have made this country a success.

I am very grateful for the opportunity I was given as a White House Fellow. My pride in our country has been strengthened; my horizons have been expanded; and my hope for our future increased.

INDEX OF AUTHORS

H. Joseph Ellen II
President, TriSolar Associates
100 Estrella
Ajo, AZ 85321
—Energy System Specialist
Page 59

H. A. Fitzhugh
Winrock International
Route 3
Morrilton, AR 72110
—Animal Scientist
Page 145

Temple Grandin
Department of Animal Science
University of Illinois
1207 W. Gregory Drive
Urbana, IL 61801
—Livestock Handling Specialist
Page 390

Cecile K. Hetzel
Stephens College
P.O. Box 2071
Columbia, MO 65215
—Horsewoman
Pages 35, 113

Dixon D. Hubbard
USDA-SEA-Extension
Room 5051 - South Building
14th & Independence
Washington, D.C. 20250
—Animal Scientist
Pages 25, 91

Ralph C. Knowles
803 Venice Drive
Silver Springs, MD 20904
—Veterinarian
Page 373

Qin Li-Rang
Professor
Head, Animal Science and Veterinary Medicine Department
Huachung Agricultural College
Wuhan, Peoples Republic of China
—Veterinarian
Pages 295, 305, 311

Matthew Mackay-Smith
Delaware Equine Clinic
RFD 2
Cochranville, PA 19330
—Veterinarian and Journalist
Pages 122, 327, 332

William C. McMullan
Large Animal Medicine & Surgery
Texas A&M University
College Station, TX 77843
—Veterinarian
Pages 216, 257, 285, 290

Doyle G. Meadows
Vice-President, Marketing and Development
Lookout Mountain Ranch, Inc.
3100 S. 33rd Ave., Route 9
Tulsa, OK 74107
—Horse Scientist
Pages 223, 226, 230, 236

Mark M. Miller
Al-Marah Micanopy
Route 1, Box 115
Micanopy, FL 32667
—Horseman
Pages 379, 430

Thomas Monin
Large Animal Clinic
College of Veterinary Medicine
Oklahoma State University
Stillwater, OK 74078
—Veterinarian
Pages 240, 247

Michael Osborne
Vice-President
North Ridge Farm
4153 Spurr Road
Lexington, KY 40511
—Veterinarian and Horseman
Pages 215, 235, 361, 402

Robert V. Price
Project Leader, Western Livestock Marketing Inf. Project
2490 West 26th, Room 240
Denver, CO 80211
—Agricultural Economist
Pages 49, 418

William A. Scheller
President, Scheller & Assoc., Inc.
917 Stuart Building
Lincoln, NE 68508
—Consulting Engineer Energy
Cycle, Inc.
Page 64

R. D. Scoggins
256 Large Animal Clinic
University of Illinois
Urbana, IL 61801
—Veterinarian
Pages 433, 430

Ben A. Scott
Marketing Consultant
Scottland Farm
5505 Schwartzmiller Road
Lake Stevens, WA 98258
—Horseman
Pages 363, 367

Arthur L. Snell
President, Snell Systems, Inc.
P.O. Box 17769
San Antonio, TX 78217
—Fencing Systems Specialist
Page 382

Joe Staheli
Al-Marah Arabians
4101 North Bear Canyon Road
Tucson, AZ 85749
—Horse Trainer
Pages 345, 348, 350

John M. Sweeten
Extension Agricultural
Engineer
Texas A&M University
College Station, TX 77843
—Animal Waste Management
Specialist
Page 73

J. W. Turner
Head, Animal Science Department
Louisiana State University
Baton Rouge, LA 70803
—Beef Cattle Geneticist
Page 162

R. L. Willham
Professor
Iowa State University of
Science and Technology
Ames, IA 50011
—Beef Cattle Geneticist
Page 201

Other Winrock International Studies Published by Westview Press

Beef Cattle Science Handbook, Volume 19, edited by Frank H. Baker.

Dairy Science Handbook, Volume 15, edited by Frank H. Baker.

Sheep and Goat Handbook, Volume 3, edited by Frank H. Baker.

Stud Managers' Handbook, Volume 18, edited by Frank H. Baker.

Future Dimensions of World Food and Population, edited by Richard G. Woods.

Hair Sheep of Western Africa and the Americas, edited by H. A. Fitzhugh and G. Eric Bradford.

Other Books of Interest from Westview Press

Animal Health: Health, Disease and Welfare of Farm Livestock, David Sainsbury.

Calf Husbandry, Health and Welfare, John Webster.

Carcase Evaluation in Livestock Breeding, Production, and Marketing, A. J. Kempster, A. Cuthbertson, and G. Harrington.

Energy Impacts Upon Future Livestock Production, Gerald M. Ward.

Livestock Behavior: A Practical Guide, Ron Kilgour and Clive Dalton.

The Public Role in the Dairy Economy: Why and How Governments Intervene in the Milk Business, Alden C. Manchester.

Other Books of Interest
from Winrock International

Bibliography on Crop-Animal Systems, H. A. Fitzhugh and R. Hart.

Bibliography of International Literature on Goats, E. A Henderson and H. A. Fitzhugh.

Case Studies on Crop-Animal Systems, CATIE, CARDI, and Winrock International.

Management of Southern U.S. Farms for Livestock Grazing and Timber Production on Forested Farmlands and Associated Pasture and Rangelands, E. Byington, D. Child, N. Byrd, H. Dietz, S. Henderson, H. Pearson, and F. Horn.

Potential of the World's Forages for Ruminant Animal Production, Second Edition, edited by R. Dennis Child and Evert K. Byington.

Research on Crop-Animal Systems, edited by H. A. Fitzhugh, R. D. Hart, R. A. Moreno, P. O. Osuji, M. E. Ruiz, and L. Singh.

The Role of Ruminants in Support of Man, H. A. Fitzhugh, H. J. Hodgson, O. J. Scoville, Thanh D. Nguyen, and T. C. Byerly.

Ruminant Products: More Than Meat and Milk, R. E. McDowell.

The World Livestock Product, Feedstuff, and Food Grain System, R. O. Wheeler, G. L. Cramer, K. B. Young, and E. Ospina.

Available directly from Winrock International, Petit Jean Mountain, Morrilton, Arkansas 72110.